PLANT FORM AND VEGETATION STRUCTURE

PLANT FORM AND VEGETATION STRUCTURE

Adaptation, plasticity and relation to herbivory

edited by M.J.A. Werger, P.J.M. van der Aart,
H.J. During & J.T.A. Verhoeven

SPB Academic Publishing bv, The Hague, 1988

CIP-DATA KONINKLIJKE BIBLIOTHEEK, DEN HAAG

Plant

Plant form and vegetation structure : adaptation,
plasticity and relation to herbivory / ed. by M.J.A.
Werger ... [et al.]. – The Hague : SPB Academic
Publishing. – Ill.
With index, ref.
ISBN 90-5103-019-3
SISO 582 UDC 58 NUGI 824
Subject heading: botany.

© 1988 SPB Academic Publishing bv, P.O. Box 97747
2509 GC The Hague, The Netherlands

All rights reserved. No part of this book may be translated or reproduced
in any form by print, photoprint, microfilm, or any other means without
the prior written permission of the publisher.

ISBN 90-5103-019-3

CONTENTS

Preface	ix
Barkman, J.J., Some reflections on plant architecture and its ecological implications. A personal view demonstrated on two species of *Quercus*	1
Barkman, J.J., New systems of plant growth forms and phenological plant types	9
Komárková, V. and McKendrick, J.D., Patterns in vascular plant growth forms in arctic communities and environment at Atkasook, Alaska	45
Epp, G.A. and Aarssen, L.W., Attributes of competitive ability in herbaceous plants	71
De Broeck, L.J., Germination and establishment of dicotyledons in grasslands	77
Wright, R.A. and Mueller-Dombois, D., Relationships among shrub population structure, species associations, seedling root form and early volcanic succession, Hawaii	87
Figueroa, M.E. and Castellanos, E.M., Vertical structure of *Spartina maritima* and *Spartina densiflora* in mediterranean marshes	105
Lotz, L.A.P. and Olff, H., Variation in biomass and architecture of plants due to small-scale environmental heterogeneity	109
Hutchings, M.J. and Slade, A.J., Aspects of the structure of clonal perennial herbs	121
Hirose, T., Nitrogen availability, optimal shoot/root ratios and plant growth	135
Hara, T. and Haraguchi, A., Phenotypic responses of plants to environmental conditions	147
Grace, J.C., Effect of foliage distribution within tree crowns on intercepted radiant energy and photosynthesis	153
Kellomäki, S., Dynamics of the branch population in the canopy of young Scots Pine stands based on modular growth	161
Werger, M.J.A. and Hirose, T., Effects of light climate and nitrogen partitioning on the canopy structure of stands of dicotyledonous herbaceous vegetation	171
Ojea, I., Pereiras, J. and Basanta, M., Vertical distribution of photosynthetic and non-photosynthetic phytomass in *Ulex europaeus*	183
Bongers, F. and Popma, J., Is exposure-related variation in leaf characteristics of tropical rain forest species adaptive?	191
Grace, J., The functional significance of short stature in montane vegetation	201
Shmida, A. and Burgess, T.L., Plant growth-form strategies and vegetation types in arid environments	211
Whigham, D.F. and O'Neill, J., The importance of predation and small scale disturbance to two woodland herb species	243
MacLean, D.A., Effects of spruce budworm outbreaks on vegetation, structure, and succession of Balsam Fir forests on Cape Breton Island, Canada	253

Brown, V.K., Gange, A.C. and Gibson, C.W.D., Insect herbivory and vegetational structure — 263

Jefferies, R.L., Pattern and process in arctic coastal vegetation in response to foraging by Lesser Snow Geese — 281

Whicker, A.D. and Detling, J.K., Modification of vegetation structure and ecosystem processes by North American grassland mammals — 301

Sala, O.E., The effect of herbivory on vegetation structure — 317

Ellenbroek, G.A. and Werger, M.J.A., Grazing, canopy structure and production of floodplain grassland at Kafue Flats, Zambia — 331

McNaughton, S.J. and Sabuni, G.A., Large African mammals as regulators of vegetation structure — 339

Index of keywords — 355

Jan J. Barkman and Juniperus communis (photo Erik Hardeman)

PREFACE

Through the ages scholars have attempted to correlate morphological features of plants and stands of vegetation with environmental variables, both biotic and abiotic. They often managed to provide satisfactory elucidations of local patterns. Exploration and travel stimulated such attempts to describe the main patterns of morphological and environmental variation and explain their 'epharmony', *i.e.*, their presumed functional meaning in the context of their specific environment. During the 19th century these efforts culminated in authoritative and comprehensive reviews of global patterns in vegetational form. It also led to the birth of ecology as a science in its own right.

Around the beginning of the 20th century the ecophysiological approach to the study of plant form and vegetation structure emerged as a separate discipline next to the existing purely descriptive approach of correlating patterns in morphological and environmental variables. While the descriptive approach continued to contribute substantially to an increasingly valuable ecological data base and succeeded in deductively generating explanatory interpretations from simple correlative observations, the ecophysiological approach was practised more sporadically and made rather slow progress. Experimental testing of causal hypotheses and quantification of ecologically relevant processes remained virtually impossible. Only after the second world war the great breakthrough with this approach could be made. This was the result of the development of sophisticated, precise and handy instrumentation suitable for measurements under controlled and field conditions. It was also enhanced by the development of computers allowing the gathering and analysis of vast amounts of ecological data and the formulation and testing of formal models. The availability of these facilities strongly reinforced the interest of scientists in trying to understand the functional aspects of plant form and their evolutionary relevance.

Though ecologists in the descriptive and ecophysiological traditions approached the questions related to the relevance of plant form and vegetation structure in the environmental context quite differently, their aims were rather similar: to investigate to what degree specific morphological features ensured an efficient functioning of the plant in its environment. This required a quantitative evaluation of the features and processes involved in a common 'currency' and led to attempts to evaluate cost-benefit analyses. For 'currency', 'dry matter' or 'energy content' were usually taken. Such analyses of expenditure and returns in terms of matter or energy proved a solid base for evaluation and allowed to bridge the gap between the descriptive and ecophysiological approaches. By considering the morphological constraints on the physiological processes in the plant and the energetic trade-offs associated with these processes in an ecological context it became possible to analyse morphological and physiological traits on a quantitative basis and to develop and experimentally test specific hypotheses. Several contributions in Givnish (1986) convincingly show the power of this approach.

Professor Jan J. Barkman, an outstanding and imaginative ecologist from the

Netherlands, used the descriptive approach. He is a keen, versatile and knowledgeable scientist with a wide interest in nature, an amazing amount of energy, an unsatiable appetite for scientific debate, and a long record of active research on and intensive concern with a variety of ecological topics. He made his marks in his phytosociological and environmental studies of epiphytic cryptogamous vegetation (Barkman 1958), terrestrial bryophyte vegetation and the fungal composition of several ecosystems; he founded a mycosociological school which enjoys international recognition; he published on the taxonomy of fungi; he decisively contributed to the formulation, establishment and acceptance of an international nomenclatural code for vegetation types; he became an authoritative microclimatologist, and above all he pushed, stimulated and modelled interest in ecological research into plant form and vegetation structure. Meanwhile he taught plant ecology and vegetation science, first at Leiden University, later simultaneously at the Universities of Wageningen and Utrecht, while at the same time he was director of the Field Station for Biological Research (of Wageningen University) at Wijster. In 1987 Jan Barkman retired from these functions and because of his association with Utrecht University since 1973, the Department of Plant Ecology organized in his honor an International Symposium on Vegetational Structure from 14 to 18 July 1987.

Jan Barkman illustrated his fascination with the many facets of vegetation structure in a personal introduction, published as the first chapter in this book. There he discusses the ecological relevance of morphological differences between two species of *Quercus* at several scales, from whole tree architecture to leaf anatomy. He also contributes his long-prepared new system of plant growth forms for North-Western Europe. This system is an elaborate hierarchical classification of all plant species (including bryophytes and macrolichens) based on the morphology of their above-ground parts. The ecological relevance of this system is shown in the contribution by V. Komárková and J.D. McKendrick.

Though he has a special interest in the nature of discontinuities, Jan Barkman's concern with plant form and vegetation structure does not primarily result from a passion for classification. It first of all stems from a fundamental interest in the evolutionary meaning of plant morphology and the ecological relevance of vegetation structure. This becomes immediately clear from the beginning pages of his well-known review on the synusial approaches to classification (Barkman 1973). There he explained that vegetation often consists of several groups of plants with a similar life or growth form and that the species in such a group often show similar reactions upon limiting environmental factors and similar patterns of habitat exploration. From this conception it is only a short way to the concept of guild which is presently so popular in community ecology. Also his review on studies of vegetation structure (Barkman 1979) instructively shows his interest in its functional relevance.

A major part of Jan Barkman's own research was devoted to a study of the ecology of juniper scrub, open scrublands with an intense pattern of patches of species of distinct growth forms correlated to a pattern of strong environmental heterogeneity, created and maintained by differences in litter fall, light climate and other microclimatological factors. Next to detailed environmental measurements sustained over long periods of time in order to closely inventory this intense pattern, he tested his ideas about the underlying causes using a variety of experiments including manipulation of litter and of the microclimate, and transplantations (*e.g.*, Barkman *et al.*, 1977). The study of this interesting ecosystem with its extremes in environment and corresponding constraints on plant performance differing over very short distances stimulated him to co-author a textbook on microclimatology. In this book a strong emphasis is put on the biological interpretation of patterns in microclimate in relation to vegetation structure (Barkman and Stoutjesdijk 1987). It is for this reason that the logo on this book depicts a juniper shrub supported by the initials of Jan J. Barkman.

The many contributions offered at the symposium in Utrecht (Woudschoten) showed that vegetation structure is conceived in a variety of meanings. Many papers addressed the implications of plant form and vegetational architecture for the functioning of individuals and populations in their environment, or dealt with the relations between structure and carbon and nutrient cycling; others discussed the horizontal patterns in plant communities and their floristic structure (composition). This allowed us to assemble the papers in three separate volumes grouped around these major themes. Accordingly, a volume edited by During *et al.* (1988) contains the papers on pattern and diversity, a volume edited by Verhoeven *et al.* (1988) those dealing with the effects of vegetation structure on production, decomposition and nutrient cycling, while the present volume focuses on the functional interpretation of plant form and the architecture of vegetation.

Many papers collected in this volume explore the plasticity in or the adaptive value of morphological aspects of growth forms and growth strategies of plant species in response to the environmental constraints of their habitat, as well as their value for the competitive vigor of the plant species in the context of the plant community. Other papers deal with the morphological consequences and ecological significance of the modes of dry matter partitioning in plants and stands of vegetation based on the view that trade-offs result in optimal allocation of matter for maximal returns through the energy capturing processes. A final set of papers investigates the consequences of herbivory on plant morphology and stand architecture under an array of environmental conditions and types of herbivores and in systems which strongly differ in the length and intensity of their grazing histories.

Together the papers show the recent insights gained in understanding the ecological significance of plant form and vegetation structure. The next step should be the rigorous application of cost-income analyses the results of which can be used to formulate quantitative predictions about plant form and the structure of the vegetation as functions of environmental variation and physiological flexibility and potential.

In presenting this volume we want to acknowledge the dedication, care and energy given by Nicolette van Splunder, Ad Vianen and George van der Vliet to the organization of the symposium. They greatly helped to make the symposium the success it was. We also acknowledge the financial support obtained from the
- College van Bestuur, Utrecht University,
- Bestuur van de Faculteit Biologie, Utrecht University,

- Ministerie van Landbouw en Visserij, Directie Natuur, Milieu en Faunabeheer,
- Koninklijke Nederlandse Akademie van Wetenschappen,
- K.L.M. Royal Dutch Airlines.

We greatly appreciate the hospitality of Mayor and Aldermen of the city of Zeist who hosted a splendid reception in Zeist Castle for all participants.

We thank dr Tom J. Givnish, dr John Grace, and dr Sam J. McNaughton, for convening the sessions in which the papers in this book were presented, and we are grateful to the publisher for the cooperation and facilities he offered during the production phase of this book and its companion volumes.

The Editors

References

Barkman, J.J. 1958. Phytosociology and ecology of cryptogamic epiphytes. Van Gorcum, Assen. 628 pp.

Barkman, J.J. 1973. Synusial approaches to classification. In: R.H. Whittaker (ed.), Ordination and classification of vegetation. Handbook of Vegetation Science 5: 437–491, Junk, The Hague.

Barkman, J.J. 1979. The investigation of vegetation texture and structure. In: M.J.A. Werger (ed.), The study of vegetation, pp. 123–160. Junk, The Hague.

Barkman, J.J., Masselink, A.K. and De Vries, B.L. 1977. Über das Mikroklima in Wacholderfluren. In: H. Dierschke (ed.), Vegetation und Klima, pp. 35–81. Cramer, Vaduz.

Barkman, J.J. and Stoutjesdijk, Ph. 1987. Mikroklimaat, vegetatie en fauna. Pudoc, Wageningen. 223 pp.

During, H.J., Werger, M.J.A. and Willems, J.H. (eds) 1988. Diversity and pattern in plant communities. SPB Academic Publishing bv, The Hague. 285 pp.

Givnish, T.J. (ed.) 1986. On the economy of plant form and function. Cambridge University Press, London. 717 pp.

Verhoeven, J.T.A., Heil, G.W. and Werger, M.J.A. (eds) 1988. Vegetation structure in relation to carbon and nutrient economy. SPB Academic Publishing bv, The Hague. 206 pp.

SOME REFLECTIONS ON PLANT ARCHITECTURE AND ITS ECOLOGICAL IMPLICATIONS
A personal view demonstrated on two species of *Quercus*

J.J. BARKMAN
Kampsweg 29, 9418 PD Wijster, The Netherlands

Abstract

The effect of plant anatomy and morphology on the life strategy and ecology of plants and on their undergrowth through modifications of microclimate, litter and soil are illustrated by the writer's observations on *Quercus robur* and *Q. rubra*, two broad-leaved deciduous climax trees of the temperate zone. They differ in bark roughness, inclination of branches and leaves, leaf size, leaf form and leaf thickness and in their reaction to sunlight (modification of sun and shade leaves, S and N in the table). Undergrowth is probably affected by differences in transmitted light, throughfall and litter density, and as a consequence by differences in moisture and oxygen content of litter and humus. Epiphytes are affected by the lack of bark fissures and the presence of a rain track on the boles of *Q. rubra*, and by a rain free underside in *Q. robur*.

Introduction

To think in terms of vegetation texture and structure requires a different approach than the floristic approach of classical plant sociology. This way of thinking has already been practised in the Soviet Union in the twenties and has been much stimulated in the past 10 to 20 years especially in the United States. Yet, I believe that there is still a wide field of knowledge to be explored. I like to illustrate this point of view with an example from my own experience.

I chose to make a comparison between *Quercus robur* L. and *Quercus rubra* L., to point out some of their differences in anatomy and architecture and to discuss the bearing of these differences upon the ecology of the two species and of the plant communities dominated by them.

Why *Quercus robur* and *rubra*? Trees in general have a major influence upon their undergrowth and their soil flora and fauna. They offer more microhabitats and niches to other organisms than shrubs or herbs. On purpose I chose two allied species in order to show how even species of one genus, occupying similar habitats in similar climates, may differ in many structural characters. *Quercus robur* and *Q. rubra* are both deciduous broad-leaved trees of climax communities on poor sandy soils in cool temperate climates of the Northern hemisphere. Although the former is a European and the latter an American species, they are now able to compete since the introduction of *Q. rubra* in Europe.

Plant morphology and autecology

One of the first differences that struck me, but which I failed to find mentioned in the literature, is that of their leaf inclination. The leaves of *Q. robur* are placed

Table 1. Growth form characteristics of Quercus robur and Quercus rubra.

	Quercus robur	Quercus rubra
Leaf inclination	spherical +90° to −90°	decumbent −10° to −50°
Absorption of solar radiation (% of maximum)	6/18 h: 62% 12 h: 62% whole day: 62%	49% 70% 56%
Leaf size (cm²) Range Average	mesophyllous 6.8−80 37	macrophyllous 34−270 125
Lobes	rounded	acuminate
Growth form of tree Branches Bark Stem flow (% of total precipitation) Rain track	Quercid horizontal rough 1.1 none	Fagid erectopatent smooth 5.3 present

spherical (all inclinations between vertical erect and vertically drooping), those of *Q. rubra* decumbent (drooping at angles between −10°C and −50°C) (Barkman 1979). In deep shade both species tend to have more horizontal leaves. This of course has consequences for the interception of sunlight. Assuming − which is probably correct − that the orientation (compass direction) of the leaves is random in both species (except at wood edges), I have calculated the amount of intercepted direct radiation as a percentage of the maximum insolation, *i.e.*, in a plane at a right angle to the sun. At the latitude of The Netherlands (52°N) unshaded leaves of *Q. robur* receive 62% of this maximum radiation throughout the day in the main growing season (May through July). For *Q. rubra* the figures are 49% at 6 and 18 h. (local natural time) and 70% at noon, with a daily average of 56% (Table 1). We may therefore conclude that the total amount of intercepted direct light is about equal, but *Q. robur* gets more light in early morning and late afternoon, while *Q. rubra* gets more in the middle of the day. Yet, *Q. robur* is probably better off, since light intensity at noon is far in excess of what is required, whereas it may be limiting early in the morning and late in the afternoon. Also stomata may, on hot days, be closed at noon.

This analysis is not complete, however. It refers only to the upper canopy. The upper (sun) leaves of *Q. robur* are thicker (av. 186μ) than those of *Q. rubra* (av. 88μ), so the latter transmit more light to lower leaves, but this light, being mainly infrared and green, cannot be used for photosynthesis. However, we have seen that *Quercus robur* also absorbs more sunlight in the early morning and late afternoon, so transmission of direct, unfiltered light by a crown of *Quercus rubra* is higher in these critical periods, which must be favorable for the lower canopy and the undergrowth. This holds true for bright days only. Diffuse light (cloudy days) is more effectively intercepted by decumbent than by spherical leaves. In cloudy weather, therefore, the upper leaves of *Quercus rubra* may have an advantage over those of *Quercus robur*, whereas the lower leaves of the canopy will receive less photosynthetic light in the former species.

Another difference is leaf size: *Q. robur* has mesophyllous leaves (leaf size one-sided averages 37 cm^2), *Q. rubra* is macrophyllous (av. 125 cm^2). Although the variation in size is enormous in both species (*cf*. Table 1), *Q. robur* has distinctly smaller leaves. As was clearly pointed out by Horn (1971), small leaves create small sun and shade spots and therefore a higher frequency of light intensity oscillations, which is favorable for photosynthesis (Sharkey *et al.* 1986). In this respect, therefore, the lower canopy of *Q. robur* may be expected to exploit sun light more effectively than *Q. rubra*.

Leaf size also affects leaf temperature and rate of transpiration. Big leaves become hotter in the sun than small leaves, which increases transpiration. However, big leaves also have a thicker boundary layer (outer diffusion layer) and a smaller edge effect, both of which reduce transpiration (Gates 1980). Which of the opposite effects outbalances the other, depends on the diffusion resistance of the leaf interior (Barkman and Stoutjesdijk 1986). As the leaves of both oak trees are moderately thin and their stomata densities high (Table 2), the interior diffusion resistance is probably low and any variation in outer resistance will therefore have a big influence on total resistance. For this reason the smaller leaves of *Q. robur* should have a higher transpiration rate, but this is probably compensated for by the temperature effect of the smaller leaves and the slightly thicker leaves of *Q. robur*. Therefore, I speculate that *Q. rubra* leaves are likely to transpire stronger in the sun than those of *Q. robur*.

In Table 2 I have presented the results of a comparative analysis of sun and shade leaves of both species, taken from the southern and the northern sides of crowns at southerly resp. northerly exposed wood edges. Obviously the shade leaves are larger and thinner in both trees, the size difference being most pronounced in *Q. rubra*, the thickness in *Q. robur*. This applies both to the thickness of the mesophyll areas (vein islets) and to that of the veins.

Dry weight is relatively higher in sun leaves of both oaks, in particular dry weight relative to leaf area (degree of sclerophylly). The sclerophylly is particularly high in sun leaves of *Q. robur* and also the difference with the shade leaves is most pronounced here. The same holds true for the degree of succulence (maximum water content in relation to leaf area). In *Q. rubra* sun leaves are characterized by much higher densities of stomata and veins.

Obviously the two oak species cope in different ways with the possibility of excessive heating and water loss in sun exposed leaves. *Q. robur* shows mainly increased thickness of its mesophyll and leaf veins and higher degrees of sclerophylly (leaf firmness) and succulence (water storage), *Q. rubra* shows strongly reduced leaf size and increased total vein length and number of stomata per cm^2. The greater vein length increases the potential rate of water supply, but this effect is counterbalanced by increased transpiration as a result of the greater stomata density. So it seems that *Q. robur* is better adapted to spells of drought than *Q. rubra*. This is in accordance with their natural habitat ranges.

In the literature few data are available on sun and shade leaves of oaks in their natural (forest) habitat. Schramm (1912, cited by Büsgen and Münch, 1927) counted 468 stomata per mm^2 in shade leaves of *Quercus petraea*, 810 per mm^2

Table 2. Leaf characteristics of *Quercus robur* and *Quercus rubra*. S = southern side of crown (sunny); N = northern side of crown (shady).

	Quercus robur		Quercus rubra	
	S	N	S	N
Leaf size (cm², one-sided)				
Range	24–48	34–61	84–115	231–269
Average	36.4	47.0	98.2	251.1
Fresh weight (mg)	864	580	1795	3421
Dry weight (mg)	340	211	744	1315
Dry weight/total weight	0.39	0.36	0.41	0.38
Dry weight/leaf area[1]	0.47	0.23	0.38	0.26
Water content/leaf area[2]	0.72	0.39	0.54	0.42
Leaf thickness	186µ	88µ	112µ	92µ
ø finest veins	35–75µ	20–50µ	15–40µ	9–40µ
ø veins 1st order	430–670µ	290–480µ	640–800µ	560–750µ
Total vein length (cm/cm²)	43.0	46.2	72.5	48.7
Number of stomata per mm²	353	328	546	250

[1] gr/dm² (2-sided) = degree of sclerophylly
[2] gr/dm² (2-sided) = degree of succulence

in sun leaves of that species. Eliáš (1978) published data on *Quercus cerris* in a dry oak-hornbeamforest in southwest Slovakia. The climate is more continental than in The Netherlands. Dry weight as a fraction of total weight was found to be 0.47 in sun leaves, 0.41 in shade leaves. Dry weights divided by leaf area were 0.66 and 0.34 respectively, degrees of succulence 0.66 and 0.49. So all parameters appeared to have higher values in sun leaves than in shade leaves, as is the case in our two oak species, but all values were higher than ours, especially the degree of sclerophylly of sun leaves. This is in accordance with the drier site and the drier climate in Slovakia. In oak trees stomata densities were measured by Eliáš only in shade leaves of *Q. cerris* (255 per mm²) and in sun leaves of *Q. petraea* (578 per mm²), values that correspond fairly well with our *Q. rubra* data. A higher number of stomata in sun leaves as against shade leaves was also found in *Carpinus betulus* (sun leaves 175 per mm², shade leaves 85 per mm²) and *Acer campestre* (388 resp. 189). This phenomenon seems to be general (*cf.* Givnish, 1987 and literature cited there). So, in this respect it seems that *Q. rubra* behaves normal, *Q. robur* does not.

Finally we have to mention the difference in leaf shape: rounded lobes in *Q. robur*, as against sharply acuminate lobes in *Q. rubra*. These somewhat resemble the drip tips of tropical rain forest trees. Should they have the same (assumed) function, it would mean that *Q. rubra* can get rid of excessive rain water more easily than *Q. robur*. Again this might suggest that *Q. rubra* is better adapted to moist climates.

Synecology

So far we have discussed the possible significance of plant morphology and anatomy for the autecology of the species. We shall now consider the possible implications for the synecology of their forest types.

Because of the smaller leaves and the spherical inclination a wood of *Q. robur* is able to develop a higher leaf area index. This means a much higher number of leaves per ground surface area. Both smaller leaves and higher LAI increase the interceptive capacity of tree crowns for rain water. This capacity is also increased by the higher number of leaves (leaf axils), particularly if the leaves are erect or erectopatent (Mitscherlich 1971) which is the case for part of the leaves of *Q. robur* (being spherical) and for none of *Q. rubra*. As a consequence interception will be higher in *Q. robur*. Moreover, rain will more easily drip off the acute leaf lobes of *Q. rubra*. Throughfall will therefore be less under *Q. robur*. In about 50 years old stands it amounts to 65% in *Q. robur*, 70% in *Q. rubra* (Brechtel in Mitscherlich 1971). These are summer values. In winter the percentages are 76 and 85 respectively. The differences in winter are probably related to crown architecture and bark relief.

A fundamental difference between these oaks is their crown architecture: *Quercus robur* belongs to the growth form which I call Quercids, *i.e.*, a type of trees with mainly horizontal branches. Rain water is neither conducted towards the stem nor to the crown periphery. *Q. rubra* belongs to the Fagids: the branches are erectopatent, the crown is of the centripetal type. Rain water is conducted towards the stem and a stem flow is developed, the more so since the smooth bark does not retain water or hamper run off, as does the very rough bark of *Q. robur* (Barkman 1958, Weihe 1968). In 50 years old stands of *Q. rubra* stem flow is 5.3% of total precipitation, in *Q. robur* only 1.1% (Brechtel l.c.). Apparently bark relief is very important, for, according to the same author, stem flow in young (17 years old) *Q. robur*, where the bark is still fairly smooth, amounts to 9.5% as against 12.6% in *Q. rubra*. The differences in epiphytic vegetation between the species are largely due to this hydrological factor, for water capacity and vapor capacity of the bark (*i.e.*, the maximum amount of water that air dry bark can absorb from a saturated atmosphere) are almost equal in *Q. robur* and *Q. rubra* (Barkman 1958).

As rain water flows around the stem, it is more or less evenly distributed in *Q. rubra*, with a maximum in a narrow strip on the underside (Fig. 1). This rain track sometimes has a special vegetation (*Protococcus* sp.) in *Q. rubra*. In *Quercus robur* a rain track is not only always lacking, but the whole underside (almost half of the circumference of the tree bole and the main branches) is dry, due to the rough bark. This dry underside is often covered with ombrophobous, leprose lichens (Caliciaceae, *Lepraria* spp.), belonging to the Calicion hyperelli (Barkman 1958).

The upper side of the stem and main branches, too, harbours different epiphytes. The smooth bark of *Q. rubra* has a scarce cover of crustaceous lichens and appressed foliaceous lichens, such as *Parmeliopsis ambigua* and *Hypogymnia physodes*. The rough bark of *Q. robur* is the favorite habitat of foliaceous and especially fruticose and semifruticose lichens, such as *Pseudevernia furfuracea*,

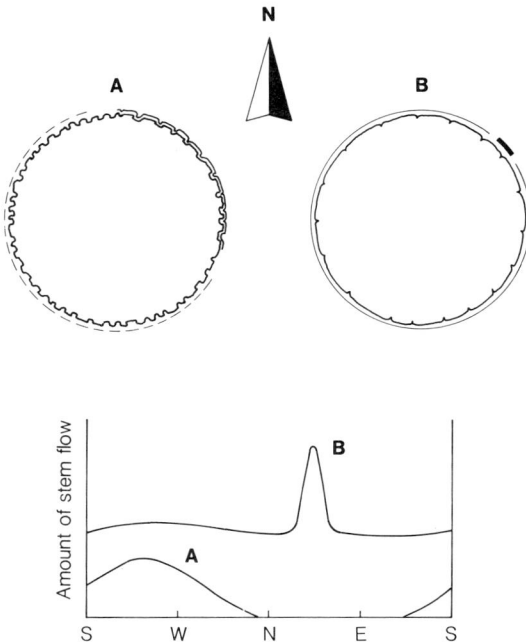

Fig. 1. Distribution of some cryptogamic epiphyte communities and variation in the amount of stem flow over trunks of *Quercus robur* (A) and *Q. rubra* (B) with regard to orientation. Key to communities in upper figures: A – dashed line, Parmelietalia physodo-tubulosae; drawn line, Calicion hyperelli. B – thin line, Lecanorion + Graphidion; thick line, Protococcetum viridis.

Evernia prunastri and *Usnea* ssp. Such lichens increase the absorptive capacity for rain water (Molchanov 1966), so that even less rain water can flow to the lower side. In this way the differences in vegetation between upper and lower side are reinforced by the vegetation itself (Barkman 1958).

The supposedly higher throughfall in *Q. rubra* woods must be favorable for the herbs and mosses, but the lower light intensity in cloudy weather and the larger sun and shade spots in sunny weather are adverse factors. The small, undulate or crisped dead leaves of *Q. robur* create a loose litter. This litter is well-aerated, but dries out quickly. It is easily blown away by the wind, which is favorable for terrestrial bryophytes. The large and heavy, flat leaves of *Q. rubra*, however, suffocate all mosses. As a matter of fact, the moss layer, poorly developed under *Q. robur*, is completely absent in most woods of *Q. rubra*. Exceptionally, however, it is well-developed, namely in small woods of *Q. rubra* on undulating ground, open to the wind. Its floristic composition is then identical with that of the Dicrano-Quercetum roboris. So, the absence of a moss layer under *Q. rubra* seems to be due to physical properties, not to chemical properties of its litter. The compact litter of *Q. rubra* also creates holes of saturated air, which is known to be favorable for many small animals, especially Isopoda. The litter will not dry out as quickly as in *Q. robur*, and this is likely to be favorable for microbial activity. For the same reason, however, inhabitants of *Q. rubra* litter might suffer occasionally from oxygen shortage.

Conclusions

In this paper a number of facts are given concerning the structure of two oak species. The consequences for their ecology and for the ecology of their associated flora and fauna are hypothesized on the basis of general theory, derived from ecophysiology and microclimatology. Some of these hypotheses have so far been confirmed by field observations. The others have yet to be checked.

Quercus robur and *Q. rubra* differ in tree architecture (inclination of the main branches), bark relief, leaf size, leaf form and thickness, vein density, stomata density and water content per leaf area. They also differ in their response to light, heat and desiccation (sun and shade leaves). This has consequences both for their autecology and for the synecology of their forest stands. The latter is affected through differences in light transmission, rain throughfall, stem flow and the physical structure of the litter. These differences affect both the epiphytes and the terrestrial bryophytes, probably also the soil inhabiting flora and fauna.

Obviously the differences in plant structure of dominant species may affect their communities profoundly and in various ways. This field of research is, in the writer's opinion, most successfully explored by synthesizing facts and theories from plant anatomy, morphology and physiology, microclimatology, vegetation science (including community structure) and animal ecology.

References

Barkman, J.J. 1958. Phytosociology and Ecology of Cryptogamic Epiphytes. Van Gorcum Publ., Assen. 627 pp.
Barkman, J.J. 1979. The investigation of vegetation texture and structure: In: M.J.A. Werger (ed.), The Study of Vegetation, pp. 125–160. Dr W. Junk Publishers, The Hague.
Barkman, J.J. and Stoutjesdijk, Ph. 1986. Microklimaat, Vegetatie en Fauna. Pudoc, Wageningen. 223 pp.
Büsgen, M. and Münch, E. 1927. Bau und Leben unserer Waldbäume. 3. Aufl. G. Fischer, Jena.
Eliáš, P. 1978. Some ecophysiological features in leaves of plants in oak-hornbeam forest. Folia Geobot. Phytotax. 13: 29–42.
Gates, D.M. 1980. Biophysical Ecology. Springer, New York. 611 pp.
Givnish, T.J. 1987. Comparative studies of leaf form: assessing the relative roles of selective pressures and phylogenetic constraints. New Phytol. 106 (suppl.): 131–160.
Horn, H.S. 1971. The Adaptive Geometry of Trees. Princeton Univ. Press, Princeton.
Mitscherlich, G. 1971. Wald, Wachstum und Umwelt, 2. Band: Waldklima und Wasserhaushalt. Sauerl. Verlag Frankfurt/Main. 365 pp.
Molchanov, A.A. 1966. The hydrological role of forests. Transl. from Russian. Israel Progr. Sci. Transl. Jerusalem.
Sharkey, T.D., Seemann, J.R., Pearcy, R.W. 1986. Contribution of metabolites of photosynthesis in postillumination CO_2 assimilation in response to lightflecks. Plant Physiol. 82: 1063–1068.
Weihe, J. 1968. Niederschlagszurückhaltung durch Wald. Allg. Forstzeitung 23: 522–525.

NEW SYSTEMS OF PLANT GROWTH FORMS AND PHENOLOGICAL PLANT TYPES

J.J. BARKMAN*
Kampsweg 29, 9418 PD Wijster, The Netherlands

Abstract

Historical surveys are given of growth forms and phenological plant types in Europe, followed by the presentation of a new system of growth forms and a new system of phenological types.

Both systems refer to terrestrial and fresh water vascular plants, bryophytes and lichens of Europe. Characters related to life strategy, life cycle, hibernation level, leaf size, leaf consistency and leaf inclination are excluded and should be studied separately. The growth form system is based on gross morphology (architecture) of the plants as it affects vegetation structure. It is free of hypotheses of environmental adaptation. New features in this sytem are: the distinction of water plants swimming 'above the surface', the distinction of Equisetoids (leaves none or reduced), the group of arcuate plants, the division of creeping plants according to the number of layers of leaves, the emphasis on leaf distribution and ramification of stems in herbs, the classification of shrubs and trees according to the inclination of branches. The system is partly mono-, partly pluridimensional.

The phenological system is two-dimensional, the main division being based on vegetative periodicity with parallel subgroups based on generative periodicity.

1. Introduction

With the increasing interest in vegetational texture and structure there is an increasing need for detailed systems of plant growth forms and phenological types. Vegetation texture can be described by elementary parameters such as height, biomass, phyllomass, leaf area index, average leaf size or average leaf inclination, but also by more differentiated features such as spectra of elementary parameters, for instance spectra of leaf size classes. Finally spectra of complex features can be made, such as growth forms, the units of which are usually based on combinations of characters (Barkman 1979). In this sequence the amount of information increases, but it also becomes more difficult and arbitrary to quantify the results and to base cluster analyses of stands and community types on these data. Also the relations between ecological factors and elementary parameters are more obvious and easy to interpret than between the former and growth forms. Yet, an analysis of the features alone that make up a growth form, gives insufficient insight into plant strategies, since these features interact and it is their interaction that determines a plant's chances of survival. Also it is the combination of such features as degree of ramification, length of internodes, branch angles and leaf distribution, that determine plant architecture and consequently vegetational structure. It is therefore recommended to study both the simple and the more complex parameters of vegetational texture and structure.

* Comm. no. 343 of the Biological Station, Wijster

Plant form and vegetation structure, pp. 9–44
edited by M.J.A. Werger, P.J.M. van der Aart, H.J. During and J.T.A. Verhoeven
©1988 SPB Academic Publishing, The Hague, The Netherlands

The complex characters can be divided into functional and morphological ones, although I am quite aware of the fact that most functional characters have a morphological basis and many morphological characters have an ecological function. As functional I consider such features as type of photosynthesis (C3, C4 or CAM), mode of pollination, mode of dissemination, periodicity, mode of hibernation (life forms of Raunkiaer), adaptation to lack and excess of water (hydrotypes of Iversen; also the *Atmophyta, Amphiphyta,* and *Ombrophyta* (Barkman 1958) belong here). Growth forms, however, belong to the morphological characters.

Ten years ago I have made unpublished systems of growth forms and of phenological plant types. At that time I was not familiar with existing growth form systems except that of Braun-Blanquet and the systems published for bryophytes, lichens and algae. The proposed systems were based on 50 years of field experience throughout Europe. The two systems mainly pertain to the Netherlands and were submitted to a number of Dutch colleagues for criticism. I want to thank those that reacted, in particular drs. S. Hennekens, drs. P. van der Knaap, drs. J. Schaminee and dr. S. Segal, who gave very thorough and detailed comments. Some of my students and my colleague prof. M.J.A. Werger tested the systems in the field and gave some valuable suggestions.

I tried the systems myself in the Netherlands, the Austrian Alps and the arctic tundra of Spitsbergen (Barkman 1987) as well as, incidentally, in the tropical mountain flora of Mexico (Popocatepetl), the Sonoran desert of Arizona, the chaparral of South California and the pine barrens of New Jersey. Except in the Sonoran desert where many new growth forms of succulents would have to be added, the systems worked well everywhere with the addition of only very few extra growth form types.

The comments of colleagues and students and my own testing in the field resulted in the systems presented here. They are not meant to be worldwide systems, since I have no knowledge of tropical vegetation nor of Asia and the Southern hemisphere. The systems deal only with macroscopic plants, excluding macrofungi and sea kelps. Epiphytes and epiliths are also disregarded. Consequently they are meant in the first place to be systems for macroscopic terrestrial and fresh water vegetation of the Holarctic, mainly Europe.

Nomenclature of vascular plants follows (for particular reasons of systematic opinion) mainly Heukels-Van Ooststroom 1977, Flora van Nederland, 19th ed., of bryophytes mainly Margadant – During, 1982, Beknopte Flora van Nederlandse Blad- en Levermossen, of lichens Wirth, 1980, Flechtenflora.

2. Growth forms and life forms

2.1. Introduction

Growth forms and life forms are often either confounded or considered synonyms in literature. Even Braun-Blanquet wrote in 1964: '. . . growth form and life form are almost identical concepts'. Gimingham and Robertson (1950) stressed the need for distinction with regard to bryophytes, but they only published a growth form system. Barkman (1958) elaborated both life forms and growth forms for epiphytic bryophytes and lichens and in 1979 he stressed the need for the same distinction in vascular plants.

Life forms are considered types of plants having the same kind of morphological and/or physiological adaptation to a certain ecological factor. In literature they are often confined to Iversen's hydrotypes and Raunkiaer's life forms or even to the latter only, but there is actually no reason not to extend the concept, for instance, to dissemination and pollination types, K- and r-strategists and to the ruderals, stress tolerators and competitors of Grime.

Growth forms are types of plants with the same gross morphology (architecture). The concept is therefore free of any hypotheses about adaptation. This does not mean that the parameters of architecture have no ecological meaning. Recent investigations (for instance Givnish 1987) have shown more and more morphological characters to be of adaptive value that were not considered as such before (although in some cases this adaptive value is, in my opinion, still rather hypothetical). Even if a morphological character should be of little ecological importance to the species in question, it can have a great impact on the plant community, for instance the inclination of branches in trees, resulting in centrifugal and centripetal crown types with different distribution patterns of precipitation on the forest floor. However, I agree with Du Rietz (1931) that growth forms should not be based on (established or supposed) adaptations to the environment. What we need is characters that describe best the architecture of plants and communities and if they should have an adaptive value, this must and certainly will be proved.

2.2. Historical development of the growth form concept

Since growth forms and life forms have so often been mixed up, combined or treated as synonyms, it is not possible to sketch the history of growth forms without discussing life forms. The first system of growth forms was by Theophrastos of Eresos (371–286 B.C.) who distinguished trees, shrubs, halfshrubs and herbs. Alexander von Humboldt (1806) distinguished 16 (later 19) 'Hauptformen' (main forms), named after representative plants, such as the cactus form, the palm form, the banana form. These were pure growth forms, too. Grisebach (1872) extended the system to 54 (later 60) 'Vegatationsformen' (vegetation forms), grouped into 7 main types, viz. woody plants (30 types), succulents (3 types), climbers (3 types), epiphytes (2 types), forbs (8 types), grasses (5 types) and 'cell plants' (*i.e.*, cryptogams, 2 types): bryophytes and terrestrial lichens. The trees were subdivided according to leaf form only. Notice the neglect of branching architecture, of water plants, of growth forms within bryophytes and lichens.

Drude (1886) stressed the importance of ecological adaptations, particularly to climatic factors. He criticized vigorously the systems by Von Humboldt and Grisebach as being mixtures of 'morphological' forms (*i.e.*, based on constitutional characters) and 'biological' forms (*i.e.*, based on adaptation characters). Only the latter should be used. He distinguished 55 'physiognomic life forms' (later he called them growth forms), divided into three classes, (1) aërophytes (terrestrial vascular plants), (2) hydrophytes (vascular water plants) and (3) 'cell plants' (mosses and thallophytes, also divided into aërophytes and hydrophytes). His divisions are partly taxonomic (monocots, dicots and Gymnosperms were strictly separated), contrary to his own principle. In 1913 he admitted this inconsistency, but maintained it. Trees were subdivided into evergreen and deciduous and further according to leaf size and leaf consistency. He was the first to combine Poaceae and Cyperaceae into graminoids which he subdivided according to growth in tussocks or mats and presence of stolons. Leafless Juncaceae and Cyperaceae were distinguished as a separated form (*cf.* our Equisetoids). Forbs were mainly classified according to: (1) climbing habit, (2) duration of life, (3) stem basis woody, (4) basal rosettes, (5) creeping or with runners, (6) cushion growth, (7) morphology of rhizomes.

A functional classification was published by Warming (1884) based on (1) lifetime of (a) the whole plant, (b) tillers, (2) number of flowering periods (hapaxanths and pollacanths), (3) vegetative expansion power (limited or unlimited growth of rhizomes, runners and tillers), (4) degree and mode of branching of rhizomes, (5) mode of hibernation. As characters like form, position and way of branching of stems, leaf-stem ratio, leaf distribution along the stem and stem consistency (woody or herbaceous) were disregarded, many groups were heterogeneous from an architectural point of view, including for instance (1) *Taraxacum* and *Chelidonium* in one group, (2) *Bryonia* and *Phyteuma* ditto, (3) *Comarum* and *Sedum*, even (4) *Pragmites, Lathyrus pratensis, Asperula odorata* and *Urtica*, and (5) *Potamogeton* and *Juncus*. His 1908 classification took more morphological features into account, resulting in such groups as creeping plants, rosette plants, cushion plants, succulents, half shrubs, dwarf shrubs, shrubs and trees. Yet the system was inconsistent, including pure growth forms, life forms (Heterophytes, *i.e.*, heterotrophic plants; hapaxanths) and ecological groups (water plants). In his Danish edition (1908) he called them life forms (livsformer), in his English edition (1909) growth forms. Some groups were small, others large. For instance muscoïd plants, lichenoïd plants and trees were not further subdivided.

In the meantime Raunkiaer's life form systems had been published (Raunkiaer 1907). In the first version the height above the ground of hibernating buds was so dominant a criterium that the group of Phanerophytes (trees) also included epiphytic vascular plants, growing high in the crowns, and the group of Cryptophytes included both geophytes and all water plants. In 1908 the life forms were subdivided into growth forms, for instance Chamaephytes into (1) half shrubs (erect), (2) 'passive' chamaephytes (first erect, then bending to the ground by their own weight), (3) 'active' chamaephytes (*i.e.*, with prostrate stems), (4) cushion plants. An important distinction was made within the Hemicryptophytes. According to leaf distribution three groups were distinguished, viz. with (1) only stem leaves, (2) both stem and basal rosette leaves ('half rosette plants'), (3) only basal rosettes. All three had parallel subgroups with and without runners (stolons), but for some obscure reason only the rosette plants were first subdivided into sympodial and monopodial plants.

Warming and Raunkiaer were both convinced neolamarckians, considering life forms a direct product of the environment. Yet they were convinced that the most conspicuous physiognomic plant forms are based on 'indifferent' (we should now say: non-adaptive or selection-neutral) characters. They both therefore admit that their life forms, being based on 'epharmonic' (adaptive) characters, are not suitable for describing the vegetation morphology.

In 1914 Warming revised his system, distinguishing between life form and 'Grundform' (basal form). In 1918 he distinguished between solitary and tussock forming plants, and between aboveground leafless runners (stolons), underground suckers (so-called 'suboles'), rhizomes, mesocorms (short erect rhizomes), rhizodes (intermediate between suboles and rhizomes), and aboveground and underground horizontal tillers bearing normal leaves. His 1919 classification of herbs was entirely focussed on these organs. He adopted Raunkiaer's three types of stem leaf, half rosette and rosette plants, but severely criticized him for only paying attention to climatic adaptations. Life forms should also mirror adaptations to soil conditions. In his last system (1923) he also distinguished such 'life forms' as rock plants and halophytes. He was the first to see the difference in growth form between *Calluna* (erect) and *Empetrum* (ascending).

Gams (1918) introduced the valuable and fundamental distinction between attached (adnate), rooting (radicant) and free floating or swimming (errant) plants, also called ephaptomenon, rhizumenon and planomenon. For the rest he followed Raunkiaer but was 'plus royaliste que le roi' using his criterion of the hibernation level in the extreme: therophytes and geophytes were therefore united, mosses and lichens incorporated in the Chamaephytes, etc. Raunkiaer's leaf distribution types were named Thyrsophylla, Basithyrsophylla and Basiphylla respectively. His subdivision of the Chamaephyta is very interesting, viz. according to water uptake: (a) mainly from the air (Bryochamaephytes), (b) from air and soil (cushion plants), (c) mainly from the soil (Euchamaephytes).

Clements (1920) distinguished between growth forms (*i.e.*, our life forms) and 'vegetation forms' (*i.e.*, our growth forms). The latter were defined as those types that determine the vegetation structure, but his main types were not in accordance with this definition, namely annuals, biannuals, perennial herbs and perennial woody plants. The first two were not further subdivided, the last two included the traditional forms of Drude and Warming.

The most extreme reaction to the epharmonic systems of Warming and Raunkiaer was the purely physiognomic system (pure growth forms) by Du Rietz (1921), although he named his types life forms. In fact he rejected all strict life form systems, since he considered the adaptive nature of plant characters to be purely hypothetical. His names are rather complicated, for instance aciculimagnolignides (coniferous trees!) and his classes are rather large. They also include ecological groups (terriherbosa vs. aquiherbosa) and taxonomic groups, viz. the groups lichens, algae and fungi. Similar impossible names were created by Rübel (1930), such as Belonidomikrophanerophytes, Makrostelechohemikryptophytes and supracruste sempervirentiherbides.

Braun-Blanquet (1928) followed Raunkiaer's system, but elaborated it in much more detail. For instance Hemicryptophytes were divided in thallose and rooting and the latter in caespitosa, rosulata, scaposa and scandentia. Rooting Chamaephytes were divided in reptantia, succulenta, pulvinata, graminoidea, velantia ('Spaliersträucher') and suffrutescentia. In this way a mixed system was created with life forms as the main classes and growth forms as subordinate units. No wonder Braun-Blanquet had to create many parallel subgroups.

Skottsberg (1926), working mainly in Antarctica and the Pacific, was of the opinion that everybody should make his own system adapted to his study area. Cockayne (1919), working in New Zealand, analyzed each character separately in addition to general (habit) growth forms.

In 1931 Du Rietz, too, stressed the difficulty of giving priority to certain characters when classifying growth forms. He therefore proposed to use several growth form systems. He again called them life forms, but his definition is extremely vague now: 'a life form is a general term for any class of plants, based on other points of view than idiotaxonomy'. Which would mean that plants with yellow flowers, poisonous plants, garden plants and edible fungi are life form classes! In fact Du Rietz only mentioned as examples pollination types, dissemination types, ecological groups, physiological types and 'life forms that affect the physiognomy of the vegetation' (*cf.* Clements 1920). These were classified into six types:
1. Main life forms, based on general physiognomy of the plant when fully developed in high season.
2. Growth forms, based on the architecture of stems and branches (*cf.* Clements' vegetation forms).
3. Periodicity life forms, based on vegetative phenology.
4. Bud height life forms (Raunkiaer's life forms).
5. Bud type life forms (naked or variously protected buds).
6. Leaf life forms (based on form, size, lifetime, consistency etc. of the leaves).

He added that a seventh system, based on root morphology would be highly desirable but not yet possible because of insufficient data*. Main life forms (1) were: woody plants, half woody plants and herbs, the last-named divided into epiphytes, stem parasites, lianas and the rest, which was subdivided into four height classes. Growth forms were based upon Warming's (1918) classification of subterranean parts. Unfortunately Du Rietz introduced a lot of new terms, such as subolegeocorms, leptoorthogeocorms etc. and he used Warming's terms rhizome and mesocorm in a different sense which is highly confusing. His division of dwarf shrubs is more detailed than any other before and very interesting. He distinguished between (1) sedentary and (2) travelling dwarf shrubs, *i.e.*, without respectively with creeping stems or stolons. The sedentary are divided into erect and ascending. He was the first to notice that *Empetrum nigrum* can behave as a 'sublianoid', climbing tree stumps and young trees or hanging down from rock faces. His tree classification is also a real improvement. The main division is into rosette trees (many palms), leafless succulents (tree cactusses) and normal trees (incorrectly called 'long-shoot trees', because many of them have both long and short shoots). This category is subdivided into (α) sedentary (no runners), (β) stoloniferous (rooting stolons, such as *Picea abies*), (γ) roottravelling (vegetative expansion by suckers, such as *Populus tremula*). His interesting 1931 publication deals in detail with woody plants only, saying 'to be continued', but a continuation was never published.

Gimingham (1951) published a system of nine growth forms for herbs in Scot-

---

* Russian investigators have always paid much attention to underground morphology. Vuesotzsky (1915, fide Fekete and Szujko-Lacza 1970) was the first to establish a life-form system based partly on subterranean plant organs and he was followed by Kazakevich (1922) and Salit's extensive studies on grass root systems (Salit 1950, 1952). From 1950 onwards attention in the Soviet Union was focussed, however, on growth forms of aerial parts (school of Serebriakov c.s.).

tish sea dunes, in which for the first time attention was paid to branching of stems. Horikawa and Miyawaki (1954), studying weeds in Japan, adopted his system but added a parallel growth form system of five underground growth forms (root types).

Of quite a different nature was the system by Dansereau (1951; Dansereau and Arros 1959). He distinguished only six main types and called them life forms, but in reality they were mainly growth forms: epiphytes, trees, shrubs, lianas, herbs, cryptogams (mosses and lichens). Each main type was divided according to four independent criteria: (1) 'function', *i.e.*, periodicity of green organs, (2) leaf form and size, (3) leaf consistency, (4) size of plant (three height classes for each 'life form'). Obviously the individual combinations cannot be used as growth forms, their theoretical number being 1728, their actual number probably at least 500.

Schmid (1956) studied phenotypic growth forms of the adult plant. He demanded that growth forms be based on vegetative characters that have an immediate ecological relevance, but his primary division coincides with large taxonomical groups such as hepatics, mosses, conifers, monocots and dicots. He divided plants according to the degree and extent of lignification into axylic, oligoxylic (slightly lignified giant herbs), hemixylic (soft, light wood), holoxylic, perxylic (extremely hard wood) and meroxylic (partly lignified halfshrubs). His second subdivision is based on the relations between seasonal (deciduous) and perennial axes, his third subdivision on the length of internodia, the type of ramification (acrotonous, mesotonous, basitonous and hypogaeic) and the degree of ramification (simple: ramose or multiple: ramulose; as well as: unicaul, pluricaul and multicaul, according to the number of main stems, sprouting from the base). He does not mention specific growth forms, so in fact his system is multidimensional or factorial, like Dansereau's.

Poplavskaja (1948) was the first author to give a system of growth forms of water plants. She distinguished three groups, the hydatophytes (whole life cycle completely submerged), 'submerged' aerohydatophytes (inflorescences above water) and 'floating' aerohydatophytes (ditto, but vegetative parts swimming at the surface). Hejny (1957, 1960) united the first two groups into the euhydatophytes, called the third group hydatoaerophytes and added a heterogeneous group of amphibious plants, the tenagophytes.

Luther (1949), dealing with waterplants, adopted Gams' tripartition, renamed them haptophytes, rhizophytes and planophytes and divided the latter into pleustophytes (large floating plants) and planktophytes (microscopic floating plants). The pleustophytes were subdivided according to their growth level (bottom, water surface or in between), into bentho-, meso- and acropleustophytes, disregarding the fact that *Stratiotes* commutes between bottom and surface and that there is an important difference between the swimming *Lemna* and *Hydrocharis* and the 'Hoover crafts' *Salvinia* and *Azolla* hovering over the surface.

Den Hartog (1955, 1959) elaborated a classification for haptophytic (epilithic) water algae, Den Hartog and Segal (1964) for the non-haptophytic aquatic cormophytes. They distinguished 11 growth forms, six rhizophytic and five pleustophytic. In addition to criteria used for land plants (leaf distribution along the stems, stolons) they used four criteria specific to aquatic plants: (1) leaf dimorphy, (2) leaf form (finely dissected or undivided), (3) leaf position (submerged, floating or emergent), and (4) position of flowers (above or under water). Segal (1965) followed this system, adding one growth form (Trapids).

Van der Maarel (1966) distinguished eight growth forms among herbaceous dune plants in the Netherlands, mostly adopted from Raunkiaer and Braun-Blanquet. New were the forms Caespitosa compacta and Tapeta.

Schmithüsen (1961) distinguished 30 growth form classes, based mainly on Raunkiaer and Braun-Blanquet. An important refinement was the subdivision of the Hemicryptophyta scandentia in (1) with winding stems, (2) with tendrils, (3) scrambling-ascending plants. The Chamaephyta were not divided into pure growth forms like the other life forms of Raunkiaer, but partly into ecological types, such as hygromorphous forbs, mesomorphous forbs, xeromorphous tussock grasses. The Geophytes were subdivided according to the nature of the storage organs (root gemmae, rhizomes, bulbs, and tubers). The system is therefore somewhat inconsistent.

Mühlberg (1967) treated growth forms of grasses in Germany and distinguished between (1) dense tussocks, (2) loose tussocks, (3) a transitional type, (4) stoloniferous grasses, (5) rhizomatous grasses and (6) annual grasses. They correspond more or less to our Graminoidea caespitosa (1), G. hemicaespitosa (2) and (3), G. reptantia and stolonifera (4), G. rhizomatosa (5) and G. solitaria (6).

Segal (1969), in his thesis on wall vegetation, stressed the fact that the more detailed the growth form subdivisions of Raunkiaer's life forms become, the more parallel groups must be created. It is therefore better, he argues, to keep Raunkiaer's system pure and simple, limiting it to life forms and to establish a separate growth form system. Segal's system of wall inhabiting herbs comprises eight growth forms which, contrary to all previous authors except Von Humboldt, were named after representative genera. His primary criterium is 'sociability' by which he means number of tillers per plant. Other criteria are: direction of stems (erect, prostrate, ascending), presence of basal rosettes and presence of stolons.

Horn (1971), referring to the foliage of the crowns, distinguished between 'multilayer' and 'monolayer' trees, which he related to pioneer and climax stages respectively.

Londo (1971), who studied pioneer vegetation in wet dune valleys, devoted a critical discussion to life forms in which he made valuable general remarks that are also applicable to growth forms. In his opinion the simultaneous use of many criteria creates an unnecessarily complicated system. He prefers to have a few parallel systems each with a limited number of parameters. He also stressed the difference between systems that are directly applicable in the field and those that are not. The former are based on characters visible in any season, for instance trees, shrubs, herbs, the latter on characters often only visible in the unfavorable season, like Raunkiaer's life forms. As a matter of fact, one should say that observation in two seasons is needed, because geophytes and summer therophytes are not visible in winter. Londo preferred the former system but inconsistently included hapaxanths, which are not recognizable in the field as such. He also distinguished between phenotypic and genotypic systems. One species can belong to different life forms in different habitats. According to Londo *Calamagrostis epigeios* is a geophyte when occurring in open pioneer vegetation, a hemicryptophyte in dense populations. We may add that the same applies to growth forms (non-climbing and climbing forms of *Hedera helix, Lonicera periclymenum* and *Solanum dulcamara*) and also to stages of development: a one year old *Digitalis* is a rosette plant, a beech seedling has exactly the same growth form (and almost the same form!) as the orchid *Listera cordata*, a fern prothallium as a thallose liverwort. Genotypic life forms are defined as the potential life form, but Londo does not define what he means by potential: under optimal conditions (and what is optimal?) or: under all conditions (*i.e.*, the sum of all its potencies). In the latter case one still has a difficulty in classifying variable species and one should create a separate category for *Calamagrostis epigeios* next to geophytes and hemicryptophytes. Londo preferred genotypical systems, except in very local situations, because one avoids the difficulty, he said, of having to place one species in different categories (but *cf.* our remark on *Calamagrostis*), of determining its life form over and over again in each new situation, and of restricted possibilities of extrapolation to other regions. But he admitted that for the description of vegetation structure phenotypical systems are better. After all he decided not to use a life form system but a (genotypical) growth form system although in my opinion some of his types are life forms. As he worked in a highly dynamic pioneer situation, it is logical that he used dynamic characters, especially the life span and the capacity of occupying bare ground. Revising the system of Warming (1884, 1914) he distinguished (1) therophytes, (2) hapaxanths, (3) 'solitary' species (no stolons, no tussocks), (4) dense tussocks, (5) short suckers or rhizomes, vegetative expansion up to 5–10 cm per year, (6) long suckers, expansion faster, (7) and (8) as (5) and (6) but with aboveground stolons or runners. He also distinguished five height classes.

Müller-Dombois and Ellenberg (1974) defined plant life forms as 'a growth form which displays an obvious relationship to important environmental factors'. They discussed life forms only, but many subgroups are in fact pure growth forms, for instance scapose (single stemmed) and caespitose (multistemmed) herbs, tuft trees and bottle trees. The first separation level is, as it should be when dealing with life forms, into autotrophic, semiautotrophic and heterotrophic plants, the second is – curiously enough – into 'cormophytes' and 'thallophytes', but what they meant is: vascular and non-vascular plants. As this corresponds roughly with homoiohydric and poikilohydric plants which is an important life strategy character (but some desert *Selaginella*'s are poikilohydric), the division is defensible, although the argument is not used. The third division level is in (1) self-supporting plants, (2) plants supported by others (such as lianas and epiphytes), and (3) 'plants whose photosynthetic organs are structurally supported by water in aquatic habitats' (p. 144). This seems a very logical subdivision and much better than the pure habitat classification in land and water plants. Unfortunately in the appendix A where the system is elaborated the third group is restricted to errant water plants, so that rooting water plants that cannot support themselves find no place in the system. They are wrongly included in different groups of the 'self-supporting' plants. This group is divided according to Raunkiaer and these life forms are subdivided multidimensionally, for instance phanerophytes (1) as to habit in: normal woody plants, tuft trees, bottle trees, tall succulents and herbaceous phanerophytes, (2) in trees, shrubs and creeping trees (Krummholz), (3) in three to four height classes, (4) in evergreen and deciduous, (5) according to the crown shape, (6) according to the relative length (height) of the crown, (7) according to leaf size, (8) according to leaf shape, (9) according to root morphology (as far as visible aboveground!), (10) according to thickness and relief of the bark, (11) according to presence/absence of thorns and spines, (12) according to position of inflorescences. Each combination is indicated by a number with digits for each parameter. This may be useful for a computer but a reader can hardly figure out which growth form is meant by 1.122.216.435.021.2. In my opinion it is probably best to consider all characters separately, also because the theoretical number of growth forms of only 'normal trees and shrubs' is already 29 million and their actual number in nature at least 10.000. Within the Chamaephytes and Hemicryptophytes attention is paid to woody/herbaceous/succulent; caespitose/reptant/ pulvinate/scapose (*i.e.*, solitary)/'aquatic'; evergreen/cold-deciduous/drought-deciduous; rosette/semi-rosette/without rosette. Aquatic hemicryptophytes are divided into caespitose (*Isoetes*), reptant (*Pilularia*) and scapose (*Lobelia*) forms. New distinctions are for instance the aphyllous plants (in different groups), the partly woody chamaephytes, the globose and flat cushion-plants, the stem-, leaf- and root-succulents, the division of therophytes in spring-green, summer-green, rain-green and winter-green annuals. The greatest value of the system is the great number of textural characters used and the world wide field knowledge of the authors emanating from the text.

Descoings (1975) classified tropical grasses in about the same way as Mühlberg (1967), but he subdivided each type into basiphyllous (rosulate) and cauliphyllous (scapose) and the latter into those with branched and unbranched culms.

Parsons (1976), dealing with shrubs in mediterranean climates (Chile, California), followed an original method. He distinguished seven growth form characters and investigated their degree of association in various species. The cluster analysis yielded nine clusters (growth forms) of shrubs, namely: (1) narrow-leaved, (2) sclerophyllous broad-leaved evergreen, (3) mesomorphic-leaved, drought-deciduous, (4) spiny, drought-deciduous, green-stemmed, (5) winter-deciduous, (6) leafless, green-stemmed, (7) pungent-leaved, mesomorphic, (8) weedy-habit shrubs of various forms, (9) succulent rosettes with spines. Even in my limited experience it is easy to find examples of 1, 2, 3, 4, 6 and 7 in the mediterranean part of southern France.

In 1978 Hallé, Oldeman and Tomlinson published a monumental work on the growth forms of tropical trees, which they called 'architectural models'. It is by far the most detailed work on tree growth forms ever published and should interest us, because growth forms of temperate trees can only be understood from studies in the tropics where trees evolved and where their form diversity is much greater. The system comprises 23 models which are seen as stepping stones in an 'architectural continuum'. Nomenclature of the types is after well-known botanists and therefore does not convey any message whatsoever about the nature of the model involved. When comparing the system to ours, however, we must stress that the architectural models are not synonymous with growth forms in our sense. To quote the authors, 'architecture is not to be confused with shape or physiognomy and cannot be equated with growth habit'. The architectural model is defined as a growth program determining the successive architectural phases. And each phase is the visible, morphological expression of the genetic blueprint of a tree at any one time. The system is therefore a genotypical and a dynamic one. It is based on fundamental morphological characters and it takes into account the whole life cycle of trees and not the adult form only. According to the authors the criteria of their system are: (1) life-span or meristems, (2) sexual vs. vegetative differentiation, (3) plagiotropy vs. orthotropy, (4) rhythmic (episodic) vs. continuous growth, resulting in different leaf arrangements: in the former case short internodes with reduced leaves alternate with long internodes with large leaves, in the latter case equal sized leaves are evenly distributed along the shoots, (5) chronology of branch development. If we examine their identification key it becomes apparent that the number of elementary features used is greater: (1) trunk simple/branched, (2) inflorescence terminal/lateral, (3) growth continuous/rhythmic, (4) vegetative axes equivalent/trunks and branches different, (5) trunk ramifies underground or near the ground/aboveground, (6) branching dichotomous/monopodial/sympodial monochasial/sympodial di- or trichasial, (7) branches orthotropic/plagiotropic/mixed, (8) axes erect/secondarily bending/primarily bending.

Among herbs in old field succession Schiefer (1981) distinguished (in addition to the three leaf distribution types) caespitose, reptant, scandent, stoloniferous, rhizomatose, bulbose and root gemmae types of herbs and found interesting correlations between growth forms and successional behavior of plants. In an interesting publication on the texture of dry grasslands and rock vegetation in the Swiss Mittelland Styner and Hegg (1984) distinguished 21 growth forms of herbaceous plants, based on: length of internodia, branching angle, consistency of axes, leaf arrangement and leaf form, and contour of the whole plant. Many of the forms correspond with my growth forms presented here. Feoli and Scimone (1984) used both growth forms and life forms as well as a number of single textural characters in a correlation analysis and a vegetation gradient analysis. Seven growth forms were distinguished, based on Parsons (1976) and Barkman (1979).

2.3. The proposed growth form system

2.3.1. Introduction and general remarks

This system is based mainly on direct observations in nature and refers to the actual (phenotypic) form of adult plants with emphasis on aboveground parts. Often plants have different forms than those mentioned in floras.

The system is not taxonomical: algae and lichens or lichens and hepatics, even bryophytes and vascular plants are sometimes united in one growth form (*cf.* growth forms 4, 8, 9, 18, 19).

The system is not purely morphological (in the classical sense): cladodia, phyllocladia and phyllodia are treated as leaves; plants with very small leaves are treated as leafless. Petiole and rhachis of *Pteridium* are treated as stems. Among water plants Rhizophytes that root very loosely in the sapropelium with hardly functional roots or float freely during part of the year, have been considered planophytes (errant plants). This is more in accordance with the reality of vegetation structure and plant physiognomy.

The system is not physiological, it is not a life form system: stem parasites with haustoria are treated as adnate plants. No distinction is made between auto- and heterotrophic, between annual and perennial, between hapaxanths and pollacanths, between the life forms of Raunkiaer. These classifications, all of them very valuable, should be used next to my system. The difference between evergreen and seasonally green is a matter of periodicity and therefore elaborated in my phenological system. But it does affect physiognomy and structure in the unfavorable season, so I have indicated the evergreens, if present, among the examples of every growth form, by printing them in italics.

The system is not ecological. The reader will not find types like halophytes, rock plants or even water plants. Only those water plants have been treated separately that have special morphological characters: namely the free swimming or floating (no roots) and those that are so flaccid (no support tissues) that they collapse outside the water. Therefore the Isoetids have been included in the Primulids, a growth form which mainly comprises land plants.

The system is a compromise between a practical and a logical system. A purely practical system would not be logical, a strictly logical system not practical. I have tried to delimit growth forms that are not too difficult to recognise in the field and that are of interest for the vegetation structure as a whole. The system is therefore not always consistent: in some growth form groups some characters are more important for subdivisions than in other groups. I am aware of not always having been successful. The difference between solitary plants and those with long rhizomes is important for the species in question, but it is not visible without digging in the ground and for vegetation structure it does not matter whether plants form dense colonies through sprouting rhizomes or by mass germination. So one might unite growth forms 61 and 62, 65 and 66, 70 and 71, 73 and 74. As a rule, however, these growth forms differ in their development: rhizomatous plants start as single sprouts and develop into colonies by vegetative expansion, whereas mass germination of solitary plants creates colonies to begin with, and in time these colonies tend to diminish their number of individuals by self thinning.

The time aspect creates a tricky problem. On the one hand we cannot consider a tree as a herb growth form, just because its seedlings are small and herbaceous, on the other hand, if we consider all juvenile trees in an abandoned grassland as tree growth forms, the resulting spectrum would suggest a woodland instead of a grassland. Just like the floristic analysis of vegetation the structural analysis should be carried out in the season of optimum development, which may differ from species to species and therefore may imply the necessity of two or more analyses in one year. This holds true particularly for plant communities with pronounced seasonal aspects like Carpinion forests. Even so the difficulty remains that juvenile trees and shrubs do not reach the adult stage in one year and in some communities they die young and do not reach it at all (trees above the timber line, pine trees in raised bogs, *Sambucus nigra* in juniper scrub on poor sandy soil). For an adequate description of the actual vegetation there is no other possibility than to describe the plants in their status in high season in the year of observation. For biannual rosette herbs this means that a steady state population always consists of a mixture of Primulids (1st year rosettes) and Verbascids (2nd year plants) or even Epipactids (if the rosette leaves have withered at the time of flowering).

One of my practical considerations was that the system should give a survey and that neither a very small nor a very large number of growth forms would be practical. For a similar practical reason no growth forms have been established for a single species or small genus, unless this form was fundamentally different.

I have tried to solve the dilemma of the uni- vs. multidimensional systems as follows. Life forms, periodicity types, leaf size classes, leaf consistency classes and leaf inclination classes have been disregarded. In my opinion they should be studied separately and the same applies to leaf hairiness, spines, thorns and aromatic substances. Nevertheless, it was sometimes difficult to choose between primary and secondary characters. So for graminoids, erect forbs and trees more than one system, in fact a more-dimensional system is proposed. For those who do not want to go into too much detail, larger groups than the individual growth forms can be used. They have been mentioned in the text with descriptive (English or Latin) names.

The growth forms, however, are consistently named after representative genera, or if impossible to find, after representative species (for instance the Junceffusids after *Juncus effusus*). Of course there is the possibility that in some area unknown to me, there exist species of a certain name giving genus that belong to a different growth form. Even now I am aware of the fact that a genus is not always ideally chosen: not all *Puccinellia*'s belong to the Puccinelliids, not all *Polygonatums* to the Polygonatids, not all *Batrachiums* to the Batrachiids, but in these cases I could not find a better name. And it would have been impossible to find short adequate descriptive terms for all growth forms mentioned here. I certainly did not want to create such long names as designated by Du Rietz and Rübel.

The species and genera mentioned under the heading of the growth forms, are only examples. Their enumeration is not complete! For nomenclature of species see p. 10.

The main differences with existing systems are the following:
1. For water plants mainly the system of Segal (1965) was followed, but not all water plants are treated as such (Isoetids).
2. A distinction is made between real swimming water plants and those that hover just over the surface or stick well out of it, not only because the water surface is a very crucial limit between habitats and an important habitat itself, but also because these plants are different morphologically and because the physiognomy of the vegetation is strongly affected by it.
3. Hydrocharids and Lemnids have been united, and so have Myriophyllids and Charids, and also Potamids and Vallisneriids.
4. For cryptogamic errant water plants new (sub)types have been added (Cladopodiellids and Enteromorphids).
5. Cryptogamic growth forms mainly follow Barkman (1958), but new are Gracilids, Cladinids, Thamnolids, Archidiids, Rhodobryids and Plagiotheciids.
6. Rosette forming lichens and hepatics have been united.
7. For stem parasites two growth forms have been established (Cuscutids and Viscids).
8. The group of aphyllous herbs and those with reduced leaves (Equisetoids) is worked out in more detail than in literature.
9. Graminoids are classified both according to leaf distribution and to growth habit, as well as to height classes.

10. The creeping herbs are divided in two groups (Illecebrids and Thymids).
11. A new group of Arcuatae is distinguished. This interesting group of plants was discussed by Givnish (1986b) as the 'arching growth form' or group of 'leaners'.
12. Within the forbs attention is paid to ramification of stems (growth forms 44 and 45, 66 and 67, 68 and 69, 71 and 72).
13. The dwarf shrubs have been divided into three different growth forms.
14. Shrubs and trees are primarily classified according to the spatial position of the main branches. Crown form and density are secondary.

Since our system of tree forms is so completely different from other systems, including the most recent and detailed one by Hallé *et al.* (1978), some explanation is needed. In the first place the latter although very detailed, is not sufficiently detailed for Europe where trees mostly belong to only two types (Rauh and Troll). More important is, that the system is not very practical in the field because one has to know the whole life cycle of a tree. The authors admit: 'Recognition of the architectural model of a tree is often difficult' (p. 75). They consider the life-span of meristems 'the single most important functional character which determines models'. It determines whether vegetative growth is sympodial or monopodial, but they also state that 'a linear sympodium may be physiognomically indistinguishable from a monopodium'. Apparently the fundamental morphology is more important to them than the visual result. It is therefore illogical from their point of view to disregard the nature of inflorescences with the argument 'whether it may be described as a spike, umbel, panicle, cincinnus, thyrse etc. is of no architectural consequence'. If, however, only vegetative morphology is considered, it is hard to understand why they have also ignored characters like long shoots vs. short shoots, leavy and leafless axes, axis modifications such as tendrils, etc., 'none of which directly determine architecture'. I think they do affect both architecture and physiognomy. Fischer (1986) pointed out that different architectural models can produce the same overall foliage presentation, for instance tiered crowns or pagoda trees, whereas one model can produce quite different crown forms.

Looking at each of the eight characters used for the key produced by the authors, it appears that (1) and (4) do not play a role in Europe and probably the Holarctis since all of our trees are branched and all have branches morphologically different from the trunks. As to (6), the dichotomous and the di- and trichasial ramifications do not occur in these regions whereas the other two types can give rise to exactly the same growth form, their difference often being completely obscured by secondary thickening growth. I have considered the possibility that the difference between monopodial and sympodial growth might have an ecological meaning: one might suppose that trees whose lateral branches always take over the direction and function of the main stem, can more easily regenerate from lateral buds after cutting (decapitation of the main stem) than monopodial trees, but the opposite seems the case. The sympodial *Fagus sylvatica* (Troll's model) cannot stand continued coppicing, whereas the monopodial *Quercus robur* (Rauh's model) can stand it excellently!

The characters (2) and (3) do not influence tree growth form. As to (5) underground ramification does not affect aboveground growth form, but may be considered an important additional criterium (*Populus tremula!*, Du Rietz's root-travelling trees). If branching is at soil level, I consider the plants shrubs and treat them in separate growth forms. As to character (8), this is important in temperate shrubs, but in my system only in so far as the adult form has arched branches, not if primarily bent branches become erect. I have distinguished this type as Sambucids and Rosids among the shrubs. Among trees it is included in the growth form of the Tiliids.

That leaves us character (7), which could be used as a subdivision of my growth forms. However, strictly (primarily) plagiotropic (*i.e.*, distichous) branches are rare and among the few European examples are the trees *Fagus* and *Ulmus* and the shrub *Corylus*. Secondarily plagiotropic branches (by diaphototropic bending of petioles and twigs) are common in lower, shaded branches of most trees. They are more pronounced in some (*e.g.*, *Robinia*, *Picea*) than in others (*e.g.*, *Quercus robur*, *Larix*), but there are all possible intergradations between plagiotropic and orthotropic even in one tree individual and we cannot possibly base a non-arbitrary classification upon this feature.

Short shoots occur in Europe in *Fagus, Betula, Pinus*, and *Larix*, but only in the latter two it affects the growth habit, leading to a leaf arrangement in bundles. Similarly we can distinguish short shoot shrubs with leaves in bundles: *Prunus spinosa, Berberis, Crataegus, Ribes uva-crispa* (all of them thorny shrubs!).

Tree physiognomy is determined mainly by height, crown form, crown density, branching, deciduousness and leaf size. Since I decided to treat deciduousness and leaf size of all plants in separate systems for the very reason that they are so important, I have not used them here. Height of adult trees is not so much varying from species to species in Europe as is the case in the tropics or in the case of native European grasses, so height classes would not give much information. In every structural vegetation analysis of forests the exact height of all tree species should, for that matter, be indicated anyhow. Crown form is used here but not as a primary criterium because it is somewhat variable and different for isolated trees and forest trees. Height of crown base is even worse a criterium: it largely depends on how far up the forester has lopped the tree trunk. Way of branching is rather typical for species and not much variable. It is an important character of physiognomy in winter and essential for distribution of rain water (pattern of throughfall) throughout the year.

I do hope this system will be tried and I hope to get criticism from my colleagues abroad in order to improve it. To facilitate its use, I here give some larger growth form groups with the numbers of the pertinent individual growth forms.

Water plants 1–7, 34–37, 45 p.p., 46 p.p., 62 p.p.
Woody plants 41, 42, 75–88.
Prostrate plants 16, 17, 28, 29, 32, 43, 47, 56, 75.
Decumbent plants 26, 30, 31, 57, 76, 78.
Carpets (Tapeta): extensive low carpets with a closed, flat canopy 5, 6, 7, 36, 37 p.p., 56 p. min. p., 57 p. max. p., 63 p. min. p., 61 p. max. p., 65 p. max. p., 70 p.p., 73, 75 p.p., 76 p. max. p.
Stoloniferous plants 48, 59, 63, 73, 78 p.p., 84 p.p. (*Picea abies*).
Rhizomatous plants (long rhizomes, mats) 49–51, 61, 65, 70, 73, 81 p.p. (*Prunus spinosa, P. avium, Hippophaë*), 87 p.p. (*Populus tremula*).

Tufted plants (short rhizomes, caespitose) 46, 52, 53, 55, 60, 64, 68, 69.
Isolated plants 3, 4, 5, 6, 33, 45, 54, 62, 66, 67, 71, 72, 74, 84–88.
Unbranched plants 13, 17, 20, 21, 22, 24, 25, 45, 46, 58 p.p., 59 p.p., 60 p.p., 61 p.p., 62, 68, 70, 71, 73, 74.
Strongly branched plants 14, 15, 16, 23, 26–31, 33, 44, 67, 69, 72, 76–88.
Rosette plant (rosulata) 5b, 7, 35b, B2a, 59–62.
Semirosette plants (scaposorosulata) B2b, 63–67.
Stem leaf plants (scaposa) 4, 6, 33, 34, 35a, 36, 38–45, B2c, 68–72, 75–88.
Top rosette plants (epirosulata) 25, B2d, 73, 74. This type is also widespread among tropical trees (tree ferns, palms, cycads).

2.3.2. The system
This is a system of growth forms of fresh water and terrestrial, non-epiphytic macrophytes of Europe, with some examples from other regions. Names of evergreen plants have been printed in italics.
   I. Errant plants (planophytes)
  II. Adnate plants (haptophytes)
 III. Radicant plants (rhizophytes)

I. ERRANT PLANTS

A. PLANKTOPHYTES.
This growth form, comprising air plankton, water plankton and edaphon ('soil plankton'), will not be discussed here, being microscopical.
B. PLEUSTOPHYTES.
Large, swimming and floating water plants.
  1. **Chaetophorids.**
     Spherical algal colonies, floating or sometimes loosely resting on the bottom.
     Chaetophora elegans.
  2. **Enteromorphids.**
     Floating or swimming, filamentous and tubular algae without stems and leaves.
     Enteromorpha, Cladophora, Vaucheria p.p.
  3. **Ricciellids.**
     Submerged, floating, small flat plants without leaves with short or no stems, but with cladodia or thalli.
     Ricciella, Lemna trisulca.
  4. **Ceratophyllids.**
     Large submerged floating plants with leaves and long stems. May be divided into two subgroups:
     4a. **Ceratophyllids s.s.**
         Finely dissected leaves. Fructification in the hydrophase.
         Ceratophyllum, Utricularia.
     4b. **Cladopodiellids.**
         Small simple leaves. Fructification only when more or less desiccated.
         Cladopodiella fluitans p.p., *Sphagnum cuspidatum* p.p., *Sphagnum crassicladum var. obesum*, Drepanocladus fluitans p.p., *Leptodictyum*

riparium p.p., *Fontinalis antipyretica* p.p.
5. **Lemnids.**
Hydrophytes swimming on the very water surface, mostly without air cavities (except Trapa, Lemna gibba and Ricciocarpus). May be divided in two subgroups:
  5a. **Lemnids s.s.**
Plants small, without leaves. One cladodium or thallus per plant. Lemna minor, L. gibba, Spirodela, Wolffia, *Ricciocarpus*.
  5b. **Hydrocharids.**
Plants large, with several genuine leaves in a rosette.
Hydrocharis, Trapa.
6. **Salviniids.**
Swimming, small hydrophytes with air cavities, without rosettes. Leaves horizontal, hovering just over the water surface.
Azolla, Salvinia.
7. **Stratiotids.**
Swimming, large hydrophytes with air cavities and rosettes of emergent, patent leaves.
Stratiotes, Pistia, Eichhornia.

II. ADNATE PLANTS

8. **Leprariids.**
Leprose lichens and unicellular algae, both forming a thin, undifferentiated, pulverulent, water repellent cover.
Lepraria, Crocynia, Protococcus.
9. **Palmogloeids.**
Gelatinous algae and lichens, embedded in mucilage, glued to the substratum, much shrinking when dry.
Palmogloea, Nostoc, Collema.
10. **Prasiolids.**
Non-gelatinous, filamentous algae.
Zygogonium, Vaucheria p.p., *Rhizoclonium, Prasiola, Hormidium, Stichococcus.*
11. **Bacidiids.**
Non-leprose crustaceous lichens.
Lecidea granulosa, L. uliginosa, Diploschistes, Bacidia, Fulgensia, Toninia, Psora decipiens, Baeomyces, Pycnothelia papillaria, Stereocaulon condensatum, Squamarina, Psoroma, Ochrolechia p.p.
12. **Psorids** (named after its most typical epiphytic representative Psora ostreata).
Ground thallus well developed, foliose, consisting of small scales.
Cladonia subg. Cenomyce p. max. p.
May be divided into three subgroups:
  12a. Without or with strongly reduced podetia.
Cladonia strepsilis, C. foliacea, C. incrassata, C. caespiticia.
  12b. Podetia well developed, with cups.
Cladonia verticillata, C. pityrea, C. pyxidata, C. chlorophaea, C. fimbriata, C. coccifera, C. pleurota, C. squamosa p.p. *C. crispata* p.p.
  12c. Podetia well developed, without cups, cylindrical, unbranched.

C. floerkeana p.p., *C. macilenta, C. bacillaris, C. glauca, C. cornuta, C. coniocraea*.

13. **Gracilids**.
Long, erect, not or sparingly branched podetia with little or no ground thallus.
Cladonia gracilis. C. subulata, C. crispata p.p.

14. **Cladinids**.
Erect, stongly branched, fruticose lichens without ground thallus, when adult. Branches cylindrical.
Cladonia furcata, C. rangiformis, C. zopfii, C. uncialis, Cladonia subg. *Cladina, Cornicularia, Stereocaulon paschale*.

15. **Cetrariids**.
Ditto, but branches flattened.
Cetraria islandica, C. nivalis.

16. **Usneids**.
Prostrate, strongly branched fruticose lichens with cylindrical branches.
Usnea articulata (terrestrial form), *Alectoria* p.p. (terrestrial species and terrestrial forms of epiphytic species).

17. **Thamnolids**.
Ditto, but unbranched.
Thamnolia.

18. **Ricciids**.
Small foliose thalli (lichens and hepatics) with narrow lobes.
Riccia, *Blasia, Riccardia, Targionia*, Anthoceros, Phaeoceros, *Hypogymnia*.

19. **Peltigerids**.
Large foliose thalli (lichens and hepatics) with broad lobes.
Nephroma, Peltigera, Platismatia, Marchantia, Conocephalum, Lunularia, Preissia, Pellia.

20. **Archidiids**.
Mostly annual, very short (1–3 mm), acrocarpous bryophytes, often growing isolated or in loose tufts.
Archidium, Nanomitrium, Pleuridium, Pseudephemerum, Ephemerum, Haplomitrium.

21. **Bryids**.
Low (3–20 mm), but dense turf of perennial mosses with erect, unbranched or slightly and sympodially branched stems. *Bryum, Pohlia, Orthodontium, Aulacomnium androgynum, Mnium hornum, Fissidens, Ceratodon, Ditrichum, Dicranella, Campylopus, Orthodicranum, Rhacomitrium* p.p., *Funaria, Pottia, Tortula, Barbula, Didymodon, Oligotrichum, Pogonatum, Atrichum, Polytrichum juniperinum* and *piliferum*, etc.

22. **Polytrichids**.
Ditto, but tall turfs (3–10 (–50) cm).
Polytrichum commune, P. formosum, P. longisetum, P. alpestre, Dicranum, Aulacomnium palustre, A. turgidum, Bartramia, Philonotis fontana, Plagiomnium cinclidioides.

23. **Sphagnids**.
Tall turfs of erect, strongly and monopodially branched stems. Branches in fascicles, many short branches at the stem top.

Sphagnum, except the floating forms and *Sphagnum pylaei*.
24. **Leucobryids**.
Unbranched, erect mosses, forming dense semiglobose cushions.
Leucobryum.
Note: the growth form of small cushions occurs only among epiphytes and epilithes.
25. **Rhodobryids**.
Erect unbranced mosses with leaves in an apical rosette.
Rhodobryum, (Rhizomnium punctatum).
26. **Anomodontids**.
Monopodially branched, decumbent mosses with a creeping stem and erect branches.
Anomodon, Calliergon stramineum, Rhytidiadelphus squarrosus, sometimes *Brachythecium albicans, Campylium stellatum, Tomenthypnum, Drepanocladus uncinatus* p.p., *Orthocaulis, Odontoschisma denudatum*.
27. **Climaciids**.
Erect, dendroid or arched stems, mainly branched and leafy near the top.
Climacium, Porotrichum, Isothecium, Mnium undulatum, Scleropodium caespitosum, Cirriphyllium crassinervium.
28. **Amblystegiids**.
Prostrate, not or monopodially branched bryophytes, forming low, compact mats. Leafy stems not complanate.
Lepidozia, Kurzia, Cephaloziella, Gymnocolea, Cephalozia, Cladopodiella p.p., *Leskea, Amblystegium, Brachythecium velutinum, Hygrohypnum, Campylium protensum, Rhynchostegium, Rhynchostegiella, Isopterygium elegans, Hypnum cupressiforme* s.s. (*cf*. 30).
29. **Plagiotheciids**.
Ditto, but leafy stems complanate (plagiotropic).
Calypogeia, Isopaches, Lophozia s.s., *Mylia, Nardia, Plectocolea, Lophocolea* p.p., *Diplophyllum, Scapania curta* and *compacta, Odontoschisma, Plagiomnium, Plagiothecium*.
30. **Pleuroziids**.
Monopodially branched bryophytes with ascending stems, forming loose wefts. Branching irregular. May be subdivided in two types:
30a. **Pleuroziids s.s.**.
Leafy stems not complanate.
Ptilidium ciliare, Hypnum lacunosum, jutlandicum and *imponens, Brachythecium rutabulum* p.p., *B. salebrosum, Camptothecium, Drepanocladus, Scorpidium, Calliergon giganteum* and *cordifolium, Pleurozium, Rhytidium, Loeskeobryum, Rhytidiadelphus loreus* and *triquetrus*.
30b. **Bazzaniids**.
Leafy stems complanate.
Bazzania, Barbilophozia, Lophocolea cuspidata and *bidentata* p.p., *Chilocyphus polyanthus, Plagiochila, Scapania nemorosa, Frullania tamarisci*.

31. **Thuidiids.**
As 30a, but stems regularly pinnate or bipinnate.
Trichocolea, Cirriphyllium piliferum, Eurhynchium praelongum and *striatum, Pseudoscleropodium, Entodon, Cratoneurum, Ptilium, Ctenidium, Abietinella, Thuidium, Hylocomium splendens.*
32. **Cuscutids.**
Rootless and leafless stem holoparasites, with haustoria and threadlike stems. Cuscuta.
33. **Viscids.**
Rootless stem hemiparasites with haustoria and branched, leafy stems. *Viscum*. (A parallel growth form with reduced leaves is found in North America: *Phoradendron*).

III. RADICANT PLANTS.

A. NON SELF SUPPORTING PLANTS.
A1. Plants supported by water.
Rooting water plants, that collapse outside the water.

34. **Myriophyllids.**
Only submerged leaves, all along the stem, linear or finely dissected into linear, filiform segments.
Myriophyllum, Ranunculus fluitans, circinatus and *trichophyllus*, Hottonia p.p., *Chara*.
35. **Potamids.**
Only submerged leaves, undivided, flat, often ribbonlike. This form might be divided in two growth forms:
35a. **Potamids s.s..**
Leaves all along the stem.
Potamogeton p. max. p., *Elodea, Hydrilla, Callitriche autumnalis, Ruppia, Zannichellia,* Najas.
35b. **Vallisneriids.**
Leaves in a basal rosette.
Vallisneria, Zostera.
36. **Nupharids.**
Both submerged and swimming leaves present, usually different in shape or size. This form may be divided in two:
36a. **Nupharids s.s..**
Submerged leaves simple, flat.
Nuphar, *Potamogeton natans, polygonifolius* and *nodosus, Callitriche* p.p., *Luronium natans* p.p., Sagittaria p.p.
36b. **Batrachiids.**
Submerged leaves finally dissected, leaves strongly dimorphous.
Ranunculus (Batrachium) aquatilis and *baudotii.*
37. **Nymphaeids.**
Only swimming leaves.
Nymphaea, Nymphaeoides, *Aponogeton, Ranunculus hederaceus, Sparganium minimum* and *angustifolium*, swimming forms of *Polygonum amphi-*

bium, Glyceria fluitans and Agrostis stolonifera. In dense populations in shallow water and between reed Nymphaea develops emergent leaves (cf. growth form 7).

A2. Scandentia. Plants supported by other plants.
A2.1 Climbing herbs.

38. **Convolvulids.**
Winding herbs.
Calystegia sepium, Convolvulus arvensis p.p., Humulus, Tamus, Solanum dulcamara, Polygonum convolvulus and dumetorum.
39. **Bryonids.**
Tendril climbers.
Bryonia, Lathyrus, Vicia, *Corydalis claviculata*.
40. **Aparinids.**
Spread-climbers ('Spreizklimmer') with slack, scrambling-ascending stems, spreading over and between other plants like a veil, often sticking to them by spines, barbs or stiff leaves.
Asparagus officinalis, Cucubalus baccifer, Fumaria officinalis, *Rubia peregrina*, Galium aparine, G. cruciata, G. mollugo, Brachypodium ramosum p.p., Stellaria nemorum p.p., Lamium maculatum p.p., Pteridium aquilinum p.p. The last-named may reach a height of 1–2 m as a free individual, of 4 m as a climber. Lamium maculatum may climb up shrubs up to 1.7 m.

A2.2 Woody climbers and lianas.

41. **Smilacids.**
Winding or twining woody plants.
Smilax, Lonicera periclymenum, *L. etrusca, L. implexa*, Clematis vitalba, Vitis, Parthenocissus.
42. **Hederids.**
Root climbers. Stems climbing with adhesive roots.
Hedera. A parallel case to that of the herbaceous Aparinids is that of the woody spread-climbers (f.i. *Asparagus acutifolius*). The dwarf shrub *Empetrum nigrum* sometimes behaves as such, reaching its maximum height when climbing juniper shrubs, namely 1.80 m. It might be worth while to create a separate growth form for these woody spread-climbers.

B. SELF SUPPORTING PLANTS.
B1. Equisetoids.
Herbaceous plants with strongly reduced leaves. Photosynthesis none or by stems.

43. **Equiscirpoids.**
Prostrate or ascending herbs.
Equisetum scirpoides, E. variegatum p.p., E. arvense var. riparium, Salicornia prostrata.
44. **Equisylvatids.**
Erect, strongly branched herbs.
Equisetum sylvaticum, arvense, pratense, palustre, telmateia, Salicornia p. max. p.

45. **Eleocharids.**
Erect, unbranched herbs. Solitary or growing in mats.
Equisetum hyemale and *fluviatile*, *Eleocharis palustris, Scirpus lacustris, Juncus filiformis, J. balticus, J. trifidus*, Salicornia stricta p.p.
Here also belong the heterotrophic, chlorophylless root parasites and holosaprophytes: Orobanche, Neottia, Coralliorhiza, Limodorum, Lathraea, Monotropa.

46. **Junceffusids.**
Erect, unbranched herbs, growing in tufts.
Eleocharis multicaulis, Scirpus caespitosus, Juncus effusus, subuliflorus and *inflexus*.
Note. Parallel forms with reduced leaves and green stems occur among woody plants, for instance in the genus *Ephedra* (dwarf shrub). *Ephedra distachya* and *Arthrocnemum* are leafless parallel forms to the Ericids, *E. fragilis* to the Empetrids. *E. fragilis ssp. campylopoda* is a spread-climber just as *Empetrum nigrum* sometimes is.

B2. Graminoids.
Herbs. Stems not branched or only at the very base, usually with many vegetative shoots with long, narrow leaves, often with intercalary growth. Two divisions are given here, (1) according to leaf distribution, (2) according to stem and rhizome characters. First system:

B2.a. Graminoidea rosulata.
Leaves basal, stems (culms) leafless.
Juncus squarrosus, Carex pulicaris, Nardus.

B2.b. Graminoidea scaposorosulata.
Many large basal leaves. Stem leaves present. Most Poaceae and Cyperaceae belong here. Carex pendula, Molinia, *Milium*, Melica, etc. etc.

B2.c. Graminoidea scaposa.
Only stem leaves, evenly distributed.
Phragmites, *Phalaris*, Calamagrostis, *Glyceria maxima*, Poa nemoralis, *Agrostis stolonifera, Anthoxanthum*, Carex brizoides.

B2.d. Graminoidea epirosulata.
Leaves in a rosette at the stem top.
Cyperus.

Second classification.
B2.1. Graminoidea reptantia.

All stems prostrate. One growth form only.
47. **Puccinelliids.**
Puccinellia rupestris, P. phryganodes, Poa annua p.p. (in strongly trampled places), *Agrostis stolonifera* p.p., Phragmites australis p. min. p. (as a pioneer on bare, fairly dry soil).

B2.2. Graminoidea stolonifera.
With stolons or runners. In smooth mats.

48. Hierochloids.
10–40 cm high Graminoids.
Hierochloe, Agrostis canina var. turfosa, *A. stolonifera, Glyceria fluitans, Juncus bulbosus var. fluitans.*

B2.3. Graminoidea rhizomatosa.
With long rhizomes. In smooth mats.

49. Trisetids.
Low graminoids ('Untergräser'), 10–50 cm high.
Agrostis tenuis, *A. vinealis, Anthoxanthum odoratum,* Briza, *Carex arenaria, dioica, distans, extensa, hostiana, limosa,* nigra, ovalis, *panicea, pulicaris, trinervis,* Cynodon, Cynosurus, Danthonia, *Eleocharis acicularis, Eriophorum angustifolium, gracile* and *scheuchzeri, Lolium perenne, Poa pratensis,* Rhynchospora, Pilularia, *Triglochin,* Trisetum.

50. Arrhenaterids.
Tall graminoids ('Obergräser'), 50–200 (–500) cm high.

50a. Arrhenaterids s.s..
50–100 cm high.
Agrostis gigantea, Alopecurus pratensis, *Ammophila,* Arrhenaterum, *Carex acuta,* acutiformis, *aquatilis,* disticha, *lasiocarpa, riparia,* vesicaria, *Elytrigia repens,* Festuca pratensis, Helictotrichon, *Holcus, Juncus subnodulosus, Phleum pratense.*

50b. Phragmitids.
1–2 m or even taller.
Bromus ramosus, Calamagrostis, *Cladium mariscus, Elymus arenarius, Festuca arundinacea, F. gigantea, Glyceria maxima, Phalaris,* Phragmites, Arundo, Typha, *Scirpus maritimus.*

B2.4. Graminoidea hemicaespitosa.
Both short and long rhizomes present. Therefore growing in uneven, lumpy mats.

51. Deschampsiids.
Deschampsia flexuosa, Festuca rubra, Koeleria, Brachypodium pinnatum.

B2.5. Graminoidea caespitosa.
With short rhizomes only. Stems many, close together. In mostly isolated tussocks.

52. Nardids.
Vegetative part 5–40 cm tall.
Corynephorus, Festuca tenuifolia, *Juncus squarrosus, Carex demissa, C. serotina, Schoenus, Eriophorum vaginatum,* Nardus, Narthecium, *Carex sylvatica.*
In detailed studies a distinction can be made between low (5–20 cm) Nardids (*cf.* the first five examples given here) and medium high Nardids (20–40 cm *cf.* the last five examples). I have called them previously Corynephorids and Schoenids respectively.

53. Moliniids.
Vegetative part 40–120 cm high.

Carex caespitosa, hudsonii, paniculata, pendula, *pseudocyperus, Dactylis glomerata*, Deschampsia caespitosa, *Luzula sylvatica*, Molinia.

B2.6. Graminoidea solitaria.
No rhizomes. Stems solitary. Most annual graminoids belong here.

54. Airids.
Aira, *Apera*, Avenua fatua, Bromus p. max. p., Catapodium, Cyperus fuscus, Echinochloa, Hordeum p.p., *Isolepis setacea*, Juncus bufonius, Parapholis, Phleum arenarium, Vulpia.

B3. Herbae.
Real herbs (forbs). Leaves well developed, with apical growth. No culms.

B3.1. Pulvinatae.
Stems very short, contracted into compact cushions. Although the basal architecture of this group is not fundamentally different from the caespitose herbs, the physiognomy of the typical alpine, arctic and desert cushion plants and the structure of their plant communities is so special, that it seems worth while to place them in a separate group. A subdivision does not seem adequate, although some are parallel forms to the Arabids (*e.g.*, Saxifraga caespitosa), others to the Linariids (*e.g.*, Silene acaulis), again others are epirosulate (Androsace p.p.). Owing to the compact structure, however, this does not show in the general appearance.

55. Androsaceids.
Androsace div. spp., *Cerastium uniflorum, C. regelii, Diapensia, Eritrichium nanum, Minuartia sedoides, Petrocallis, Rhizobotrya, Sempervivum* p.p., *Saxifraga* div. spp., *Silene acaulis* and *exscapa*.

B3.2. Decumbentes.
Stems long, prostrate or decumbent.

56. Illecebrids.
Stems prostrate, leaves in one plane.
Anagallis tenella, Calystegia soldanella, Convolvulus arvensis p.p., Coronopus squamatus, Corrigiola, *Glechoma, Hedera* (creeping form), *Herniaria, Hydrocotyle, Hypericum humifusum*, Illecebrum, *Linnaea, Lysimachia nemorum* and *nummularia, Montia fontana*, Peplis portula, Polygonum aviculare (on open ground), *Sagina procumbens, Rubus fruticosus* p.p., *Trifolium repens*, Wahlenbergia.

57. Thymids.
Stems decumbent, leaves in several planes.
Chrysosplenium oppositifolium, Honckenya peploides, Lycopodium clavatum, L. inundatum, Lonicera periclymenum (non climbing form), *Polygala serpyllifolia*, Polygonum persicaria p.p., *Saxifraga oppositifolia, Stellaria holostea, Thymus serpyllum, Veronica filiformis*, hederaefolia and *officinalis, Vinca*.
Note. Intermediate forms between 56 and 57 can be found.

B3.3. Arcuatae.
Main stem (or aboveground axis) arching.

58. Polygonatids.
Polygonatum (excl. P. verticillatum), Uvularia, Smilacina, Streptopus, Gentiana asclepiadea, Pteridium aquilinum, Rubus idaeus, Polygonum cuspidatum and sachalinense.

B3.4. Erectae.
Stems, if present, erect. Like the graminoids the erect forbs can be divided into Rosulatae (59–62), Scaposorosulatae (63–67), Scaposae (68–72) and Epirosulatae (73–74), but also into Stoloniferae (59, 63), Caespitosae (60, 64, 68, 69), Rhizomatosae (61, 65, 70, 73) and Solitariae (62, 66, 67, 71, 72, 74). The two systems are more or less equivalent, but – contrary to the graminoids – we have chosen here the former system as the primary division: whilst among graminoids the difference between rosulatae and scaposorosulatae is not very clear, this difference is more clearcut among forbs. Also the former quadripartition of the graminoids is unbalanced, since the great majority belong to only one growth form (scaposorosulatae). A drawback of the second system among forbs is the practical difficulty to decide in the field whether plants are isolated or connected by rhizomes. In the following enumeration they have always been placed next to each other (*e.g.*, 61 and 62), so that they can easily be combined, if desired.

B3.4.1. Erectae rosulatae.
Basal leaves only.

59. Fragariids.
With runners (stolons).
Fragaria, Hieracium subg. Pilosella, Potentilla anserina, Tussilago, *Viola odorata.*

60. Bellids.
With short branched rhizomes or radical gemmae. Basal rosettes crowded, caespitose.
Bellis, Goodyera, Hypochaeris p.p., *Leontodon autumnalis.*

61. Oxalids.
With long rhizomes, not caespitose, often forming mats. No rosettes, only basal leaves. Botrychium, Calla, Convallaria, Marsilia, *Menyanthes, Oxalis*, Ophioglossum, Petasites, *Polypodium*, Thelypteris (excl. T. limbosperma), *Viola palustris.*

62. Primulids.
Plants solitary, stems single or absent.
Allium ursinum, Arnoseris, *Asplenium*, Athyrium, *Blechnum, Ceterach*, Cystopteris, Drosera, Dryopteris s.s., Erophila, Isoetes, *Littorella, Lobelia*, Narcissus, Parnassia, *Phyllites*, Pinguicula, *Plantago* p.p., Platanthera, *Polystichum*, Primula, *Pyrola*, Taraxacum, Teesdalia, Thelypteris limbosperma.

B3.4.2. Erectae scaposorosulatae.
Both stem leaves and a differentiated basal rosette present.

63. **Ajugids.**
    With runners (stolons).
    Ajuga, Antennaria, Chrysosplenium alternifolium, Galeobdolon, Potentilla anglica, reptans and *verna*, Ranunculus repens, Saxifraga flagellaris.
64. **Arabids.**
    Basal rosettes crowded, caespitose.
    Arabis hirsuta, Carlina vulgaris, Echium vulgare, Sanguisorba minor, Stachys officinalis.
65. **Aegopodiids.**
    With long rhizomes.
    Adoxa, Aegopodium, Campanula trachelium, Cirsium arvense, *Potentilla palustris, Pulmonaria.*
66. **Verbascids.**
    Plants solitary. Leafy stem part not or sparingly branched.
    Arnica, *Capsella, Cardamine,* Cirsium palustre, *Digitalis* p.p., *Lychnis, Oenothera,* Orchis, Pedicularis sylvatica, *Rumex* p. max. p., *Samolus,* Silene otites, *Verbascum.*
67. **Ranunculids.**
    Plants solitary. Leafy stems strongly branched.
    Aconitum, Actaea, Angelica, *Anthriscus,* Aquilegia, Caltha, Campanula rapunculus, Crepis, Delphinium, Geranium div. spp., Heracleum, *Hieracium lachenalii, Lapsana,* Meum, Mycelis, Pedicularis palustris, Ranunculus subg. Ranunculus, Sisymbrium, Stellaria nemorum p.p.

B3.4.3. Erectae scaposae.
Only stem leaves present (at least basal leaves not differentiated).
68. **Linariids.**
    Many stems from a caespitose base, not or little branched.
    Agrimonia eupatoria, Centaurea serotina, *Hypericum pulchrum,* Linaria vulgaris, Lythrum salicaria.
69. **Minuartiids.**
    Ditto, but stems strongly branched.
    Minuartia verna, Campanula rotundifolia, *Jasione montana* p.p.
70. **Asperulids.**
    With long rhizomes.
    Asperula odorata, Cornus suecica, Polygonatum verticillatum.
71. **Epipactids.**
    Solitary. Leafy stems not or sparingly branched.
    Chenopodium album, Chamaerium, Epipactis, Euphrasia p.p., Saponaria, Fritillaria, *Hippuris, Lysimachia thyrsiflora,* Ophrys, Rhinanthus p.p., Urtica, Polygonum aviculare (in tall hayfields), Veratrum, Lilium martagon.
72. **Gypsophilids.**
    Solitary. Leafy stems strongly branched.
    Arenaria, Cakile, Centaurea p.p., Cichorium, Corydalis lutea, *Dianthus* p.p., Euphrasia p.p., Galeopsis, *Gypsophila, Hieracium umbellatum,* Melilotus, *Odontites,* Radiola, Rhinanthus p.p., Salsola, Scrophularia, Viola arvensis, *canina* and tricolor.

B3.4.4. Erectae epirosulatae.
Leaves only in a rosette at the top of the vegetative stem.

73. **Trientalids.**
   With long rhizomes. Plants forming mats.
   Anemone, *Mercurialis perennis*, Paris, *Pachysandra*, Podophyllum peltatum, Trientalis.
74. **Eranthids.**
   Solitary, sprouting from a short rhizome, bulb or tuber.
   Colchicum, Eranthis, Listera.

B4. Dwarf shrubs.
Many-stemmed, woody plants, less than 1 m high. This height limit is of course arbitrary, but we want to keep up the distinction between shrubs and dwarf shrubs, not only because the terms are so familiar and widely used, but also because the distinction has proved to be practical.

75. **Loiseleureids.**
   Creeping dwarf shrubs with prostrate stems, less than 5 (10) cm high.
   Loiseleurea, Oxycoccus palustris and *microcarpus, Cassiope hypnoides*, Salix retusa, herbacea and polaris and Arctous alpina.
76. **Empetrids.**
   Ascending dwarf shrubs with decumbent stems, usually 10–30 cm high.
   Empetrum (but *cf.* 42), *Calluna* p.p., *Arctostaphylos, Oxycoccus macrocarpus, Gaultheria, Dryas, Juniperus nana*, Salix repens p.p., S. reticulata, Ononis repens, Genista pilosa.
77. **Myrtillids.**
   Erect dwarf shrubs, 10–50 (–100) cm high.
   Calluna p.p., *Erica* p. max. p., *Ledum, Andromeda, Phyllodoce, Cassiope tetragona, Vaccinium vitis-idaea, Thymus vulgaris, Rosmarinus, Ruscus, Vaccinium myrtillus* (evergreen, but deciduous), V. uliginosum, Betula nana, Genista anglica, G. tinctoria, Ononis spinosa, Rosa pimpinellifolia, Daphne, *Suaeda fruticosa*.
   Note. Extreme forms of *Calluna vulgaris fo. decumbens* belong to the Loiseleureids, less extreme forms represent a dwarf shrub type, analogous to the Rosids (80).

B5. Shrubs.
Many-stemmed woody plants, 1–5 (–10) m high.

78. **Mugids.**
   Creeping shrubs with decumbent stems. *Pinus mugo ssp. mugo* ('Legföhre'), *Juniperus communis* type III (prostrate form = var. intermedia), *Quercus coccifera*, Alnus viridis, also the creeping oaks of Wistman's Wood (Dartmoor, Devonshire) and the dwarf beeches at the forest limit in the Vosges.
79. **Rosids.**
   Scrambling-trailing shrubs with arching stems, often covering other plants and rooting at the nodes.
   Rosa p. max. p., *Rubus fruticosus s.l.* p. max. p.

80. **Sambucids.**
Erect shrubs with arching branches, not rooting at the nodes.
Lycium (evergreen, but deciduous), *Sambucus nigra, S. racemosa, Berberis, Crataegus, Prunus padus, Salix cinerea* p.p., *S. aurita* p.p., *S. caprea*.
81. **Cornids.**
Erect shrubs with straight, divergent, erectopatent branches.
Ligustrum, Euonymus (deciduous), *Ulex europaeus* (deciduous), *Cornus sanguinea, Cornus mas* (deciduous), Rhamnus, Frangula, Prunus spinosa, P. avium, Hippophae, Salix p. max. p., Viburnum, Ribes, Corylus, Lonicera xylosteum. Coppiced trees of Quercus, Betula, Alnus, Ulmus, Fraxinus, Carpinus may be assigned to this or to a separate growth form.
82. **Cornoflorids.**
Erect shrubs with straight horizontal branches. Not in Europe.
Cornus florida, C. nuttallii (N. America). Many examples in African savannahs.
83. **Sarothamnids.**
Erect fastigiate shrubs with stiff, erect branches.
Juniperus communis type II (fastigiate type), *Erica arborea, E. scoparia* p.p., *Sarothamnus* (deciduous), *Spartium* (deciduous), Myrica.

B6. **Trees.**
Woody plants with a single erect trunk.

84. **Piceids.**
Middle and lower branches hanging down. Crown of the centrifugal type.
Picea, Abies, Pseudotsuga.
85. **Quercids.**
Most branches horizontal.
Pinus p. max. p., *Cedrus*, Larix, Quercus p. max. p., Alnus glutinosa and incana.
86. **Pinoflexids.**
Branches near the trunk horizontal, then curving upwards.
Pinus flexilis (N. America).
87. **Fagids.**
Branches erectopatent. Centripetal crown type.
Cypressus, Juniperus communis type I (columnar type), *Arbutus*, Fagus, Ulmus, Populus, Fraxinus.
88. **Tiliids.**
Branches arcuate, erectopatent at base, drooping near apex.
Tilia, Aesculus, Castanea, Betula, *Ilex aquifolium* p.p., Pyrus, Malus, weeping forms (cultivars) of Fagus and Salix.
According to the crown form we may classify European trees as follows:
a. Narrow cylindrical. *Cypressus, Juniperus communis type I*, Populus canadensis, *Pinus sylvestris var. lapponica*.
b. Pyramidal. *Picea, Abies, Pseudotsuga*, Larix, Populus p.p.
c. Ovate. Salix fragilis, S. pentandra, Betula, Alnus, Fraxinus, Acer p.p., *Juniperus type II* (fastigiate type).
d. Dome-shaped. Fagus, Quercus, Castanea, Aesculus, Platanus, Tilia, Acer pseudoplatanus, Malus, Pyrus, Sorbus, *Pinus cembra, Pinus sylvestris var. scotica*.

e. Umbrella shaped. *Pinus pinea* and many other trees from mediterranean and dry subtropical regions.

According to the average crown density European trees can be classified as follows:
1. Thin crowns.
    Fraxinus, Salix p.p., Betula, Larix, *Pinus halepensis*, Robinia, Populus tremula.
2. Moderately dense crowns.
    Pinus sylvestris, Quercus robur and petraea, Tilia, Acer, Alnus.
3. Dense crowns.
    Pinus cembra, Picea, Pseudotsuga, Abies, Juniperus, Taxus, Cypressus, Quercus ilex, Ilex, Fagus, Carpinus, Castanea, Ulmus.

According to vegetative propagation we may divide trees into:
a. With runners (stoloniferous). *Picea abies* p.p.
b. With suckers (root travelling). Populus tremula.
c. Without vegetative means of propagation. Most trees.

Considering the relations between these growth forms and vegetative periodicity, we can state the following.

Of course all growth forms of lichens and bryophytes (8–31), except the few annual bryophytes, are evergreen.

The graminoids except the solitary (54) are mainly evergreen, the summergreen species being conspicuously concentrated in the tall grasses (50, 53).

Among water plants most free floating and swimming representatives (1–7), except of course the bryophytes, are summergreen, whereas most of the rooting ones (34–37) are evergreen.

Most climbing herbs (38–40) are summergreen, most Equisetoids (43–46) are either evergreen or nevergreen (heterotrophic).

Most prostrate and decumbent herbs (43, 47, 56, 57) are evergreen, but the Arcuatae are all deciduous.

The stoloniferous erect herbs (59, 63) are mostly evergreen and so are all pulvinate herbs (55), but most caespitose herbs (60, 64, 68, 69), which have fundamentally the same architecture as the cushion plants, are summergreen.

Among the solitary semirosette herbs (erectae scaposorosulatae solitariae) the strongly branched (67) are mainly summergreen, whereas the majority of the unbranched (66) are evergreen.

The majority of the dwarf shrubs (75–77) is evergreen, most shrubs (78–83), however, are deciduous.

Trees with dense crowns, finally, are mainly evergreen, those with thin crowns mainly deciduous.

It remains to be investigated whether there is any ecological meaning to this character clustering.

3. Phenological plant types

3.1. History

Contrary to growth and life forms, little attention has been paid to phenological plant types, although some are implicit in certain life forms (geophytes, therophytes, chamaephytes, hapaxanths) or explicitly mentioned as subtypes of life forms, *e.g.*, summer and winter annuals, evergreen and deciduous phanerophytes. Periodicity has mainly been studied on the individual species level and represented by beautiful graphs on a time axis. Sometimes this has been done for all species of a plant community and on this base phenophases (seasonal aspects) have been distinguished within such communities. In literature one can find many interesting observations on the phenology of individual species and its variation with habitat (*e.g.*, Westhoff 1949), but general classifications of all species into phenological types are hard to find. However, I think it is worth while to classify plants according to their type of periodicity, so that generalizations are possible, correlations with other plant characters and habitat factors are possible and spectra can be drawn for stands and community types.

The first and only detailed system was published by Massart (1910) in his famous classical work on the vegetation of Belgium. He distinguished 12 types (A–L), but they were only presented in the forms of a figure, not mentioned or explained in the text. Although time of flowering was indicated in the figure, this was not a criterium for the distinction of his types, for (1) he stated that the flowering time only refers to the first example of each type, and (2) this time obviously varies widely within each type. He mentioned for instance *Taraxacum officinale* and *Gentiana ciliata* as belonging to the same phenological type, and similarly *Fagus sylvatica* and *Parnassia palustris*.

If we try to extricate the characters of his types from the figures, we arrive at the following system (the description and the references to Raunkiaer's life forms are entirely mine, the capitals refer to Massart's symbols).
1. Winter green.
    Winter annuals.   A
2. Winter- and spring green. Geophytes. Perennial plants, budding in autumn, flowering in spring, dying aboveground in summer.   B
3. Spring green. Ditto, but leaves do not appear until spring.   C
4. Summer green.
    a. Hibernating as seeds: summer annuals.   J
    b. Hibernating as rhizomes, bulbs or tubers: Geophytes.   L
    c. Hibernating above ground, but without leaves. Chamaephytes and Phanerophytes.   K
5. Evergreen.
    a. Biannual hapaxanths.   E
    b. Perennial species.
        b1. Stems and leaves living several years.
            b1.1 Green parts at soil surface. Hemicryptophytes.   F
            b1.2 Green parts above the soil. Chamaephytes.
                b1.2.1 Flowering in spring.   D
                b1.2.2 Flowering in summer.   G
        b2. Stems biannual, leaves annual. Chamaephytes.   I
        b2. Stems and leaves living half a year. Summer and winter generations (the latter are vegetative) alternate to the effect that green leaves are present all year round.   H

This remarkably detailed system of a brilliant observer is obviously based partly on phenology, partly on life form. Unfortunately it has been largely ignored by later authors.

Diels (1918), studying forest herbs in Central Europe, distinguished only three phenological types: (1) 'aperiodic' (*i.e.*, evergreen), (2) with summer rest, (3) with winter rest. Du Rietz (1931) stressed the need of a system of periodicity life forms, but did not work it out. Dansereau (1951) distinguished between evergreen, evergreen-leafless, semideciduous and wholly deciduous.

Ellenberg (1974) and Müller-Dombois and Ellenberg (1974) made a distinction between evergreen, winter green, spring green (early summer green) and rain green. Their winter greens are green throughout the year, but the leaves last one year and are replaced in spring. The rain greens are ephemeral desert plants, the vegetative periods of which are irregular and coincide with rainy spells. They may last only a few weeks and do not occur each year. This type is very rare in Europe, and is not dealt with here.

Bakker *et al.* (1966) worked out a most valuable and detailed system of hapaxanthic vascular plants of the Netherlands. They distinguished ephemerals, annuals and biannuals and divided each of them according to (1) the stage of hibernation (as seeds, root buds, leaf rosettes or 'summer dress'), (2) the presence/absence of dormancy and (3) vernalization requirements. The two lastnamed features also cause differences in flowering time. The system is partly physiological and experiments under controlled conditions are needed to tell some of their 11 periodicity types apart. In my system I have indicated where their types (A 1–3, B 1–5, C 1–3) should be placed, but they are not synonymous with my types which are larger, also comprising pollacanthic perennials.

3.2. The proposed system

In my system I have disregarded Raunkiaer's life forms, which can much better be studied separately, being no phenological types proper. Also I have ignored the life span of leaves, stems and whole plants, inasfar as they do not affect seasonal aspects of the community, but I have paid much more attention to flowering periodicity and its relation to vegetative periodicity. Eight types of flowering periodicity have been distinguished.

I have adopted Ellenberg's five types of vegetative periodicity, but I have added Dansereau's semideciduous type and extended this concept to a degree that it can better be called semi-evergreen.

I have split up the summer greens in spring-summer greens and real summer greens, because there is a conspicuous and ecologically important difference between plants that become green in March or April and those that bud in June. It also affects vegetation physiognomy and structure.

Thus I arrive at seven types of vegetative periodicity and eight types of generative periodicity. The classification is based primarily on the former, because they determine vegetation structure in the first place. They are subdivided into parallel flowering subtypes. Since the two classifications are largely independent (of the theoretical 56 combinations 29 have been found), it would have been cumbersome to give each combination a name.

Contrary to my growth forms these types of course cannot be determined in the field by a single observation. They must be established by observation all the year round. As types may vary within species according to habitat and macroclimate (and often to ecotype), it is indispensable to check the type of a species in extreme habitats and in climates different from Western Europe. The examples of vegetative and flowering periodicity mentioned here, refer to 'normal' habitats and 'normal' years in the lowland of Western Europe only. The examples are not exhaustive.

For the generative subdivision the following abbreviations have been used:

a = autumn flowering* (September and October)
b = (bare) flowering when the plant is leafless
e = always (ever)flowering. The flowering may be interrupted by spells of drought or frost, but the interruptions are short and irregular and flowering is possible in any season
i = (Indian summer) late summer flowering: August and September
l = late spring flowering (main period between April 15 and June 15)
n = not flowering (plants always sterile)
s = summer (midsummer) flowering: June and July
v = vernal, flowering early spring (March and April)
w = winter flowering (November to February inclusive)

The Ellenberg's symbols (1974) are:
I = immergrün (evergreen)
S = sommergrün (summer green)
V = vorsommergrün (spring green)
W = überwinternd grün (hibernating green)
They are added in parentheses.

A. Evergreen (Ell: I)
Leaves lasting several years.
1. Vegetative growth mainly in autumn (and spring). Fructification usually in winter and spring.
n, v Lichens and bryophytes, excl. Sphagnum
2. Vegatative growth mainly in spring and summer.
n   not flowering.
    Ricciella.
w   Flowering in winter.
    Hedera, Ulex europaeus, Arbutus unedo, Viscum.
v   Flowering early spring.
    Empetrum, Eriophorum vaginatum.
l   Flowering late spring.
    Most conifers, Ilex, Arctostaphylos, Andromeda, Oxycoccus, Vaccinium vitis-idaea, Vinca, Ajuga, Galeobdolon, Glechoma, Stellaria holostea, Eriophorum angustifolium.
s   Flowering midsummer.
    Asplenium ruta-muraria, Blechnum, Erica tetralix, Equisetum hyemale, Lycopodium, Polystichum, Sedum, Sempervivum, Sphagnum.
i   Flowering late summer.
    Calluna.

---

* 'Flowering' also refers to fruiting of lichens, bryophytes and pteridophytes.

B. Semievergreen
3. Green until the frost, in mild winters evergreen. Leaves annual, renewed in spring (Ell: W).
e   Always flowering. A1.
      Bellis perennis, Lamium purpureum, Senecio vulgaris, Stellaria media.
l   Flowering late spring.
      Taraxacum, Luzula pilosa.
s   Flowering in midsummer. C2.
      All perennial grasses, except Calamagrostis, Molinia, Nardus and Phragmites. Some perennial chamaephytes (Sagina procumbens, Teucrium scorodonium, etc.), a few shrubs (Ligustrum, Rubus fruticosus), water plants (Nuphar, Stratiotes, Sagittaria) and perennial rosette plants (Arnica, Pyrola), most of the biannual rosette plants (Digitalis, Oenothera, Cirsium palustre etc.) and a number of ferns (Cyrtomium, Ceterach, Dryopteris dilatata, D. filix-mas, Polypodium).
i   Late summer flowering. C1.
      Verbascum.
4. Deciduous, but only 1–2 months. Breaking into leaf in February, in mild winters in January (Ell: S).
l   Flowering late spring.
      Anthriscus etc.
s   Flowering midsummer.
      Corydalis claviculata, Lonicera periclymenum etc. The lastnamed plant is a special case, because the leaves of the naked buds stay green all winter.
5. Deciduous, but twigs remain green and keep up photosynthesis.
vb   Flowering early spring, before breaking into leaf.
      Cornus mas.
l   Flowering late spring.
      Vaccinium myrtillus, Sarothamnus, Spartium.
s   Flowering midsummer.
      Euonymus, Lycium.

C. Wintergreen (Ell: V p.p.). Drought deciduous.
6. Winter annuals, germinating in autumn, flowering at the end of the vegetative season (spring).
v   Flowering early spring. B3.
      Cerastium semidecandrum, C. tetrandum, Erophila verna.
l   Flowering late spring. A2, B4, B5.
      Aira caryophyllea and praecox, Cochlearia danica, Myosotis ramosissima, Phleum arenarium, Saxifraga tridactylites, Spergula morisonii.
7. Perennial geophytes, flowering in autumn, before the leaves develop.
ab   Scilla autumnalis.

D. Spring green (Ell: V p.p.). Shade deciduous.
8. Breaking into leaf end of the winter/early spring, dying aboveground before summer. Perennial geophytes.
w   Flowering late winter.
      Galanthus, Eranthis.

v   Flowering early spring.
    Adoxa, Anemone nemorosa and ranunculoides, Ficaria, Gagea.
l   Flowering late spring.
    Ornithogalum, Scilla verna.

E. Spring-summergreen (Ell: S p. max. p.) Cold deciduous.
9. Green from March/April till September/November.
vb Flowering before breaking into leaf.
    Acer negundo, Alnus (excl. A. viridis), Corylus, Daphne mezereum, Equisetum arvense, sylvaticum and telmateia, Hamamelis, Larix, Myrica gale, Prunus spinosa, Salix (excl. S. pentandra), Ulmus.
l   Flowering in beginning vegetative period.
    Acer pseudoplatanus, Allium ursinum, Alnus viridis, Amelanchier, Arum, Betula, Caltha, Cardamine pratensis, Carex brizoides, C. caryophyllea, C. dioica, C. pilulifera, C. praecox, C. pulicaris, Carpinus, Convallaria, Fagus, Fritillaria, Majanthemum, Menyanthes, Pulmonaria, Quercus (excl. the evergreen oaks), Salix pentandra, Sambucus racemosa, Trientalis, etc.
s   Flowering in the middle of the vegetative period. B1.
    Apera, Calamagrostis, Nardus, Narthecium, many Compositae, Cruciferae, Geraniaceae, Labiatae, Ranunculaceae and Rosaceae, among the trees f.i. Sambucus nigra, Sorbus aucuparia, Tilia.
i   Flowering at the end of the vegetative period. B2, C3.
    This large group comprises mainly Compositae and Chenopodiaceae, among ecological groups mainly weeds of root crop fields, and plants of salt marshes, dune slacks and chalk ground.
    Arctium, Aster, Atriplex, Bassia, Bidens, Blackstonia, Callitriche hermaphroditica, Campanula glomerata, Centaurea nigra, Centaurium, Chenopodium album, ficifolium, murale, polyspermum and rubrum, Cicendia, Cichorium intybus, Cirsium acaule, Corispermum marschallii, Dipsacus fullonum, Epilobium hirsutum, Erigeron, Eupatorium, Euphrasia p.p., Galeopsis bifida, Gentiana pneumonanthe, Gentianella, Halimione, Herniaria, Hieracium sabaudum and umbellatum, Hydrocotyle, Impatiens glandulifera, Inula, Knautia, Kochia, Leontodon autumnalis, Limonium, Lythrum, Najas, Odontites serotina, Origanum, Parnassia, Phragmites, Picris, Polygonum convolvulus, dumetorum, hydropiper, lapathifolium, minus and mite, Pulicaria, Rumex maritimus, Salicornia, Salsola, Saponaria, Sanguisorba officinalis, Scabiosa, Scrophularia aquatica and neesii, Senecio erucifolius, fluviatilis, fuchsii and jacobaea, Setaria glauca, Solidago, Spartina, Spiranthes spiralis, Suaeda, Succisa, Tanacetum, Teucrium botrys, chamaedrys and scordium, Verbena.
ab Flowering after the vegetative period.
    Colchicum.

F. Summergreen (Ell: S) Cold deciduous.
10. Green only from May or June.
s   Flowering in midsummer.
    Rhynchospora.
i   Flowering late summer.
    Molinia, Pteridium, Centaurium erythraea.

Note. This list is probably very incomplete and a number of species of type 9 i may have to be transferred to type 10.

References

Bakker, D. et al. 1966. Ecological research at the Plant-ecological Laboratory, State University, Groningen. Wentia 15: 1-24.
Barkman, J.J. 1958. Phytosociology and ecology of cryptogamic epiphytes. Van Gorcum, Assen.
Barkman, J.J. 1979. The investigation of vegetation texture and structure. In: M.J.A. Werger (ed.), The Study of Vegetation, pp. 125-160. Junk, The Hague.
Barkman, J.J. 1987. Preliminary investigations on the texture of high arctic tundra vegetation. In: A.H.L. Huiskes, C.W.P.M. Blom and J. Rozema (eds), Vegetation between Land and Sea, pp. 120-132. Junk, The Hague.
Braun-Blanquet, J. 1928. Pflanzensoziologie. Grundzüge der Vegetationskunde. Biol. Studienbücher 7. 1. Aufl. Berlin.
Braun-Blanquet, J. 1964. Pflanzensoziologie. 3. Aufl. Springer Wien-New York.
Clements, F.E. 1920. Plant indicators. The relation of plant communities to process and practice. Carnegie Inst. Wash. 290.
Cockayne, L. 1919. New Zealand plants and their story. 2nd ed. Wellington.
Dansereau, P. 1951. Description and recording of vegetation upon a structural basis. Ecology 32: 172-229.
Dansereau, P. and Arros, J. 1959. Essais d'application de la dimension structurale en phytosociologie. I. Quelques exemples européens. Vegetatio 9: 48-99.
Descoings, B. 1975. Les types morphologiques et biomorphologiques des espèces graminoides dans les formations herbeuses tropicales. Naturalia monspeliensia, sér. Bot. 25: 23-35.
Diels, L. 1918. Das Verhältnis von Rhytmik und Verbreitung bei den Perennen des europäischen Sommerwaldes. Ber. deutsch. Bot. Ges. 6.
Drude, O. 1886. Deutschlands Pflanzengeographie. Handb. Deut. Landes- u. Volksk. 4: 502.
Drude, O. 1913. Die Ökologie der Pflanzen. Die Wissenschaft. F. Vieweg, Braunschweig. 50.
Du Rietz, G.E. 1921. Zur methodologischen Grundlage der modernen Pflanzensoziologie. Akad. Abh. Uppsala. Holzhausen, Wien.
Du Rietz, G.E. 1931. Life-forms of terrestrial flowering plants. Acta Phytogeogr. Suec. 3: 1-95.
Ellenberg, H. 1974. Zeigerwerte der Gefässpflanzen Mitteleuropas. Ser. Geobot. 9: 5-97.
Fekete, G. and Szujkó-Lacza, J. 1970. A survey of the plant lifeform systems and the respective research approaches II. Ann. Hist.-Nat. Mus. Nat. Hung. Pars Bot. 62: 115-127.
Feoli, E. and Scimone, M. 1984. A quantitative view of textural analysis of vegetation and examples of application of some methods. Archiv. Bot. e Biogeogr. Ital. 60 (1/2): 73-94.
Fischer, J.B. 1986. Branching patterns and angles in trees. In: T.J. Givnish (ed.), On the Economy of Plant Form and Function, pp. 493-523. Cambridge Univ. Press, Cambridge.
Gams, H. 1918. Prinzipienfragen der Vegetationsforschung. Vierteljahresschr. naturf. Ges. Zürich 63: 293-493.
Gimingham, C.H. 1951. The use of life form and growth form in the analysis of community structure, as illustrated by a comparison of two dune communities. J. Ecol. 39: 396-406.
Gimingham, C.H. and Robertson, E.T. 1950. Preliminary investigations on the structure of Bryophyte communities. Trans. Brit. Bryol. Soc. 1 (4): 330-344.
Givnish, T.J. (ed.) 1986a. On the economy of plant form and function. Cambridge Univ. Press, Cambridge.
Givnish, T.J. 1986b. Biomechanical constraints on crown geometry in forest herbs. In: T.J. Givnish (ed.), On the Economy of Plant Form and Function, pp. 525-583. Cambridge Univ. Press, Cambridge.
Givnish, T.J. 1987. Comparative studies of leaf form: assessing the relative roles of selective pressures and phylogenetic constraints. New Phytol. 106 (Suppl.): 131-160.
Grisebach, A. 1872. Die Vegetation der Erde nach ihrer klimatischen Anordnung. W. Engelmann Leipzig, vol. I, II.
Hallé, F., Oldeman, R.A.A. and Tomlinson, P.B. 1978. Tropical trees and forests. An architectural analysis. Springer, Berlin-Heidelberg-New York.

Hartog, C. den 1955. A classification system for the epilithic algal communities of the Netherlands' coast. Acta Bot. Neerl. 4: 126-135.
Hartog, C. den 1959. The epilithic algal communities occurring along the coast of the Netherlands. Wentia 1: 1-241.
Hartog, C. den and Segal, S. 1964. A new classification of the water-plant communities. Acta Bot. Neerl. 13: 367-393.
Hejny, S. 1957. Ein Beitrag zur ökologischen Gliederung der Makrophyten der tschechoslowakischen Niederungsgewässer. Preslia 29: 349-368.
Hejny, S. 1960. Ökologische Characteristik der Wasser- und Sumpfpflanzen in den slowakischen Tiefebenen (Donau- und Theissgebiet). Bratislava.
Horikawa, Y. and Miyawaki, A. 1954. Studies on the growth forms of weeds as related to community structures. Jap. J. Ecol. Sendai 4: 79-88.
Horn, H.S. 1971. The adaptive geometry of trees. Princeton Univ. Press, 144 pp.
Humboldt, A. von 1806. Ideen zu einer Physiognomik der Gewächse. Cotta, Stuttgart.
Londo, G. 1971. Patroon en proces in duinvalleivegetaties langs een gegraven meer in de Kennemerduinen. Thesis. Derks, Cuyk.
Luther, H. 1949. Vorschlag zu einer ökologischen Grundeinteilung der Hydrophyten. Acta Bot. Fenn. 44: 1-15.
Maarel, E. van der 1966. Over vegetatiestructuren, -relaties en -systemen in het bijzonder in de duingraslanden van Voorne. Thesis, Univ. Utrecht.
Massart, J. 1910. Esquisse de la géographie botanique de la Belgique. Lamertin (ed.), Bruxelles.
Mühlberg, H. 1967. Die Wuchstypen der mitteldeutschen Poaceen. Hercynia 4 (1): 11-50.
Müller-Dombois, D. and Ellenberg, H. 1974. Aims and methods of vegetation ecology. John Wiley & Sons, New York-London-Sidney-Toronto.
Parsons, D.J. 1976. Vegetation structure in the mediterranean scrub community of California and Chile. J. Ecol. 64: 435-447.
Poplavskaja, G.I. 1948. Ekologija rastenij. Sov. nauka, Moscow.
Raunkiaer, C. 1907. Planterigets Livsformer og deres Betydning for Geografien. Munksgaard, Copenhagen.
Rübel, E. 1930. Pflanzengesellschaften der Erde. Huber, Bern-Berlin.
Schiefer, J. 1981. Vegetationsentwicklung und Pflegemassnahmen auf Brachflächen in Baden-Württemberg. Natur und Landschaft 56 (7/8): 263-268.
Schmid, E. 1956. Die Wuchsformen der Dikotyledonen. Ber. geobot. Forsch.-inst. Rübel f. 1955: 38-50.
Schmithüsen, J. 1961. Allgemeine Vegetationsgeographie. 2. Aufl. De Gruyter & Co., Berlin.
Segal, S. 1965. Een vegetatieonderzoek van de hogere waterplanten in Nederland. Wet. Meded. Kon. Ned. Nat. hist. Ver. 57: 1-80.
Segal, S. 1969. Ecological notes on wall vegetation. Junk, The Hague.
Skottsberg, C. 1926. Plant communities of Juan Fernandez Islands. Proc. Int. Congr. Pl. Sc. Ithaka.
Styner, E. and Hegg, O. 1984. Wuchsformen in Rasengesellschaften am Südfuss des Schweizer Juras. Tuexenia N.S. 4: 195-215.
Theophrastos of Eresos, 371-286 B.C. Historia Plantarum (Latin transl.).
Warming, E. 1884. Om Skudbygning, Overvintring og Foryngelse. Festskr. Naturh. Foren. Copenhagen.
Warming, E. 1895. Plantesamfund: Grundtraek af den Økologiske Plantegeografi. Philipsens, Copenhagen.
Warming, E. 1908. Om Planterigets Livsformer. Festskr. udg. af Univ. Copenhagen.
Warming, E. 1909. Oecology of plants: An introduction to the study of plant-communities. Oxford Univ. Press, Oxford.
Warming, E. 1918. Bemärkninger om Livsform og Standplats. Kristiania.
Warming, E. 1919. Dansk Plantevakst. 3. Skovene, 6. Bot. Tidsskr. 35, 6.
Warming, E. 1923. Ökologiens Grundformer. D. Kgl. Danske Vidensk. Selsk. Skrift. Naturv. og Math. Afd.; 8. Räkke, IV, 2.
Westhoff, V. 1949. De betekenis van de phaenologie voor het plantensociologisch onderzoek. Ned. Kruidk. Arch. 56: 24-31.

PATTERNS IN VASCULAR PLANT GROWTH FORMS IN ARCTIC COMMUNITIES AND ENVIRONMENT AT ATKASOOK, ALASKA

V. KOMÁRKOVÁ[1] and J.D. McKENDRICK[2]
[1] *American College of Switzerland, CH-1854 Leysin, Switzerland, and Institute of Arctic and Alpine Research (INSTAAR), University of Colorado, Boulder, Colorado 80309, USA;* [2] *Palmer Research Center, University of Alaska, Palmer, Alaska 99645*

Abstract

In the low arctic landscape at Atkasook, Alaska, the distribution of vascular plant growth forms and their communities is determined by natural disturbances (rivers, wind, thaw-lake cycle, cryoturbation) which diversify the landscape into a variety of landforms and control surface soil properties and material movement. These disturbances rework the landscape every several thousand years and limit the time available for the differentiation of landforms and growth form- and taxa-based communities which are relatively poorly defined. The high proportion in the vegetation of rhizomatous and other growth forms adapted to rapid vegetative reproduction, colonization of newly bared areas, and growth on unstable surfaces probably reflects this.

Introduction

The flatness, high content of water or ice in the ground, and loose, sandy substrate of the low arctic (Aleksandrova 1977) landscape at Atkasook enhance the effects of various disturbances which create bare patches at frequent intervals (Fig. 1). As growth form characteristics are subject to environmental selection, it should be easier to demonstrate the relationship between growth forms (Barkman 1979, 1988) and the environment here than in other landscapes. This paper covers the total diversity of landforms, vascular plant taxa, and mature communities undisturbed by humans in the Atkasook area, which is representative for a relatively large portion of the Arctic Coastal Plain, and which is environmentally, biotically, and historically distinctive and uniform. The analysis is thus based on a complete sample of a uniform landscape unit (a 'concrete flora' of Tolmačev 1971). The ecological range and optimum of each growth form and community can be determined and other similarly sampled areas can be compared.

Description of the study area

Atkasook (70°29'N and 157°25'W), a small village, is located on the Arctic Coastal Plain (Wahrhaftig 1965) on Meade River on the north slope of the Brooks Range, where the latitudinal gradient of increasing temperatures toward the south (Clebsch 1957; Clebsch and Shanks 1968) controls the environment. The climate at Atkasook is less humid and warmer than the climate at the Arctic Ocean coast at Barrow (100 km north of Atkasook; Komárková and Webber 1980). The nearly

Fig. 1. Wet and sandy terrain northwest of Atkasook is typical for the Alaskan Arctic Coastal Plain. Meandering rivers and thaw lakes, which are moved across the landscape by wind, rework the surface every few thousand years.

flat landscape at Atkasook is formed by sandy unconsolidated Pleistocene sediments (Black 1964; Williams *et al.* 1977). The surface of permafrost is about 10–150 cm belowground during the summer.

Streams expose sand on meander rivercuts with 10 m high cutbanks, on pointbars which border on sand dune fields, and on dry river bottoms during periods of low water. Old meanders outline the highest ancient sand dune ridges, the remnants of a sand sea which probably covered most of this area before 50,000 years BP (Carter 1981a, 1981b). Wind redistributes sand from rivers, creates dunes (Black 1951), and moves thaw lakes which travel across the landscape in the direction perpendicular to the leading wind (Britton 1967; Billings and Peterson 1980). Cryoturbation creates polygonal patches about 5 to 10 m in diameter, beaded streams, and string bogs. The landform ages range from 400 to 12,000 years. Away from the oldest ridges, each piece of landscape is probably being reworked each 3000–4000 years (Everett 1979, 1980, pers. comm.). Migrating animals, primarily caribou, redistribute nutrients (White and Trudell 1980); microtine rodents are widespread (Batzli and Jung 1980), and arctic ground squirrels enrich burrows on ridges with deep thaw (Batzli and Sobaski 1980). Herbivorous ptarmigans are active in willow stands. Human surface disturbance is limited to the immediate vicinity of Atkasook.

The area is floristically relatively rich, as a result of the absence of glaciers from the north slopes of the Brooks Range during the Quaternary (Péwé 1975). The

Table 1. Vegetation units classified in a Braun-Blanquet hierarchy. Atkasook, Alaska (608 samples, 245 vascular species).

Classes/ landscape units	Orders	Alliances	Associations
Acetosella graminifolia-Bromopsis pumpelliana spp. *arctica*			
sand dunes	1	2	8
Carex obtusata-Dryas integrifolia ssp. *integrifolia*			
consolidated sand dunes	3	5	12
Diapensia lapponica ssp. *obovata-Salix phlebophylla*			
ancient sand dune ridges	1	2	8
Ledum palustre ssp. *decumbens-Eriophorum vaginatum* ssp. *spissum*			
uplands	1	2	4
Carex tripartita-Salix rotundifolia ssp. *rotundifolia*			
snowpatches	2	6	14
Pedicularis langsdorffii ssp. *arctica-Salix planifolia* ssp. *pulchra*			
lowlands	2	6	17
Pedicularis sudetica ssp. *albolabiata-Carex aquatilis* ssp. *stans*			
marshes	2	4	11
Sparganium hyperboreum-Arctophila fulva			
lakes and streams	1	2	4
Aster sibiricus ssp. *subintegerrimus-Salix glauca* var. *glauca*			
streambanks	4	6	12
Dodecatheon frigidum-Cardamine nymanii			
springs	1	3	4
Wilhelmsia physodes-Deschampsia brevifolia			
riverbars	1	2	5
Poa rigens-Arctagrostis arundinacea			
disturbed surfaces	1	4	7
12 classes	20 orders	44 alliances	106 associations

vegetation of the Atkasook area (177 km^2) has been mapped by Komárková and Webber (1980) using subjectively established units and described according to the Braun-Blanquet approach (Westhoff and Van der Maarel 1978) by Komárková (in prep.; Table 1). The results of the Braun-Blanquet analysis contradict Chapin and Shaver (1985) who concluded that the continuum model of community organization is valid in arctic tundra. Most classes correspond to major landforms; one class includes human and animal disturbances. After disturbance, the complexity of plant communities increases through the development of complex microtopography; some of these changes have been studied by Peterson (1978) and Peterson and Billings (1978).

Methods

Visually homogeneous sample plots were selected subjectively; except on small-sized landforms, the plots were about the size of the minimum area of tundra vegetation (around 25 m^2; Tüxen 1970; Westhoff and Van der Maarel 1978). In each plot, the percentage cover of all taxa of vascular plants, bryophytes, and lichens was recorded. Environment was evaluated according to subjective environmental gradient scales (after Komárková 1979). For example, caribou disturbance scale (0-10) value is a composite evaluation of caribou trampling, feces, browsing, and clipping. Surface age was estimated according to the position of the landform in the landscape and the available radiocarbon dates. Surface soil layer was collected in most plots. One representative sample was selected for each vascular plant-dominated Braun-Blanquet association. Vascular plants were named according to Á. Löve and D. Löve (1975), Argus (1973), and Hultén (1968). Following the flora of Hultén (1968) and the system of Barkman (1983, 1987), taxa were classified into growth forms and taxa data were transformed. Physical analysis of the surface horizon of the selected soils has been carried out at INSTAAR, and chemical analysis by atomic absorption spectrophotometry by J.D.M. at the University of Alaska.

Either entire cases with missing data (multiple regression) or pairs of values with one missing were deleted from the analysis. Among numerical methods used were correlation, factor analysis, discriminant analysis, simple and multiple regression, clustering, and multidimensional scaling (see figure legends for details). Comparisons of the results of various methods served to eliminate chance correlations and to identify trends. Many relationships between variables were nonlinear, either normal or bimodal; such relationships are important but poorly explained by linear regression or correlation. They were indicated by low correlation coefficients and low R^2 values, and recognized in scattergrams which are not presented to save space. Most computing has been done using the Macintosh computer and packages Systat and Statview. Packages SPSS, Cornell Ecology, and TAXON were used on mainframes.

Limitations to geographical comparison included the different growth form systems of different authors and the lack of data for entire uniform landscape units. The comparison included several sites of the International Biological Programme (IBP) Tundra Biome (French 1981) some of which represent primarily azonal vegetation controlled by local disturbances (*e.g.*, bogs). The study of Webber *et al.* (1975) along a latitudinal gradient was not used due to its lack of quantitative data. All data were approximated to Barkman's system.

Results and discussion

Growth form system

Prior to the work of Barkman (1979), no detailed growth form system has been available for studies of vegetation texture. In the Arctic, Polozova (1978, 1981, 1983) used the life forms of Raunkiaer (1934). The arctic growth form systems of Webber *et al.* (1975), Webber (1978), Webber *et al.* (1980), and Webber and Wielgolaski (in French 1981, unpublished IBP Tundra Biome materials) are relatively incomplete and inconsistent. For example, 'single' graminoids include all non-caespitose ones, with or without rhizomes or stolons, and monocotyledonous forbs and Pteridophyta are classified as such rather than according to their form. Barkman (1979, 1983, 1988) devised a new system, based on the plants of the Netherlands, which he extended to arctic Spitsbergen (Barkman 1987). His growth forms are free of any hypothesis as to environmental adaptation (Barkman 1988). The system also avoids the taxonomical system and relies primarily on form.

Barkman's system as applied at Atkasook (Fig. 2) includes forms parallel to graminoids among forbs, and caespitose graminoids divided into looser caespitose forms and tighter tussocks. Finer divisions of the system such as stem branching, height, and stem or basal leaves do not appear among 18 growth forms which represent the higher level of more than 60 categories. Attributes which do not occur at Atkasook (thorny, succulent) have been omitted. The placement of taxa was sometimes a compromise; for example, plants with long rootstocks or long subterranean runners were grouped with rhizomatous plants. In the absence of data, it was assumed that due to severe winters the herbaceous plants are largely deciduous.

Growth forms in vegetation

Graminoids (mostly Cyperaceae) have the highest cover in most associations and in the landscape, but they account for a lower proportion of taxa (37.9% of cover, 26.6% of taxa; Tables 2 and 3). Rhizomatous graminoids are the most important growth form; together with rhizomatous forbs, they account for 39.4% of cover and 27.9% of taxa. Rhizomes were reported as a common vegetative reproduction method in cold-dominated regions where seed production and germination are less reliable (Sørensen 1941; Billings and Mooney 1968; Callaghan and Collins 1981). However, rhizomes also enable plants to rapidly adjust to being covered by fine mineral material or to colonize a newly bared, neighboring surface. For example, rhizomatous graminoids are the last remaining vascular plants to occupy centers of late-melting snowpatches with rapid material movement in Colorado alpine areas (Komárková 1979), and colonize all human-disturbed surfaces at an arctic site similar to Atkasook (Komárková 1983). At Atkasook, they are both colonizers and dominants in habitats continually disturbed by water or wind, or colonizers in newly disturbed uplands. Their predominance thus appears to reflect surface properties rather than the short growing season. Reproduction by seeds is dominant in some most severe climates (*e.g.*, in tussocks and cushions in the Antarctic), and many species reproduce by seeds at Atkasook. Grulke and Bliss (1985) did not find increased vegetative reproduction in high vs. low arctic species of saxifrages.

High importance of deciduous large (14.6% of cover, 2.5% of taxa) and dwarf shrubs (22.0% and 8.1%) indicate the relatively southern arctic location of Atkasook; shrubs are less well-adapted to low temperatures than other forms (Tieszen *et al.* 1981). At Atkasook, large deciduous shrubs do not dominate zonal uplands as they do approximately 100 km to the southwest. Evergreen dwarf shrubs are half as important (7.8% and 2.6%) as deciduous dwarf shrubs (15.0% and 6.2%). Evergreen dwarf shrubs appear to be well-adapted to cold environments: their leaves allow energy capture early in the short growing season (*e.g.*, Callaghan and Collins 1981), and because they have low metabolic rates they can function at relatively low water potentials and utilize nutrient-poor habitats (Small 1972; Tieszen *et al.* 1981). As the severity of winter increases they may become less important;

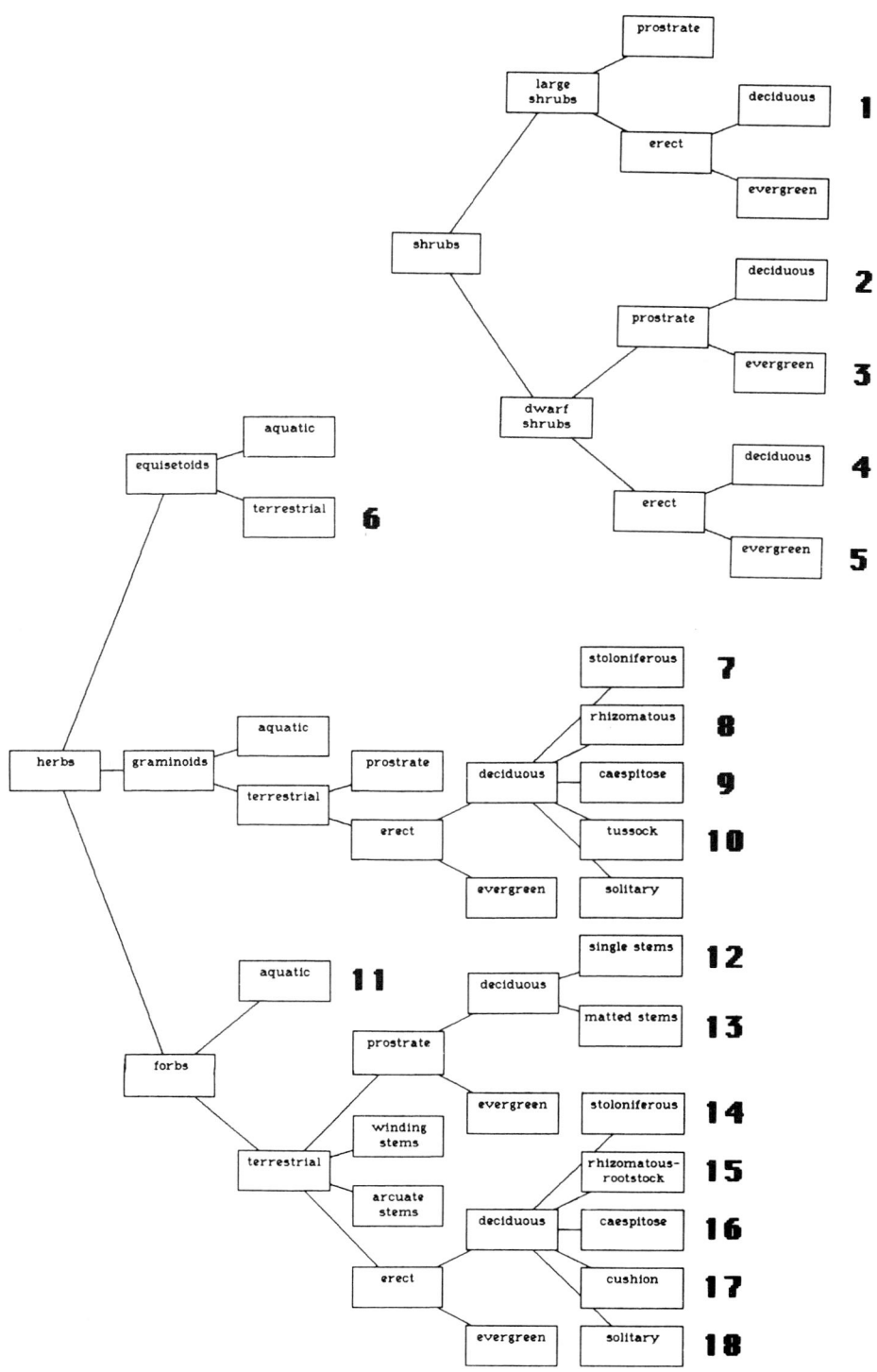

Table 2. Importance of growth form categories. Atkasook, Alaska.

Growth form	Total % cover in all associations	% of taxa
1. Large deciduous shrubs	14.6	2.5
2. Prostrate deciduous dwarf shrubs	13.0	4.2
3. Prostrate evergreen dwarf shrubs	4.6	1.3
4. Erect deciduous dwarf shrubs	2.0	1.3
5. Erect evergreen dwarf shrubs	3.2	1.3
6. Equisetoids	3.4	2.1
7. Stoloniferous graminoids	1.7	1.7
8. Rhizomatous graminoids	29.3	11.4
9. Caespitose graminoids	3.3	7.2
10. Tussock graminoids	3.6	6.3
11. Aquatics	1.2	1.3
12. Prostrate single forbs	1.6	4.2
13. Prostrate matted forbs	1.3	2.5
14. Erect stoloniferous forbs	1.3	0.3
15. Erect rhizomatous forbs	10.1	16.5
16. Erect caespitose forbs	1.1	11.0
17. Erect cushion forbs	0.8	0.8
18. Erect solitary forbs	4.9	23.3
Total	101.0	99.2

deciduous shrubs and forbs have better developed winter storage systems which facilitate rapid early-season growth (Tieszen et al. 1981) and are less vulnerable to freeze-desiccation.

The tussock graminoid growth form is predominantly made up by *Eriophorum vaginatum* ssp. *spissum*, which dominates uplands, is slowly growing and reproduces only by seeds (Chester and Shaver 1982; Gartner 1982; Gartner et al. 1986). Forbs account for less cover than graminoids and for the highest number of taxa (21.1% and 58.6%); many also reproduce by seeds. Cushions grow very slowly, conserve radiation, maintain a warmer and more humid microclimate inside, and dominate in extremely windy and dry sites with poor soils and low pH (*e.g.*, Bliss 1970; Komárková 1981); their low number and importance also indicate the relatively mild arctic conditions of Atkasook.

←

Fig. 2. Growth form system of 18 categories, after the system of Barkman (1979, 1983, 1988). Some categories were not recognized in the Atkasook flora. The highest number of categories were recognized among forbs.

The examples for individual categories include *Salix glauca* var. *glauca* (1), *Arctous alpina* ssp. *rubra* (2), *Empetrum eamesii* ssp. *hermaphroditum* (3), *Vaccinium gaultherioides* (4), *Vaccinium vitis-idaea* ssp. *minus* (5), *Hippochaete variegata* ssp. *variegata* (6), *Bromopsis pumpelliana* ssp. *arctica* (7), *Carex rupestris* ssp. *rupestris* (8), *Hierochloë alpina* (9), *Eriophorum vaginatum* ssp. *spissum* (10), *Potamogeton pectinatus* (11), *Koenigia islandica* (12), *Saxifraga oppositifolia* (13), *Saxifraga flagellaris* ssp. *setigera* (14), *Bistorta vivipara* (15), *Saxifraga caespitosa* ssp. *sileniflora* (16), *Silene acaulis* ssp. *arctica* (17), and *Taraxacum ceratophorum* (18).

Table 3. Means and standard errors of growth forms and environmental variables in the Braun-Blanquet classes. Atkasook, Alaska.

Class/landscape unit	1 Acetosella graminifolia-Bromopsis pumpelliana ssp. arctica sand dunes	2 Carex obtusata-Dryas integrifolia ssp. integrifolia consolidated sand dunes	3 Diapensia lapponica ssp. obovata-Salix phlebophylla ancient sand dune ridges	4 Ledum palustre ssp. decumbens-Eriophorum vaginatum ssp. spissum uplands	5 Carex tripartita-Salix rotundifolia ssp. rotundifolia snowpatches	6 Pedicularis langsdorffii ssp. arctica-Salix planifolia ssp. pulchra lowlands	7 Pedicularis sudetica-Carex aquatilis ssp. stans marshes	8 Spergenium hyperboreum-Arctophila fulva lakes and streams	9 Aster sibiricus subintegerrimus-Salix glauca var. glauca streambanks	10 Dodecatheon frigidum-Cardamine nymanii springs	11 Wilhelmsia physodes-Deschampsia brevifolia riverbars	12 Poa rigens-Arctagrostis arundinacea disturbed surfaces	mean for all units
Growth forms, mean percentage cover													
large deciduous shrubs	15.3/8.26	4.8/2.47	0.13/0.13	0.2/0.19	6.1/4.60	17.2/6.83	2.0/2.13	0	32.6/9.33	22.7/21.15	0.3/0.23	17.2/15.35	11.0/2.32
prostrate deciduous dwarf shrubs	13.1/7.76	31.0/8.61	10.4/7.99	16.3/1.25	14.6/5.76	8.2/3.53	1.3/1.19	0	7.6/3.04	2.7/1.77	0.25/0.25	1.3/0.93	9.8/1.83
prostrate evergreen dwarf shrubs	2.4/2.30	7.2/6.81	8.8/5.77	2.5/0.09	8.4/4.30	2.5/1.93	0	0	0.5/0.41	0	0	0.05/0.03	3.4/1.13
erect deciduous dwarf shrubs	0	0	15.7/10.28	4.32/4.32	0	0.8/0.6	0	0	0.3/0.22	0	0	0.08/0.08	1.6/0.9
erect evergreen dwarf shrubs	0.7/0.45	0	11.3/7.92	14.8/5.37	1.4/1.15	3.6/1.96	0	0	0	0	0	0.03/0.02	2.3/0.8
equisetoids	0.1/0.12	0.8/0.38	0.01/0.01	0	0.5/0.37	3.8/2.82	0	0	4.0/2.79	2.7/2.63	12.9/10.78	9.3/9.27	2.5/0.9
stoloniferous graminoids	2.0/0.78	1.4/0.68	0	0	0.42/0.42	0.4/0.29	4.3/2.35	1.50/1.50	0.8/0.51	0	5.8/4.79	0.08/0.08	1.2/0.3
rhizomatous graminoids	14.2/3.22	12.7/4.67	4.9/1.88	12.6/5.71	15.1/6.73	25.4/7.42	45.4/7.58	13.3/7.82	15.9/7.99	48.9/18.97	24.1/20.66	43.7/14.59	21.8/2.67
caespitose graminoids	4.1/0.95	2.1/1.29	4.6/1.58	1.1/1.00	6.3/2.54	0.2/0.08	0.9/0.66	0	1.5/0.56	2.8/2.71	0.9/0.44	2.2/1.15	2.4/0.47
tussock graminoids	2.5/2.21	6.9/3.30	0.4/0.27	18.4/14.12	0.13/0.11	0.4/0.23	0.2/0.19	0	4.0/3.11	0	5.2/5.01	2.4/1.62	2.7/0.84
aquatics	0	0	0	0	0	0	0	15.0/10.6	0	0	0	0	0.6/0.47
prostrate single forbs	0.5/0.10	0.5/0.20	0.05/0.03	0	4.2/3.91	0.2/0.09	0	0	1.3/0.63	0.03/0.02	1.8/0.64	2.5/0.87	1.1/0.56
prostrate matted forbs	1.8/0.50	1.14/0.38	0.6/0.24	0	0.02/0.01	0.05/0.05	0	0	3.0/1.49	0.06/0.06	0.01/0.01	1.5/1.41	0.8/0.23
erect stoloniferous forbs	0.06/0.06	0.63/0.63	0	0	0	0	0	0	0.7/0.49	0	0	0	0.2/0.09
erect rhizomatous forbs	1.6/0.64	1.6/0.53	2.8/1.00	5.8/1.87	15.8/7.15	6.2/2.06	8.4/6.55	0.8/0.73	12.9/2.47	17.0/9.10	5.4/3.17	8.2/4.74	7.6/1.39
erect caespitose forbs	0.9/0.26	1.7/0.36	0.5/0.28	0	0.06/0.04	0.2/0.09	0.11/0.11	0	1.3/0.54	0	0.02/0.01	0.7/0.42	0.6/0.10
erect cushion forbs	0.03/0.02	0.2/0.12	3.5/1.90	0	0.02/0.02	0.01/0.01	0	0	0.1/0.08	0	0	0.14/0.14	0.3/0.17
erect solitary forbs	1.4/0.06	3.7/1.77	0.5/0.32	0	2.8/1.08	0.9/0.39	4.8/1.85	0	12.6/4.13	9.8/4.20	0.14/0.11	1.5/0.9	3.5/0.70
Environmental variables													
slope°	3.4/1.74	3.9/2.65	0.9/0.52	0.8/0.48	5.0/2.05	0.20/0.20	0	0.75/0.75	13.1/3.93	5.7/3.84	4.6/3.47	1.6/0.48	3.6/0.75
elevation, m.s.m.	13.7/0.81	14.5/1.25	14.8/1.22	16.3/1.25	15.5/0.37	15.7/0.45	16.0/0.67	15.0/0	15.1/1.16	13.3/1.67	13.0/1.23	15.7/0.71	15.1/0.27
temperature, scale 0-10	3.4/0.05	3.2/0.06	2.7/0.13	2.5/0.09	1.8/0.09	2.2/0.07	2.0/0.05	1.9/0	2.6/0.19	2.2/0.25	2.1/0.03	2.6/0.09	2.4/0.06
site moisture, scale 0-10	2.3/0	2.8/0.16	3.5/0.32	4.6/0.20	5.4/0.24	5.8/0.25	7.3/0.23	8.2/0.60	4.3/0.26	6.6/1.30	4.7/0.33	3.8/0.19	4.8/0.18
wind, scale 0-10	6.9/0.38	5.6/0.38	5.1/0.50	4.3/0.10	1.6/0.18	3.1/0.30	2.8/0.27	1.2/0.38	2.7/0.35	2.2/0.43	3.0/0.21	3.4/0.36	3.5/0.19
snow duration, scale 0-10	1.7/0.05	2.9/0.44	3.2/0.56	4.4/0.25	7.7/0.14	5.5/0.23	6.4/0.07	6.6/0.12	5.9/0.38	6.1/0.79	2.4/0	4.6/0.41	5.2/0.20
surface age, scale 0-10	4.9/0.93	4.8/0.55	7.2/0.54	6.2/0.43	4.9/0.27	4.7/0.32	4.8/0.28	3.7/0.43	3.8/0.27	3.9/0.17	2.4/0	4.1/0.71	4.7/0.17
erosion, scale 0-10	8.0/0.21	6.7/0.21	4.6/0.39	2.7/0.46	5.0/0.44	4.0/0.55	5.6/0.46	7.8/0.22	7.1/0.32	7.5/0.63	8.0/0.13	5.0/0.50	5.8/0.20
cryoturbation, scale 0-10	0	1.1/0.28	2.6/0.37	3.3/0.22	1.7/0.19	3.0/0.46	3.5/0.30	0.6/0.22	1.2/0.25	2.2/0.70	0.05/0.05	1.1/0.36	1.8/0.15
stream disturbance, scale 0-10	1.1/0.42	1.4/0.31	0.6/0.31	0	0.5/0.11	1.4/0.45	0.9/0.32	5.2/1.89	2.4/0.57	3.4/1.68	7.6/0.28	0.7/0.59	1.6/0.22
lake disturbance, scale 0-10	0	0	0	0.62/0.62	0.2/0.09	7.2/5.89	2.2/1.01	4.3/2.46	0.6/0.49	0	0	0.29/0.29	0.8/0.19
gravity disturbance, scale 0-10	0.6/0.29	0.07/0.07	0	0	0.1/0.08	0.05/0.05	0	0	0.07/0.07	0	0	0	0.2/0.10
human disturbance, scale 0-10	0.8/0.09	0.9/0.12	0.8/0.08	1.0/0.04	0.9/0.12	0.67/0.08	0.3/0.05	0.2/0.03	1.1/0.19	1.2/0.58	0.2/0.03	1.1/0.23	0.8/0.05
animal disturbance, scale 0-10	1.5/0.33	1.5/0.29	1.8/0.25	2.9/0.42	1.6/0.33	1.2/0.29	0.2/0.11	0	1.2/0.27	2.4/1.44	0.18/0.18	2.0/1.05	1.3/0.13
caribou disturbance, scale 0-10	0	0	0.7/0.36	3.1/1.15	2.9/0.64	1.2/0.39	0.7/0.37	0	1.4/0.54	2.2/1.27	0	2.8/0.53	1.2/0.18
microtine disturbance, scale 0-10	0.6/0.35	0	0.3/0.22	0	0.1/0.08	0.3/0.20	0	0	1.5/0.68	0.83/0.83	0	0	0.5/0.13
arctic ground squirrel disturbance, scale 0-10	0	1.7/0.67	0.7/0.29	0	0.2/0.13	0.4/0.18	0	0	1.7/0.35	1.8/1.32	0	0.6/0.47	0.7/0.11
ptarmigan disturbance, scale 0-10	1.4/0.54	1.0/0.28	0	0	10.2/3.47	6.9/1.46	6.9/2.19	4.5/2.18	15.0/5.21	24.0/4.30	2.0/0.71	26.3/5.17	10.2/1.21
litter cover, %	40.1/6.6	55.1/5.4	11.4/4.26	13.8/3.38	0	0	0	0	0	0	23.0/22.34	0	0.92/0.90
rock cover, %	0	0	0	2.3/1.05	6.8/4.14	10.6/4.59	12.9/5.85	63.3/12.34	32.8/6.93	6.1/4.50	51.3/14.77	6.5/1.87	21.7/2.65
bare soil cover, %	61.1/12.62	29.2/4.63	7.2/2.38	0	9.3/4.99	0.7/0.50	46.4/11.37	72.0/24.10	0	18.2/12.65	0	0	9.5/2.46
water cover, %	0	0	0	0	38.3/6.20	44.0/6.73	31.7/4.41	30.7/12.73	39.3/14.87	63.5/	20.0/	31.9/8.66	42.2/3.21
maximum depth of thaw, cm	87.1/6.85	56.8/15.05	38.9/13.74	33.6/7.83	0.4/0.18	0.5/0.26	9.3/2.91	48.8/17.84	11.3/3.96	21.0/19.52	15/	5.8/3.22	3.6/1.32
water depth, cm	0	0	0	4.0/	7.2/4.06	9.6/3.67	21.3/1.25	6.3/5.75	48.1/19	12.5/4.66	2.0/0.54	14.1/4.75	7.2/1.26
surface horizon depth, cm	1.3/0.25	2.7/0.33	4.9/0.99	16.1/7.00	9.4/2.90	24.4/6.65	45.1/9.52	46.1/6.2	82.2/4.54	78.4/11.0	84.2/3.97	86.6/4.51	13.1/1.95
weight loss on ignition, %	96.6/0.67	92.9/0.93	89.5/1.74	84.0/4.33	85.6/2.96	84.4/4.92	86.3/4.56	80.1/10.05	12.3/3.37	15.0/7.17	10.3/3.22	10.3/1.89	86.6/1.27
sand, %	1.9/0.70	4.3/0.84	7.5/1.66	10.9/2.86	10.1/2.20	8.9/3.15	8.5/3.25	14.1/7.48	5.8/2.66	6.6/3.85	5.5/0.76	3.2/0.68	8.9/0.91
silt, %	1.6/0.24	2.7/0.33	3.0/0.30	5.2/1.46	4.3/1.05	6.7/2.18	5.2/1.36						4.4/0.44
clay, %													

Vascular plant growth forms in arctic communities

moisture loss at 106°C, %	0.3/0.03	0.7/0.22	0.9/0.06	1.8/0.69	1.7/0.48	3.5/0.97	5.5/1.04	0.8/0.29	1.0/0.23	2.0/0.63	0.4/0.18	2.1/0.82	1.9/0.25
maximum moisture retention capacity, %	28.8/1.70	38.2/2.29	49.8/2.19	81.7/24.03	72.0/12.81	175.5/75.0	399.5/96.38	50.7/13.16	52.5/7.10	97.8/24.1	31.6/3.14	112.9/50.04	110.5/18.30
moisture tension, 1/3 atm, %	2.9/0.49	6.1/0.89	9.8/0.68	28.5/12.51	22.2/6.96	66.1/21.05	89.2/19.25	13.3/7.42	11.3/2.33	30.6/11.35	5.3/1.70	28.2/10.32	30.1/4.77
moisture tension, 15 atm, %	2.1/0.51	5.2/0.81	8.5/0.56	19.2/10.0	16.1/4.36	54.0/15.44	80.7/17.47	5.2/2.18	7.9/2.11	22.4/9.15	2.5/0.63	24.5/10.48	24.2/3.90
available water, %	0.8/0.28	1.2/0.28	3.0/1.20	16.9/10.03	10.2/4.32	17.2/6.04	24.8/9.66	8.1/5.29	2.8/0.51	15.5/8.36	2.7/1.12	10.7/4.57	9.6/1.73
pH, paste	7.1/0.30	6.5/0.20	5.2/0.19	4.9/0.19	5.1/0.13	5.6/0.28	4.9/0.16	6.5/0.47	6.8/0.19	5.2/0.28	7.5/0.12	5.1/0.22	5.8/0.11
CaCO₃, %	0.3/0.19	0	0	0	0	0	0	0	0	0	2.8/1.96	0	0.1/0.09
total N, ppm	13.6/1.22	16.7/1.40	19.2/1.30	45.4/21.78	35.3/7.09	54.4/21.73	79.9/20.18	20.2/2.81	20.8/2.19	35.9/9.17	18.0/2.07	139.5/78.35	41.2/7.25
total N, %	0.03/0.006	0.07/0.009	0.1/0.01	0.4/0.15	0.3/0.07	0.4/0.12	0.7/0.22	0.1/0.04	0.1/0.05	0.3/0.10	0.05/0.01	0.3/0.10	0.2/0.03
NH₄, ppm	6.9/0.45	8.4/0.51	11.9/0.87	38.2/22.81	23.2/6.73	46.7/21.60	60.8/18.08	12.0/1.62	10.6/0.77	28.8/7.98	8.6/0.68	57.2/41.08	26.2/4.83
NO₃, ppm	6.7/0.81	8.3/1.02	7.3/1.20	7.2/1.29	12.1/3.44	7.7/0.44	19.1/1.68	8.1/1.68	10.2/1.93	7.1/1.21	9.4/1.59	82.3/69.14	15.0/5.15
P, ppm	9.6/2.25	7.1/1.57	9.0/1.94	9.3/3.13	8.0/1.17	8.4/1.53	6.8/2.39	9.0/2.03	8.4/1.03	13.5/7.74	9.8/1.44	109.3/72.82	15.7/5.54
K, ppm	121.6/36.15	101.1/23.90	117.5/37.84	324.3/176.53	98.4/19.95	166.4/23.56	291.9/117.61	85.5/50.86	685/20.42	167.0/110.55	26.3/4.57	132.7/42.69	137.8/16.98
Ca, ppm	337.9/68.99	578.2/94.92	488.3/83.38	2978.8/2265.18	570.1/130.05	836.3/205.44	1303.9/305.77	826.0/327.22	1115.8/221.1	957.7/440.83	541.8/131.40	1005.7/405.77	872.6/113.53
Mg, ppm	40.6/8.28	91.5/13.12	86.6/14.46	473.8/352.72	921/25.61	139.8/39.77	240.6/41.60	173.0/85.38	172.4/33.51	177.0/102.65	107.0/40.50	152.1/40.41	145.0/18.18
Na, ppm	83.6/12.68	117.3/12.93	192.3/75.25	68.5/6.31	122.3/16.43	111.7/13.83	155.6/27.10	109.8/15.10	129.4/13.33	108.3/8.25	136.0/13.90	148.7/32.40	126.1/8.36
Cu, ppm	0.3/0.04	0.7/0.09	1.0/0.14	1.5/0.97	0.8/0.11	1.8/0.59	3.2/0.70	1.1/0.52	1.9/0.46	3.3/1.63	1.3/0.38	1.9/0.64	1.4/0.15
Zn, ppm	0.6/0.09	0.9/0.14	4.0/1.90	3.1/1.70	2.1/0.28	3.5/1.82	7.6/1.95	1.8/0.84	1.7/0.35	5.6/3.00	0.7/0.09	6.2/3.33	2.9/0.45
Mn, ppm	4.3/0.65	6.7/0.95	13.8/3.14	3.4/2.20	13.2/2.70	14.2/4.74	22.2/5.97	17.2/9.37	9.1/1.44	41.6/10.06	6.8/1.62	19.8/9.18	12.9/1.41
Fe, ppm	115/1.80	43.7/10.94	301.6/57.53	356.9/63.98	278.4/58.75	506.6/188.12	1113.8/525.90	133.1/56.94	120.4/46.87	322.4/115.18	25.0/5.64	280.8/75.54	283.4/50.18
Al⁺⁺, meq 100 g⁻¹	0.1/0.06	0.4/0.08	0.5/0.15	2.4/0.52	1.6/0.98	1.6/0.65	2.6/0.56	0.3/	0.1/0.1	1.4/		0.6/0.26	1.3/0.26
H⁺, meq 100 g⁻¹	0.5/0.09	0.6/0.13	3.2/2.65	6.6/4.88	5.5/2.14	1.9/0.84	0.3/0.15	1.4/	0.2/0.2	3.3/		1.9/0.61	2.5/0.61
Ca⁺⁺, meq 100 g⁻¹	0.6/0.09	1.3/0.25	1.4/0.65	2.0/1.28	2.3/0.79	2.8/0.88	2.9/1.18	0.5/	3.5/1.95	1.7/		2.2/1.49	2.1/0.32
Mg⁺⁺, meq 100 g⁻¹	0.03/0.03	0.6/0.48	1.8/1.20	1.0/0.52	0.9/0.38	0.4/0.10	0.4/0.10		0.65/0.65	0.3/		0.7/0.50	0.6/0.12
cation exchange capacity, meq 100 g⁻¹	1.2/0.21	2.8/0.52	6.8/3.35	12.1/4.43	10.3/3.52	6.2/1.95	6.2/1.68	2.2/	4.4/2.5	6.6/		5.5/2.50	6.3/0.95
base saturation, %	49.2/6.16	64.6/8.22	56.6/19.95	31.1/14.75	35.0/7.13	49.3/7.53	46.4/7.77	22.7/	88.0/9.1	30.3/		42.9/12.39	41.4/3.48
pH, KCl	5.5/0.15	5.2/0.35	4.5/0.55	3.8/0.20	4.1/0.09	4.7/0.27	4.0/0.29	4.1/	6.5/0.25	4.0/		4.4/0.03	4.6/0.13

Table 4. Environmental variables in multiple regression equations for growth forms. Atkasook, Alaska.

Growth forms	R^2	Variables in equation						F to enter
1. Large shrubs	.452	temperature	snow	– surface age	– sand	animals		4.0
2. Prostrate deciduous dwarf shrubs	.416	surface age	– bare soil	pH	– streams	– gravity		2.5
3. Prostrate evergreen dwarf shrubs	.444	temperature	snow	– pH	– K	Mg	sand wind – total N max moisture capacity	2.0
4. Erect deciduous dwarf shrubs	.467	– microtines	cryoturbation	– Cu	Zn	– Fe		2.0
5. Erect evergreen dwarf shrubs	.285	– temperature	– erosion	sand	silt	wind		2.1
6. Equisetoids	.151	– clay	– cryoturbation	– gravity	– total N	Cu		1.1
7. Stoloniferous graminoids	.197	– water	– cryoturbation	stream	– gravity	– CaCO$_3$		1.0
8. Rhizomatous graminoids	.320	– surface age	NH$_4$	K	– silt	– cryoturbation	Mn – stream	2.0
9. Caespitose graminoids	.363	– pH	NO$_3$	P	– K	– cryoturbation	– stream – total N	2.2
10. Tussock graminoids	.201	– snow	– erosion	– Ca	silt	stream	– Mn	1.5
11. Aquatics	.328	– temperature	– snow	sand	silt	– wind	stream – CaCO$_3$	1.0
12. Prostrate single forbs	.215	– bare soil	water	NO$_3$	– P	– cryoturbation	– Fe	1.0
13. Prostrate matted forbs	.252	temperature	snow	erosion	– microtines	animals		1.0
14. Erect stoloniferous forbs	.400	temperature	– moisture	bare soil	K	– Ca	– wind – stream	1.0
15. Erect rhizomatous forbs	.469	– water	– pH	NH$_4$	– wind	gravity	– Fe	4.0
16. Erect caespitose forbs	.454	– bare soil	NO$_3$	wind	gravity	– Cu	– Fe max moisture cap.	1.6
17. Erect cushion forbs	.249	surface age	– sand	wind	– total N	– Zn		1.4
18. Erect solitary forbs	.375	pH	NO$_3$	animals	gravity	Fe		2.4

Growth forms and environment

Disturbance and surface soil variables explained a relatively large portion of growth form distribution (Table 4, Fig. 3). *Large deciduous shrubs*, which are associated with browsing caribou and ptarmigans, are dominant on unstable streamsides and relatively stable streambanks. This is the source of their bimodality with pH (5.5 and 7.5), wind, snow, organic matter, silt, clay, most nutrients, and available water. They show highest cover and height on streambanks where deep snow provides protection from wind in winter. Shrubs have only small stature and low cover in moist to wet lowlands. *Prostrate deciduous dwarf shrubs* dominate in partly consolidated habitats which may be controlled by either erosion or deposition, and are bimodal along temperature, thaw, snow, wind, pH (5.5 and 7.5), microtines, NH_4, and Na gradients. They increase with increasing surface age and thaw, and decrease with moisture, most nutrients, OM, available water, and close to lakes. They can occupy more extreme (wetter, longer snow cover) habitats than their evergreen counterparts. *Erect deciduous dwarf shrubs* are largely limited to uniform stable ancient sand dunes and zonal uplands and show no bimodality. They increase with increasing surface age, thaw, temperature, caribou, and most nutrients (except for Zn and Mn), and decrease with snow, erosion, and moisture. *Prostrate evergreen dwarf shrubs* dominate only in dry habitats (*e.g.*, consolidated sand dunes), at lower erosion and lower levels of most nutrients (except for N and Mn), and at higher organic matter and available water than prostrate deciduous dwarf shrubs. They are bimodal along temperature, snow, thaw, wind, surface age, microtines, and erosion, and strongly decrease with most increasing nutrients (except for K, Ca, Mg, and Fe). *Erect evergreen dwarf shrubs* grow on mesic, older, nutrient-poor surfaces. They show normality along many gradients. They dominate at lower values along the gradients of temperature, pH, Ca, Mg, K, Zn, sand, cryoturbation, lakes, animals, and at shallower thaw and higher NH_4, N, Cu, Na, Fe, organic matter, available water, silt, clay, and streams than erect deciduous dwarf shrubs. They decrease with increasing nutrients, and increase with surface age, caribou, Zn, Mn, and Fe.

Equisetoids have an optimum at high erosion and low surface age and show very little bimodality. They decrease with increasing cryoturbation, animals, K, Ca, Mg, organic matter, available water, Zn, Fe, and N, and increase with erosion, moisture, thaw, pH, P, base saturation, Na, silt, clay, and streams. *Stoloniferous graminoids* also occur at high erosion and low surface age and organic matter. Their relationship to snow, pH (5.2, 7.5), and streams is bimodal. They decrease with increasing animal effects, pH, thaw, and most nutrients (except for Cu, Na, and Mn), and increase with erosion and moisture. *Rhizomatous graminoids* dominate in very diverse habitats along lakes and streams and range wider than other growth forms. They have an optimum at high water, moisture, proximity of lakes, cryoturbation, organic matter, available water, nutrients, snow, erosion, and sand, shallow thaw, and low silt, clay, and pH (5.25). *Caespitose graminoids* are important in snowpatches, sand dunes, and on old ridges and show much bimodality, including the gradients of temperature, thaw, wind, pH (5.25 and 7.25), snow, surface age, erosion, caribou, microtines, animals in general, and moisture. They rapidly decrease with most increasing nutrients, organic matter,

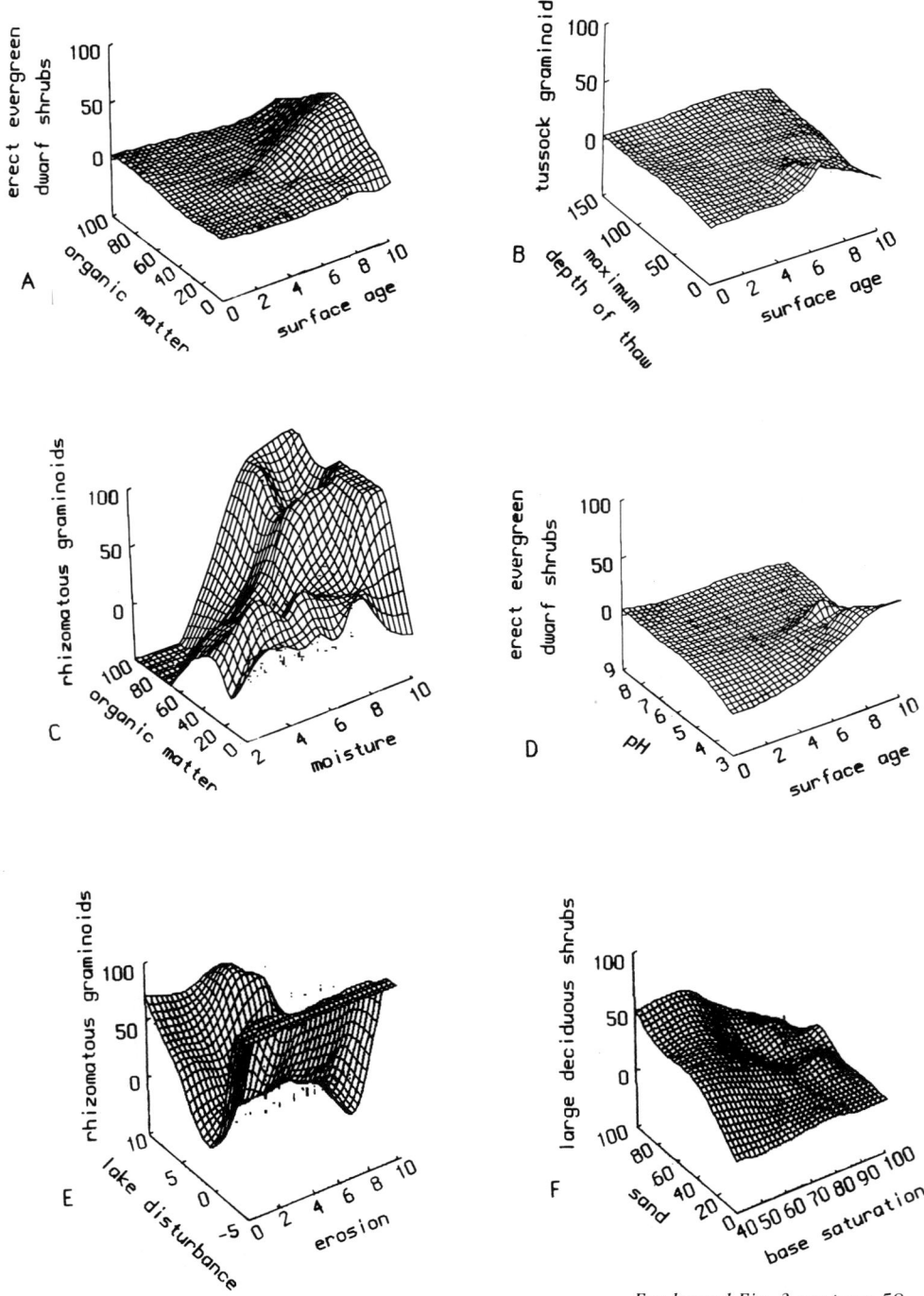

For legend Fig. 3 see page 58.

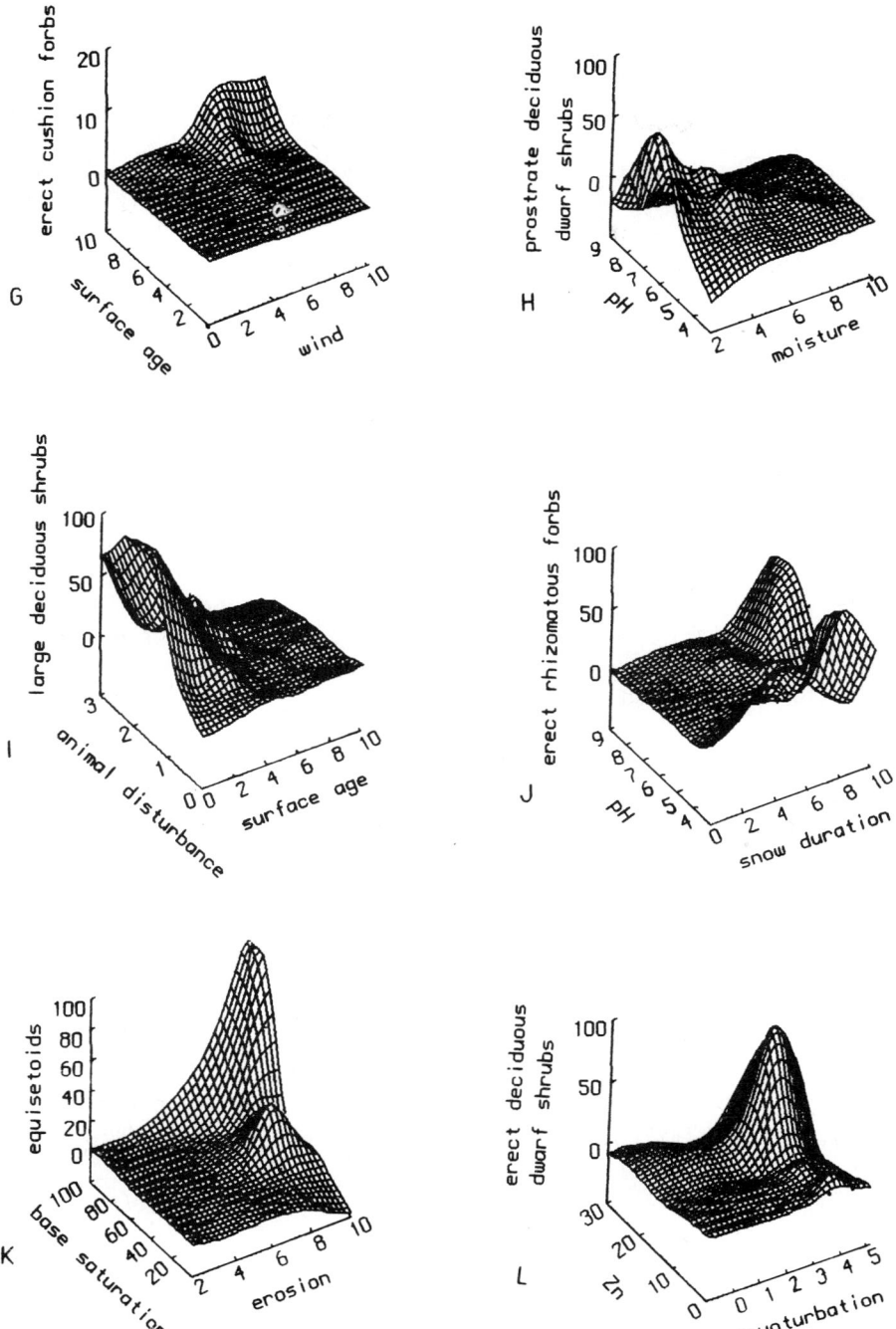

For legend Fig. 3 see page 58.

available water, and proximity of lakes. *Tussock graminoids* dominate mesic zonal uplands which are controlled by local climate. Probably because of this, unlike any other growth form, they show high normality with optima at medium values. Their occurrence on streamsides accounts for bimodalities with erosion, surface age, pH (4.5 and 7), and streams. They are associated with high values of microtine, caribou and other animal disturbance, temperature, organic matter, available water, sand, silt, clay, wind, cryoturbation, and with low values of most nutrients (except for N) and shallow thaw.

Aquatic forbs occur only in lakes and streams. Erosion and pH (7.0) are high, and organic matter, moisture, and most nutrients (except for Mg, Na, Cu, and Mn) are low. *Prostrate forbs with single stems* reach higher cover in only one snowpatch association; they are common, with an optimum at medium high moisture and erosion, long snow cover, pH 6, low temperature, wind, streams, shallow thaw, and very low nutrients and organic matter. *Prostrate forbs with matted stems* occur in small amounts in mostly warm, dry, sandy habitats with medium snow. They are bimodal along surface age, erosion, wind, thaw, microtines, Cu, Zn, Mn, and pH. Most nutrients (except for NO_3, Ca, Mg, N and P) are very low. Organic matter, cryoturbation, Fe, and available water are usually also low. *Erect stoloniferous forbs* occur in very small quantities in warm, dry, young, sandy habitats with high pH (7) and medium snow. Like stoloniferous graminoids, they occur at high erosion. Their optimum is at low cryoturbation, animals, organic matter, available water, silt and clay, very low nutrients (except for K, Cu, Ca, and Mg), and high gravity disturbance. *Erect rhizomatous forbs* have a wide range and show much normality. In marshes, they are the second most important growth form after rhizomatous graminoids with which they have similar relationship to pH and most nutrients. Their optimum is at lower erosion, wind, moisture, streams, cryoturbation, organic matter, available water, clay, silt, Cu, Zn, Mn and shallower thaw, and at higher sand, gravity disturbance, and longer snow cover. *Erect caespitose forbs* have a wide ecological range and are bimodal along snow, wind, thaw, and surface age. Their optimum is in considerably drier, sandier sites, and at higher pH (7.5) and gravity disturbance than that of caespitose graminoids. Both forms reach optima at long snow cover, high erosion, low cryoturbation, low nutrient levels, and small effects of streams, lakes, and animals. *Erect cushion forbs* are important only on windy old ridges and show very little bimodality. Their optimum is in dry, very sandy sites with short snow cover, low caribou, animals in general, erosion, pH (5), and deep thaw. Organic matter, available water, proximity of lakes and streams, silt, clay, and all nutrients (except for K,

---

Fig. 3. Smoothed locally weighted three dimensional surfaces for some growth forms and environmental variables. The units are the same as in Table 3. Erect evergreen dwarf shrubs occur only on old surfaces with high percentages of organic matter (A) and low pH (D). Tussock graminoids occur primarily on shallowly thawed old surfaces (B). Rhizomatous graminoids have an optimum at high organic matter, moisture (C) and erosion (E), and a bimodal relationship to lake disturbance (E). Large deciduous shrubs have an optimum at high sand and animal disturbance, relatively high base saturation, and low surface age (F, I). Erect cushion forbs occur on windy old surfaces (G), prostrate deciduous dwarf shrubs at high pH and low moisture (H), equisetoids at high base saturation and erosion (K), and erect deciduous dwarf shrubs at high cryoturbation and zinc (L). Erect rhizomatous forbs have bimodal relationship to pH and an optimum at long duration of snow cover (J).

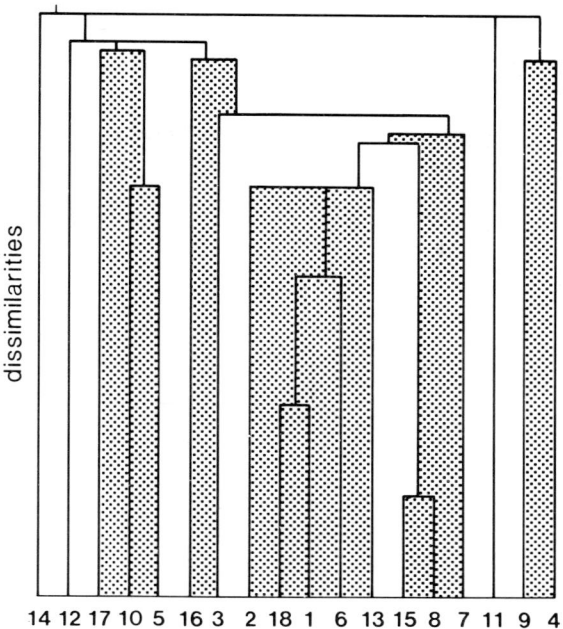

Fig. 4. Single linkage (nearest neighbor) hierarchical tree diagram from a covariance correlation matrix showing the groups of the 18 growth forms based on their percentage cover in 100 samples representing Braun-Blanquet associations. Growth form communities are shaded. See text for interpretation. Numbers 1 to 18 are growth forms as in Fig. 2.

Na, and Cu) are very low. Except for uplands and aquatic habitats, *erect solitary forbs* occur everywhere and show high normality and bimodality. Their optimum is on streambanks and in springs at relatively low surface age, organic matter, available water, and nutrients, and at high pH (7.25).

Growth form communities

Growth forms which utilize the same habitats form several growth form communities (Table 3, Figs 4 and 5) which are analogous to groups of higher plant taxa categories and therefore cannot be paralleled to the Braun-Blanquet system. Like units at all levels of the Braun-Blanquet system, growth form communities are related to landforms. The number of dominant growth forms and growth form diversity are lowest in uplands, springs, streams and lakes. Like the low taxa and association diversity in these habitats, this possibly reflects the lack of diversifying disturbances in zonal habitats and the dominance of water in the others. Marsh growth form and taxa diversity is also low. The highest growth form diversity occurs on streambanks, in lowlands, sand dunes, and in disturbed habitats, all probably diversified by several types of disturbance.

Large deciduous shrubs and erect solitary forbs lead a group which dominates streambanks and disturbed areas, both dry to mesic habitats controlled by surface disturbance. Erect rhizomatous forbs are correlated with rhizomatous graminoids

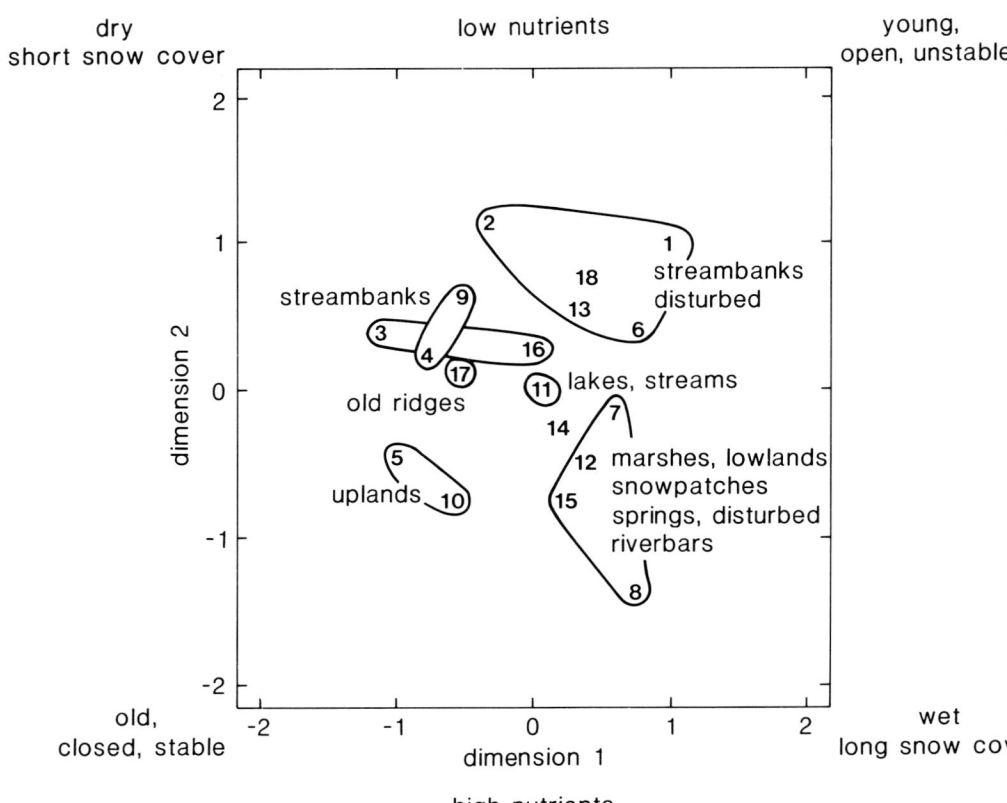

Fig. 5. A configuration of growth forms based on their percentage cover in 100 samples representing Braun-Blanquet associations, produced by monotonic multidimensional scaling using Guttman's coefficient of alienation from covariance correlation matrix. Rhizomatous graminoids are closer to rhizomatous forbs along the third dimension. The landforms on which the growth form groups occur and their controlling environmental gradients are indicated. Numbers 1 to 18 are growth forms as in Fig. 2.

in moist to wet habitats with surfaces disturbed by water, cryoturbation, animals, or humans. On undisturbed zonal uplands, tussock graminoids dominate, accompanied mainly by erect evergreen dwarf shrubs; cushions occur on old ridges, and aquatics in their unique habitat. The remaining groups which are formed at low similarity levels occur in dry, disturbed habitats near rivers. Growth forms which utilize contrasting habitats include those which mainly grow in either unstable or stable ones (shrubs vs. dwarf shrubs, rhizomatous vs. caespitose and tussock graminoids, deciduous vs. evergreen dwarf shrubs, prostrate vs. erect dwarf

Fig. 6. Single linkage (nearest neighbor) hierarchical tree diagram from a Pearson correlation matrix showing the groups of environmental variables and associated groups of growth forms based on 100 samples representing Braun-Blanquet associations. The two major clusters are shaded. See text for interpretation. Similar configuration was produced by monotonic multidimensional scaling using Guttman's coefficient of alienation from Pearson correlation matrix.

Vascular plant growth forms in arctic communities 61

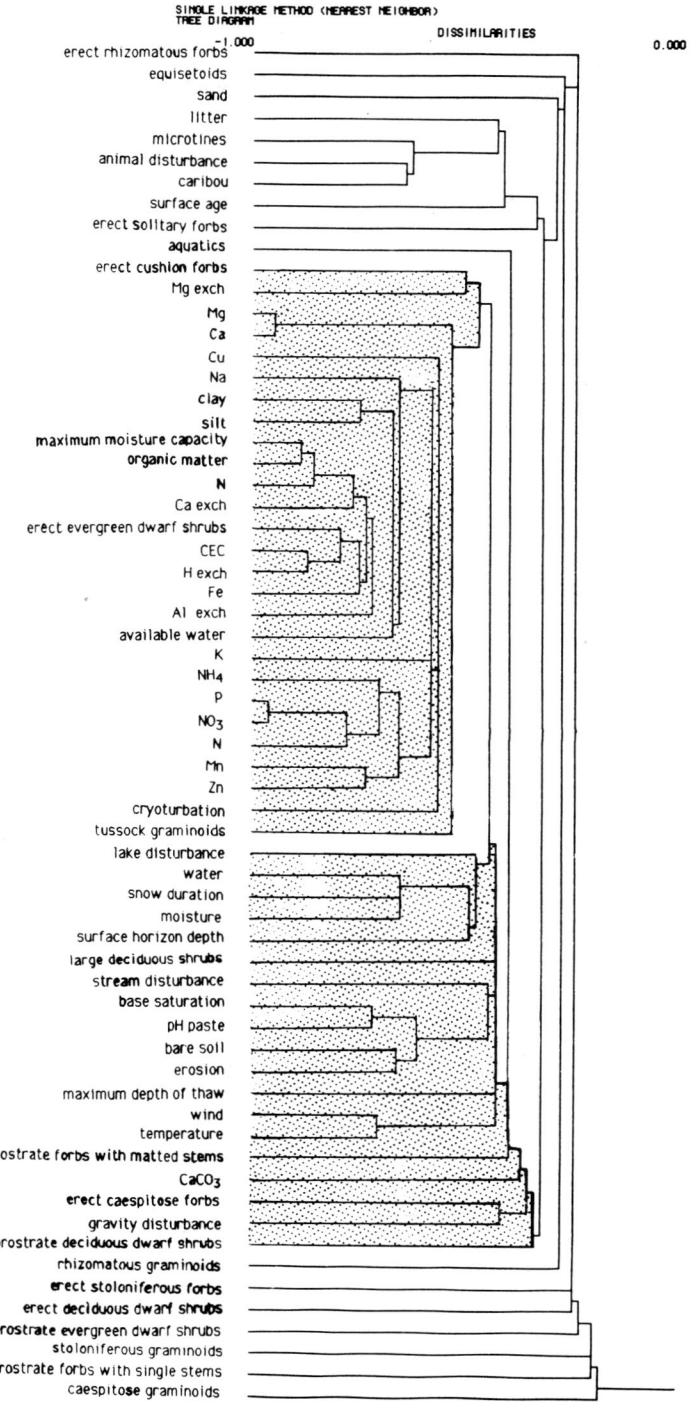

shrubs). Erect evergreen dwarf shrubs appear to be more strongly environmentally differentiated from erect deciduous dwarf shrubs than their prostrate counterparts; this probably reflects the greater microtopographical differentiation on surfaces with less material movement, occupied by the erect forms.

Growth forms which grow together can be expected to have some similarity in their physiology. For example, forbs, graminoids, and deciduous and evergreen shrubs which grow together in exposed fellfields are mostly slow-growing, the forbs form compact cushions or mats with evergreen or semi-evergreen leaves, and reserve metabolisms of the forbs and graminoids resemble those of evergreen shrubs (*e.g.*, Tieszen *et al.* 1981). Webber *et al.* (1980) reported high fidelity of growth forms for vegetation types at Barrow; however, their example, evergreen shrubs in heaths, are at Barrow at the northern edge of their area and may be limited to one habitat for that reason. Growth forms at similar level of resolution did not have high fidelity for vegetation types at Atkasook.

Distribution over landscape units and relation to disturbance

Two groups of classes/landscape units, separated primarily by material erosion or deposition, are indicated by two large variable clusters in Fig. 6 (also Table 3, Fig. 7). They include: (1) surfaces with organic matter and nutrient accumulation, low pH, and high cryoturbation, such as old uplands, marshes, and lowlands; and (2) continually disturbed, eroded surfaces with high pH, $CaCO_3$, and low organic matter and nutrients, such as young habitats along streams and lakes. Both groups of habitats contain surfaces of varying ages. In the first group, soil organic matter is the important reservoir of nutrients; plants are such reservoirs in the second group. Overall, the landscape is nutrient-poor. According to Dowding *et al.* (1981), low temperature keeps the tundra nutrient input low. The input decreases and conservatism of tundra communities with nutrients increases with increasing latitude and continentality. Plant production is limited by low availability of N and especially P in the tundra soils (Tieszen *et al.* 1981).

The oldest surfaces on ancient sand dune ridges are leached and lose light particles to strong winds; they are acidified and are poor in both mineral and organic nutrients as a result. Due to long weathering, cation exchange capacity (CEC) is average. Wind is negatively correlated with organic matter, clay, and almost all nutrients; cushions and dwarf shrubs grow close to the ground. Snowpatches in the lee of old sand dune ridges are also acidified and nutrient poor; they are eroded by creeping snow and leached by meltwater. In both habitats, Fe accumulates through long weathering and leaching; pH and Fe show a strong negative correlation. Due to low relief, very few snowpatches are late-melting. Differences in snow duration, erosion, and moisture produce several different snowpatch dominants.

Tussock graminoids and erect evergreen dwarf shrubs dominate extensive zonal uplands which are flat, mesic, relatively stable, undisturbed, and controlled by the local climate. Organic matter and nutrients accumulate under tussocks but are being incorporated into the permafrost; tussocks insulate the ground and the permafrost table rises. Individual tussocks may slowly grow for several hundreds

Vascular plant growth forms in arctic communities 63

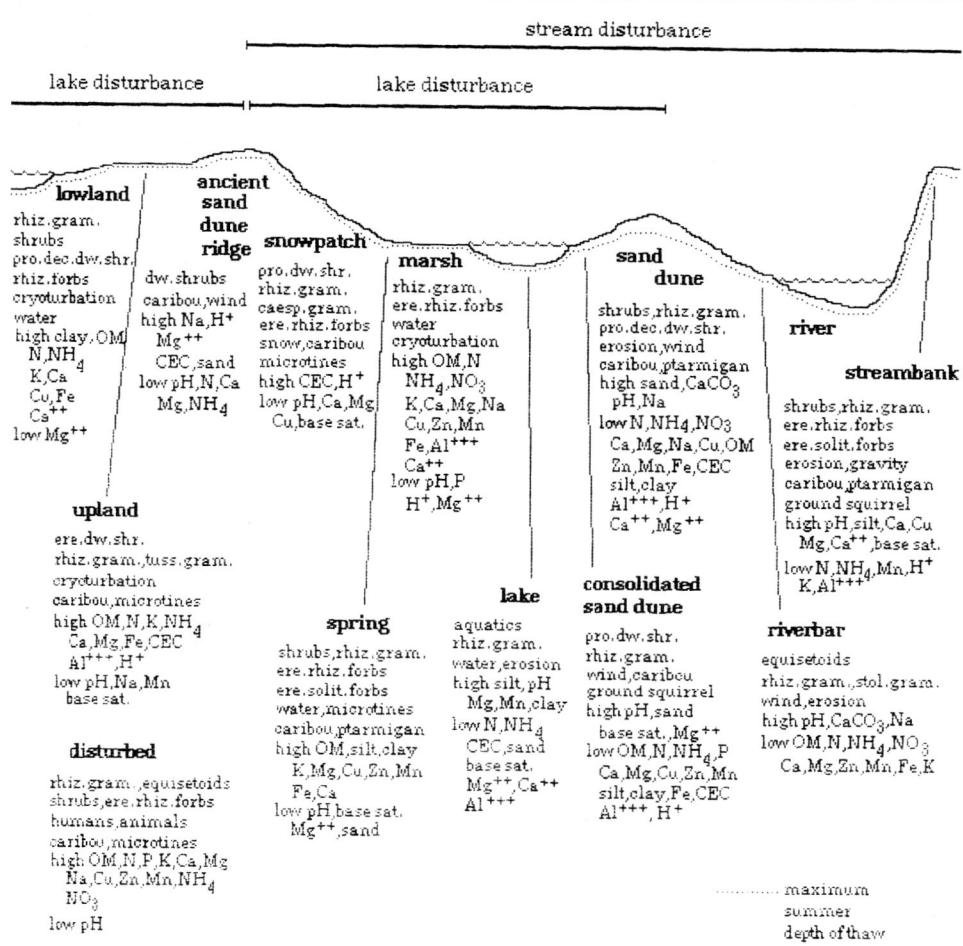

Fig. 7. Summary of the disturbance-controlled classes/landform units and their environmental variables and growth forms in the Atkasook landscape.

of years and only growth forms which fit between them and can respond to their growth can occur; erect stature may be an advantage. Soil between tussocks is often organic nutrient-poor and affected by microcryoturbation. Rhizomatous graminoids are dominant in larger spaces created by cryoturbation such as polygon troughs, which are enriched by nutrients in inflowing water (Dowding *et al.* 1981).

Streams produce most rapid landscape changes. They are, more than lakes, positively correlated with erosion, pH, base saturation, bare soil, and sand, and negatively correlated with organic matter and nutrients. Gravity disturbance (cutbanks) has similar relationships. Silt, clay, and organic matter, which increase the soil exchange complex, are carried away by streams and by wind from streambared surfaces. Silt and clay are deposited on riverbars. $CaCO_3$ occurs only in riverbars and sand dunes, which have the highest pH and strongly correlated base saturation. Thaw is the deepest in sand dunes. The rapidly moving surfaces found in these habitats favor shrubs, rhizomatous graminoids and forbs, prostrate deciduous dwarf shrubs, and equisetoids.

Marshes and lowlands occur mostly along lakes; along streams, exposed sand drains well. Moving fine particles of organic matter, silt, and clay accumulate here and cryoturbation is the strongest. Along with animal-controlled sites, uplands, and partly lowlands, marshes have high amounts of nutrients (except for P), and are positively correlated with organic matter and negatively with pH; most of N and P are bound in organic matter and not available to plants (Dowding et al. 1981). Marshes and uplands have the lowest pH and highest N in tundra soils also elsewhere (Everett et al. 1981). The dominant rhizomatous graminoids may be supported by high nutrients and selected for by high material movement.

In Fig. 6, animal disturbance is separated from the two major clusters. It controls only small areas within major landforms, where organic nutrients may be extremely high (on bird posts, and near arctic ground squirrel burrows and caribou carcasses). Here rhizomatous graminoids are dominant, and forbs and deciduous shrubs, which have higher nutrient requirements (Tieszen et al. 1981), increase. The overall effect on the landscape is probably small; obvious anti-herbivore adaptations are absent. Microtines and other animals are positively correlated with organic nutrients. Microtines may cause an increase of nutrients in snowpatches and in other nesting places; they exert strong influence only in large numbers (Batzli 1981). Caribou have a negative correlation to nutrients, probably because they (and bears) move along old ridges and streambanks where it is easier to walk and which are nutrient-poor. They graze in generally drier habitats (White et al. 1981).

Geographical comparison

The growth form composition at Atkasook (Table 5) is low arctic. Like at high arctic Edgeøya, there are no winding, climbing, or arcuate forbs, no solitary (annual) graminoids, and only very few forbs with stolons; these are growth forms of warmer latitudes. Unlike at Edgeøya, low arctic shrubs play a significant role, stoloniferous graminoids occur in disturbed habitats, and erect forbs with branched stems grow mostly under tallest willows. At Edgeøya, forbs are important in drier habitats, taking the place of the dwarf shrubs at Atkasook, and cushions play a greater role, probably due to a more severe climate. Barkman (1987) found little correspondence between Edgeøya and Barrow.

All shrubs are much less important at Barrow than at Atkasook despite the relatively short distance. The Barrow area is azonal and controlled by the severe Arctic Ocean climate. Its climate and ecosystems perhaps correspond to the high Arctic, and in Alaska represent only a very narrow coastal strip. At Barrow, marshes and, consequently, rhizomatous graminoids are more important than at Atkasook. Graminoids increased along with moisture and forbs were most important in dry sites at Barrow; at Atkasook, graminoids decrease in very wet sites, and forbs are highest in association with shrubs.

The diverse IBP sites, including alpine sites, produced (compared to Atkasook) poorly defined groups of 13 growth forms (French 1981). Some trends at these

Table 5. Comparison of vegetation growth form composition at different latitudes. Percentages of total importance/vegetation cover of vascular plants.

Growth form: Site, latitude	Large shrubs	Dwarf shrubs		Rhizomatous graminoids	Caespitose graminoids	Prostrate forbs	Erect forbs	
		deciduous	evergreen				cushion	others
Edgeøya, 78°05′N	0	0	0	23.4	5.5	33.4	6.8	30.7
Barrow, 71°18′N	1.7	4.5	1.7	46.4	7.2	7.8	0.2	30.3
Atkasook, 70°29′N	14.6	15.0	7.8	31.1	6.9	2.9	0.8	17.4

	Large and dwarf shrubs		'Single' graminoids	Caespitose graminoids	Prostrate forbs	Erect forbs	
	deciduous	evergreen				cushion	others
High Arctic							
Edgeøya, Spitsbergen, 78°05′N	0	0	23.4	5.5	33.4	6.8	30.7
Devon Island, Canada, 75°33′N	5.0	7.8	46.1	11.1	9.4	11.8	8.1
Low Arctic							
Tareja, USSR, 73°30′N	7.0	2.3	28.5	21.2	2.0	27.0	11.7
Barrow, Alaska, 71°18′N	9.8	2.2	48.3	9.4	6.0	0.3	24.0
Atkasook, Alaska, 70°29′N	29.6	7.8	31.1	6.9	2.9	0.8	17.4
Disko Island, Greenland, 69°15′N	28.4	2.3	2.6	16.2	6.0	12.6	28.6
Subarctic							
Abisko, Sweden, 68°22′N	9.3	22.4	35.1	12.0	0	0	21.1
Temperate							
Glenamoy, Ireland, 54°12′N	1.1	35.5	11.3	35.0	2.2	0.4	13.5
Low Antarctic							
South Georgia, 54-55°S	31.8	0	16.8	49.6	1.8	0	0
High Antarctic							
Signy Island, 60°S	0	0	0	91.1	0	8.9	0

Note: Based on data for all habitats available from each site. In the upper half, data for Edgeøya from Barkman (1987) and for Barrow from Webber *et al.* (1980). For Barrow, large shrubs (*S. planifolia* ssp. *pulchra*, *S. lanata* ssp. *richardsonii*?), which grow low at Barrow, were obtained from a preprint species table despite its discrepancies with the published growth form table.

In the lower half, data from Webber and Wielgolaski (in French 1981). Dwarf shrubs and large shrubs were not distinguished in the IBP data. 'Single' graminoids include rhizomatous, stoloniferous, and solitary graminoids, and caespitose graminoids include both caespitose and tussock graminoids. Prostrate forbs include only matted forbs at sites other than Atkasook and Edgeøya. Pteridophytes were omitted (they were included in the totals) because they are a taxonomical rather than a growth form category. The columns of caespitose and 'single' graminoids were apparently reversed in French (1981) and are, therefore, switched here; at Signy Island, the only graminoid (*Deschampsia antarctica* Desv.) is caespitose, and most graminoids are rhizomatous at Barrow, Alaska.

sites did not occur at Atkasook; low average cover of vascular plants, dominance of forbs in mesic meadows, and few forbs in shaded habitats (Wielgolaski et al. 1981). Shrubs decrease with increasing latitude and increasing cold in both hemispheres. Evergreen shrubs seem to be less important than deciduous shrubs at higher latitudes and more than deciduous shrubs in bogs at lower latitudes.

Graminoids are most important at all sites except for forb-rich Edgeøya and Disko Island, and rhizomatous graminoids are more important than caespitose graminoids at all arctic sites except Disko Island. This probably reflects the absence of marshes in the data from these two sites. In the Antarctic, caespitose graminoids predominate. Highly permeable, little weathered surface materials and sloping topography do not support marshes and on the Antarctic Peninsula; those few seen are dominated by bryophytes. Caespitose (tussock) graminoids may be enhanced by dry, stable surfaces and perhaps by locally severe climatic conditions; their high importance at Tareja, on the Antarctic Peninsula, and on Disko Island coincides with high importance of cushion forbs. Erect forbs other than cushions appear to be important in relatively nutrient-rich meadows which are absent at Atkasook.

Conclusions

Frequent surface disturbances create the heterogeneity of landform patches and communities in the Atkasook landscape. Each landform type has a characteristic suite of disturbances and is environmentally and vegetationally unique. These disturbances may also have an opposite, homogenizing effect when they interrupt the gradual differentiation of landforms and plant communities at regular intervals. This still produces mature communities, but only on low-relief terrain with gentle environmental and vegetational gradients. Compared to a mid-latitude, high-relief Northern Hemisphere alpine area of comparable size and plant diversity (e.g., the Indian Peaks area in the Southern Rocky Mountains; Komárková 1979), the number of Braun-Blanquet units in the Atkasook area is higher at all hierarchy levels, the units are less well-defined and, like taxa, show less clear environmental responses. Many growth forms occur in two or more habitats and their environmental response is bimodal or broadly normal. The growth form communities at the 18 growth form level are also relatively poorly defined. This agrees with some of the suggestions made in the literature; for example, Summerhayes and Elton (1928), Griggs (1934), Polunin (1934, 1948), and Savile (1960, 1961) emphasized the poor definition of vegetation units in the Arctic and related it to the instability and youth of the substrate or to reduced competition.

Latitude (temperature) and zonal climate may determine which growth forms will be present at a site and which are the dominant growth forms on zonal surfaces; for example, the importance of trees and shrubs regularly decreases with increasing latitude and severity of climate. Some growth forms (e.g., graminoid tussocks) may contribute to their dominance on zonal surfaces by influencing surface properties. The composition of growth forms at a specific site reflects the predominance of certain habitats. Some sites are marshy, with high surface movement and high importance of rhizomatous forms (Atkasook, Barrow), others are dry and stable, with high importance of cushions and caespitose or tussock graminoids (Antarctic Peninsula, South Orkney Islands).

It appears that distribution and dominance of the recognized 18 growth forms on various landforms at Atkasook is determined both by surface properties and by micro-climate, and that surface stability has more influence than the amount of organic nutrients (and usually associated pH and base saturation) in the surface soil horizon. For example, cushion forbs occur on stable, organic nutrient-poor, windy surfaces; rhizomatous graminoids are dominant in unstable habitats which may either be organic nutrient-rich or -poor; and shrubs dominate in highly disturbed, usually wind-protected, both organic nutrient-poor and -rich habitats. Caespitose and tussock graminoids dominate on usually stable surfaces that may be organic nutrient-rich or -poor. Rhizomatous forbs are important in relatively unstable, usually organic nutrient-rich habitats, and solitary forbs occur in association with shrubs. Dwarf shrubs are important in dry to mesic, relatively stable, both organic nutrient-rich and -poor habitats; deciduous dwarf shrubs in usually less stable ones and evergreen dwarf shrubs in nutrient-poor ones. Equisetoids and stoloniferous graminoids dominate in unstable and organic nutrient-poorer habitats. In some cases microtopographical patches may allow growth forms other then the dominants. This is best seen on old surfaces; for example, on uplands, evergreen dwarf shrubs dominate small surface disturbances between tussocks and rhizomatous graminoids dominate polygon troughs.

Acknowledgments

We would like to greatly thank prof. Dr. J.J. Barkman for his growth form system, his generous discussions of it, and for his help with the article. We are also grateful to L. Granato and H.A. Todd Ferwerda for programming and analysis of the Atkasook taxa data, to prof. Drs. M.P. Hainard and M.H. Clémençon for the use of facilities at the University of Lausanne, to Dr. M. Dale and D. Ross for program package TAXON, to Drs. K.R. Everett and L.D. Carter for help with landscape dynamics, to Dr. N.V. Matveyeva for Russian literature, and to G. Kamani and Drs. M. Rejmánek and B. Bressoud for help. This work was supported by the U.S. National Science Foundation grants DPP-7825748, OPP-7512949, and DPP-8611827. University of Colorado and National Center for Atmospheric Research (NCAR) provided computer funds. The travel of V.K. to the symposium was supported by the American College of Switzerland faculty development funds.

References

Aleksandrova, V.D. 1977. The Arctic and Antarctic: their division into geobotanical areas. Nauka, Leningrad (Translation by D. Löve). Cambridge University Press, New York, 1980. 247 pp.
Argus, G.W. 1973. The genus Salix in Alaska and the Yukon. National Museums of Canada, Ottawa. Publications in Botany 2. 279 pp.
Barkman, J.J. 1979. The investigation of vegetation texture and structure. In: M.J.A. Werger (ed.), The Study of Vegetation, pp. 123–160. Dr W. Junk Publishers, The Hague.
Barkman, J.J. 1983. Growth form system for the plants of the Netherlands. Unpublished manuscript. Landbouwhogeschool, Biologisch Station, Kampsweg 27, 9418 PD Wijster, The Netherlands.
Barkman, J.J. 1987. Preliminary investigations on the texture of high arctic tundra vegetation. In: A.H.L. Huiskes et al. (eds), Vegetation Between Land and Sea, pp. 120–130. Dr W. Junk Publishers, Dordrecht.
Barkman, J.J. 1988. New systems of plant growth forms and phenological plant types. In: M.J.A.

Werger, P.J.M. van der Aart, H.J. During and J.T.A. Verhoeven (eds), Plant Form and Vegetation Structure: Adaptation, plasticity and relation to herbivory, pp. 9–44. SPB Academic Publishing, The Hague.
Batzli, G.O. 1981. Populations and energetics of small mammals in tundra ecosystem. In: L.C. Bliss, O.W. Heal and J.J. Moore (eds), Tundra Ecosystems: A Comparative Analysis, pp. 377–396. Cambridge University Press, New York.
Batzli, G.O. and Jung, H.G. 1980. Nutritional ecology of microtine rodents: resource utilization near Atkasook, Alaska. Arctic and Alpine Research 12: 483–499.
Batzli, G.O. and Sobaski, S. 1980. Distribution, abundance, and foraging patterns of ground squirrels near Atkasook, Alaska. Arctic and Alpine Research 12: 501–510.
Billings, W.D. and Mooney, H.A. 1968. The ecology of arctic and alpine plants. Biological Review of the Cambridge Philosophical Society 43: 481–529.
Billings, W.D. and Peterson, K.M. 1980. Vegetational change and ice-wedge polygons through the thaw-lake cycle in arctic Alaska. Arctic and Alpine Research 12: 413–432.
Black, R.F. 1951. Eolian deposits of Alaska. Arctic 4: 89–111.
Black, R.F. 1964. Gubik Formation of Quaternary age in northern Alaska. U.S. Geological Survey Professional Paper 302-C: 59–91.
Bliss, L.C. 1970. Primary production within tundra ecosystems. In: W.A. Fuller and P.G. Kevan (eds), Productivity and Conservation in Northern Circumpolar Lands, pp. 51–79. International Union for the Conservation of Nature and Natural Resources, Morges, Switzerland. New Series, No. 16.
Britton, M.E. 1967. Vegetation of the arctic tundra. In: H.P. Hansen (ed.), Arctic Biology, pp. 67–130. Oregon State University Press, Corvallis, Oregon.
Callaghan, T.V. and Collins, N.J. 1981. Life cycles, population dynamics and the growth of tundra plants. In: L.C. Bliss, O.W. Heal and J.J. Moore (eds), Tundra Ecosystems: A Comparative Analysis, pp. 257–284. Cambridge University Press, New York.
Carter, L.D. 1981a. A Pleistocene sand sea on the Alaskan Arctic Coastal Plain. Science 211: 381–383.
Carter, L.D. 1981b. Middle Wisconsinan through Holocene climate in the Ikpikpuk River region, Alaska. University of Colorado, Institute of Arctic and Alpine Research, Arctic Workshop 10: 5–9.
Chapin, F.S. III and Shaver, G.R. 1985. Individualistic growth response of tundra plant species to environmental manipulation in the field. Ecology 66: 564–576.
Chester, A.L. and Shaver, G.R. 1982. Reproductive effort in cotton grass tundra. Holarctic Ecology 5: 200–206.
Clebsch, E.E.C. 1957. The summer season climatic and vegetational gradient between Point Barrow and Meade River, Alaska. M.S. thesis, University of Tennessee, Knoxville, Tennessee. 60 pp.
Clebsch, E.E.C. and Shanks, R.E. 1968. Summer climatic gradients and vegetation near Barrow, Alaska. Arctic 21: 161–171.
Dowding, P., Chapin III, F.S., Wielgolaski, F.E. and Kilfeather, P. 1981. Nutrients in tundra ecosystems. In: L.C. Bliss, O.W. Heal and J.J. Moore (eds), Tundra ecosystems: a comparative analysis, pp. 647–683. Cambridge University Press, New York.
Everett, K.R. 1979. Evolution of the soil landscape in the sand region of the Arctic Coastal Plain as exemplified at Atkasook, Alaska. Arctic 32: 207–223.
Everett, K.R. 1980. Distribution and variability of soils near Atkasook, Alaska. Arctic and Alpine Research 12: 433–466.
Everett, K.R., Vassiljevskaya, V.D., Brown, J. and Walker, B.D. 1981. Tundra and analogous soils. In: L.C. Bliss, O.W. Heal and J.J. Moore, J.J. (eds), Tundra Ecosystems: A Comparative Analysis, pp. 139–179. Cambridge University Press, New York.
French, D.D. 1981. Multivariate comparisons of IBP Tundra Biome site characteristics. In: L.C. Bliss, O.W. Heal and J.J. Moore (eds), Tundra Ecosystems: A Comparative Analysis, pp. 47–75. Cambridge University Press, New York.
Gartner, B.L. 1982. Controls over regeneration of tundra graminoids in a natural and man-disturbed site in arctic Alaska. M.Sc. thesis, University of Alaska, Fairbanks, Alaska.
Gartner, B.L., Chapin III, F.S. and Shaver, G.R. 1986. Reproduction of Eriophorum vaginatum by seed in Alaskan tussock tundra. J. Ecol. 74: 1–18.
Griggs, R.F. 1934. The problem of arctic vegetation. Wash. Acad. Sci. 24: 153–175.
Grulke, N.E. and Bliss, L.C. 1985. Growth forms, carbon allocation, and reproductive patterns of high arctic saxifrages. Arctic and Alpine Research 17: 241–250.
Hultén, E. 1968. Flora of Alaska and neighboring territories. Stanford University Press, Stanford, California. 1008 pp.

Komárková, V. 1979. Alpine vegetation of the Indian Peaks area, Front Range, Colorado Rocky Mountains. Flora et vegetatio mundi 7, R. Tüxen, (ed.), 2 vols. Cramer, Vaduz. 591 pp.
Komárková, V. 1981. Holarctic alpine and arctic vegetation: circumpolar relationships and floristic-sociological, high-level units. In: H. Dierschke (ed.), Syntaxonomie, pp. 451–476. Cramer, Vaduz.
Komárková, V. 1983. Recovery of plant communities and summer thaw at the 1949 Fish Creek Test Well 1, arctic Alaska. Proceedings of the Fourth International Conference on Permafrost, University of Alaska, Fairbanks, Alaska, 18 to 22 July 1983, pp. 645–650. National Academy Press, Washington, D.C.
Komárková, V. Arctic vegetation at Atkasook, Alaska. American College of Switzerland, CH-1854 Leysin, Switzerland (in preparation).
Komárková, V. and Webber, P.J. 1980. Two low arctic vegetation maps along the Meade River near Atkasook, Alaska. Arctic and Alpine Research 12: 447–472.
Löve, Á. and Löve, D. 1975. Cytotaxonomical atlas of the arctic flora. Cramer, Vaduz. 598 pp.
Peterson, K.M. 1978. Vegetational successions and other ecosystemic changes in two arctic tundras. Ph.D. thesis. Duke University, Durham, North Carolina. 323 pp.
Peterson, K.M. and Billings, W.D. 1978. Geomorphic processes and vegetational change along the Meade River sand bluffs in Northern Alaska. Arctic 31: 7–23.
Péwé, T.L. 1975. Quaternary geology of Alaska. U.S. Geological Survey Professional Paper 835. 139 pp.
Polozova, T.G. 1978. Žiznennye formy sosudistych rastenij Tajmyrskogo stacionara (Life forms of vascular plants of the Tajmyr station). In: Struktura i funkcii biogeocenozov tajmyrskoj tundry (Structure and function of the biogeocoenoses of Tajmyr tundra), pp. 114–143. Akademija nauk SSSR, Naučnyj sovet po problemam biosfery, Botaničeskij Institut im. V.L. Komarova. Nauka, Leningrad.
Polozova, T.G. 1981. Žiznennyje formy sosudistych rastenij v različnych podzonach tajmyrskoj tundry (Life forms of vascular plants in the different subzones of the Tajmyr tundra). In: Žiznennye formy: struktura spektry i evolucija (Life forms: structure spectra and evolution), pp. 265–281. Akademija nauk SSSR, Moskovskoe obščestvo ispytatelej prirody. Nauka, Leningrad.
Polozova, T.G. 1983. Sostav biomorf i nekotorye osobennosti struktury reliktovych stepnych soobščestv zapadnoj Čukotki (The life form composition and some structural features of relict steppes in western Čukotka). Botaničeskij Žurnal 68: 1503–1512.
Polunin, N. 1934. The flora of Akpatok Island, Hudson Strait. J. Bot. 72: 197–204.
Polunin, N. 1948. Botany of the Canadian Eastern Arctic, Part III. Vegetation and ecology. Nat. Mus. Can. Bull. 104. 304 pp.
Raunkiaer, C. 1934. The life forms of plants and statistical plant geography. Clarendon Press, Oxford.
Savile, D.B.O. 1960. Limitations of the competitive exclusion principle. Science 132: 1761.
Savile, D.B.O. 1961. The botany of the northwestern Queen Elizabeth Islands. Can. J. Bot. 39: 909–942.
Small, E. 1972. Photosynthetic rates in relation to nitrogen recycling as an adaptation to nutrient deficiency in peat bog plants. Can. J. Bot. 50: 2227–2233.
Sørensen, T. 1941. Temperature relationships and phenology of the northeast Greenland flowering plants. Meddelser om Grønland 125(9): 1–305.
Summerhayes, V.A. and Elton, C.S. 1928. Further contributions to the ecology of Spitzbergen. J. Ecol. 16: 13–267.
Tieszen, L.L., Lewis, M.C., Miller, P.C., Mayo, J., Chapin III, F.S. and Oechel, W. 1981. An analysis of processes of primary production in tundra growth forms. In: L.C. Bliss, O.W. Heal and J.J. Moore (eds), Tundra Ecosystems: A Comparative Analysis, pp. 285–356. Cambridge University Press, New York.
Tolmačev, A.I. 1971. Florenreichtum als Gegenstand vergleichender Forschung. Feddes Repert. 82: 561–572.
Tüxen, R. 1970. Einige Bestandes- und Typenmerkmale in der Struktur der Pflanzengesellschaften. In: R. Tüxen (ed.), Gesellschaftsmorphologie, pp. 76–98. Dr W. Junk Publishers, The Hague.
Wahrhaftig, C. 1965. Physiographic divisions of Alaska. U.S. Geological Survey Professional Paper 482. 52 pp.
Webber, P.J. 1978. Spatial and temporal variation of the vegetation and its production, Barrow, Alaska. In: L.L. Tieszen (ed.), Vegetation and Production Ecology of an Alaskan Arctic Tundra, pp. 37–112. Ecological Studies 29. Springer, New York.
Webber, P.J., Miller, P.C., Chapin III, F.S. and McCown, B.H. 1980. The vegetation: pattern and succession. In: J. Brown, P.C. Miller, L.L. Tieszen and F.L. Bunnell (eds), An arctic ecosystem: the

coastal tundra at Barrow, Alaska, pp. 186–218. Hutchinson, Dowden and Ross, Stroudsburg, Pennsylvania.

Webber, P.J., Walker, D.A. and Rowley, F. 1975. Gradient analysis of plant growth forms. In: J. Brown (ed.), Ecological and limnological reconnaissances from Prudhoe Bay into the Brooks Range, Alaska, pp. 31–35. RATE (Research on Arctic Tundra Environments) Report.

Westhoff, V. and Van der Maarel, E. 1978. The Braun-Blanquet approach. 2nd ed. In: R.H. Whittaker (ed.), Classification of Plant Communities, pp. 287–399. Dr W. Junk Publishers, The Hague.

White, R.G., Bunnell, F.L., Gaare, E., Skogland, T. and Hubert, B. 1981. Ungulates on arctic ranges. In: L.C. Bliss, O.W. Heal and J.J. Moore (eds), Tundra ecosystems: a comparative analysis, pp. 397–483. Cambridge University Press, New York.

White, R.G. and Trudell, J. 1980. Habitat preference and forage consumption by reindeer and caribou near Atkasook, Alaska. Arctic and Alpine Research 12: 511–529.

Wielgolaski, F.E., Bliss, L.C., Svoboda, J. and Doyle, G. 1981. Primary production of tundra. In: L.C. Bliss, O.W. Heal and J.J. Moore (eds), Tundra Ecosystems: A Comparative Analysis, pp. 187–225. Cambridge University Press, New York.

Williams, J.R., Yeend, W.E., Carter, L.D. and Hamilton, T.D. 1977. Preliminary surficial deposits map of National Petroleum Reserve-Alaska. U.S. Geological Survey Open File Report 77-868 (map + refs).

ATTRIBUTES OF COMPETITIVE ABILITY IN HERBACEOUS PLANTS

GARY A. EPP and LONNIE W. AARSSEN
Department of Biology, Queen's University, Kingston, Ontario, Canada K7L 3N6

Abstract

The ability of a plant to compete for resources may depend on a combination of several of its attributes relative to those of other plants. This paper reviews some of the attributes which have been shown to confer competitive ability in plants. Attention is drawn to the problems related to the use of combinations of attributes as predictors of success in competition.

Introduction

Competitive ability is generally regarded as the relative ability of an individual to acquire contested resources, and to deny resources to others. For plants, essential resources (*i.e.*, light, water, carbon dioxide, and nutrients) vary considerably in nature and form and, for the most part, must be obtained independently throughout the life of a plant. Consequently, there are numerous attributes of a plant that may confer competitive ability at all life phases.

Table 1 lists plant attributes, according to life phase, that have been demonstrated to confer competitive ability in plants. Some of these attributes are important in exploitation competition, whereas others are important in interference competition. The purpose of this paper is to review the evidence for these attributes of competitive ability in herbaceous plants. It is not, however, implied that competitive ability is restricted to only these few traits.

Competitive ability attributes

Seed to seedling phase

Perhaps the most important stage of life during which the success of an individual is determined is the seed to seedling phase (Rhodes 1986; Cavers 1983; Sugiyama and Takahashi 1985). During this critical period, attributes that enable rapid occupation of space, germination, and establishment are important (Table 1) (Ross and Harper 1972; Fowler 1984). Of these, seed weight, or the storage of large seed reserves, has been shown to influence the ability of a plant to compete in the early stages of growth (Black 1958; Kaufman and McFadden 1960; Harper and Obeid 1967; Twamley 1967; Cideciyan and Malloch 1982; Fenner 1983; Peters 1985; Wulff 1986a, 1986b). Where plants from seeds of varying weights have been

Table 1. Attributes that have been demonstrated to confer competitive ability in plants. Attributes are classified by life phase. Each attribute is also classed as an attribute important in exploitation competition (E), versus an attribute important in interference competition (I).

LIFE PHASE/ATTRIBUTE	E or I	REFERENCE
SEED – SEEDLING		
Early seed release	E	Ross and Harper (1972)
Efficient dispersal	E	Davidson and Morton (1981)
Large seed weight	E	Peters (1985)
High seed quality*	E	Parrish and Bazzaz (1985)
Early germination	E	Fowler (1984)
Rapid seedling growth	E	Weaver (1984)
JUVENILE – ADULT		
Rapid growth rate	E	Austin (1982)
Tall plant height	E	Black (1960)
Large shoot biomass	E	Black (1960)
Large root biomass	E	Ennik and Baan Hofman (1983)
Large numbers of roots	E	Sugiyama and Takahashi (1985)
Resource sharing within clones	E	Hartnett and Bazzaz (1983)
Association with symbionts	E	Barea and Azon-Aquilar (1983)
Secretion of allelopathic substances	I	Rice (1979)
Deposition of dense litter	I	Grime (1973); Werner (1975)
Tolerance of drought	E	Thomas (1984)
Tolerance of shading	E	Hutchinson (1967)
Tolerance of defoliation	E	Krans and Beard (1985)
Carrying pathogens of other species	I	Rice and Westoby (1982)
REPRODUCTIVE		
Attractive and numerous flowers	E	Levin and Anderson (1970)
Rapid pollen germination	E	Sari Gorla and Rovida (1980)
Interference with pollen of other species	I	Kanchan and Jayachandra (1980)

* seed quality is defined by the presence of nutrients essential to rapid growth and establishment of seedlings.

grown in competition with each other, plants originating from heavier seeds have out-competed those from lighter seeds (*e.g.*, Peters 1985). It has also been shown that plants growing from seeds with a higher concentration of essential nutrients are superior competitors (Parrish and Bazzaz 1985). Both the quantity and quality of seed reserves, therefore, may be important for success during competition at the seedling phase.

As plants require light to grow, the rapid occupation of space for the growing seedling is crucial. Therefore, the rate at which seeds germinate (and germination date) and subsequent seedling growth rates can be important (Ross and Harper 1972; Cook 1980; Naylor 1980; Klemow and Raynal 1981; Fowler 1984; Weaver 1984). Several studies have shown that early germinating plants on average are better able to compete than later germinating plants (Guneyli *et al.* 1969; Fowler 1984; Sugiyama and Takahashi 1985). Coupled with early germination may be the ability of the germinated seedling to grow rapidly, thereby over-topping and shading other plants (Rhodes 1968; Guneyli *et al.* 1969; Weaver 1984).

Juvenile to adult phase

As growth of the individual progresses from seedling to adult plant, morphological characters become increasingly more important (Table 1). Due to the sessile nature of growing plants, the pre-emption of resources is often increased by numerous or large plant parts (Black 1960; Ennik and Baan Hofman 1983; Sugiyama and Takahashi 1985). In *Lolium perenne* L. root mass has been found to be positively correlated with competitive ability when grown with other species (Baan Hofman and Ennik 1980, 1982; Ennik and Baan Hofman 1983; Snaydon and Howe 1986). Also, the depth of roots (Berendse 1982), and the number of roots (Sugiyama and Takahashi 1985) have been associated with high competitive ability. The proliferation and growth of various plant parts, such as roots, rhizomes, and tillers enables the plant to exploit resources by virtue of its size (Harper 1977).

However, not all competitive ability attributes are number- or size-dependent. Association with symbionts (Barea and Azcon-Aquilar 1983), and tolerance to adverse environmental conditions (Hutchinson 1967; Thomas 1984; Krans and Beard 1985) are two examples of attributes which can indirectly affect a plant's ability to procure resources and hence deny them to neighbors. These attributes, although they may result in increased number of plant parts or size, are not directly dependent on size.

During the adult phase interference attributes may be important. For example, studies have indicated that the deposition of litter by a plant may give it a competitive edge (Grime 1973; Werner 1975; Marshall and Naylor 1984). By producing a large dense mat of leaves around its base, a plant can reduce the germination and establishment of potential competitors (Grime 1979). Other interference attributes may include the secretion of allelopathic chemicals in the soil, which may harm competitors (Rice 1979), and the ability to serve as a host to pathogens that affect the growth of competitors (Rice and Westoby 1982).

Reproductive phase

Competitive ability is dependent, not only on attributes relevant to plant growth, but it is also dependent on attributes that enhance the reproductive success of a plant (Fenner 1978). During the reproductive phase of a plant's life cycle, characteristics related to competition for pollinators and seed production become important. Such attributes may include the production of numerous attractive flowers (Levin and Anderson 1970), rapid pollen germination on the stigma (Sari Gorla and Rovida 1980), and interference with pollen of other plants (Kanchan and Jayachandra 1980). In competition for pollinators, a large number of flowers maximizes the potential for successful seed production.

Attribute complexes

Attributes that confer competitive ability in plants are not always independent of each other. As plants require resources from various components of the environment (soil, air, pollinators, etc.) the ability to compete successfully will depend on a combination of many characteristics of the plant. The study of competitive ability in plants should, therefore, be concerned with combinations of several strategic attributes, or attribute complexes.

Some studies have shown that a combination of measured traits can effectively predict competitive success in plants (Morishima *et al.* 1961; Grime 1973). Grime (1973) used a competitive index calculated from four characters (plant height, growth form, relative growth rate, and persistent litter) to demonstrate the severity of competitive exclusion in six different habitats in Britain. He found that species richness decreased with the competitive ability of coexisting species in 1 m^2 quadrats. Similar results have been found in abandoned hayfields in Canada using ten plant traits (Epp and Aarssen, unpublished data).

Finale

Several problems exist with the use of plant traits as predictive measures of competitive success. Firstly, not all attributes pertinent to the competitive success of an individual can possibly be accounted for. The attribute, or attributes that best determine relative competitive ability of a species may consequently be overlooked. Secondly, the relationship between an attribute and competitive ability may not be linear. The direct measurement of such an attribute may, therefore, not accurately reflect its contribution towards the overall competitive ability of a plant. Finally, some traits which are considered as competitive ability attributes are also used to measure the relative success of species growing together. Such traits are usually related to the size, or biomass of a plant. If success is measured as percent cover, or total plant biomass, the obvious conclusion will be that the competitively superior species are those which produce large plants. Large plants, however, are not necessarily superior competitors. Attributes and measurements of success which are size dependent should, therefore, be handled with a degree of skepticism.

Despite the difficulties involved, it is important to consider combinations of attributes as determinants of competitive success, because it is with a combination of its attributes that a plant competes for resources. Competitive ability of a plant may be affected not only by its relative ability to deny resources to neighbors, but also by its relative ability to tolerate resource depletion or interference by neighbors (Aarssen 1983; Goldberg and Werner 1983). The way in which these two components of competitive ability interact is poorly understood and in need of further research.

Acknowledgments

Support was provided by the Queen' University School of Graduate Studies and Research, and by an operating grant from the Natural Sciences and Engineering Research Council of Canada.

References

Aarssen, L.W. 1983. Ecological combining ability and competitive combining ability in plants: Toward a general evolutionary theory of coexistence in systems of competition. Am. J. Nat. 122: 707–731.

Austin, M.P. 1982. Use of a relative physiological performance value in prediction of performance in multispecies mixtures from monoculture performance. J. Ecol. 70: 559–570.

Baan Hofman, T. and Ennik, G.C. 1980. Investigation into plant characters affecting the competitive ability of perennial ryegrass (*Lolium perenne* L.). Neth. J. Agric. Sci. 28: 97–109.

Barea, J.M. and Azcon-Aquilar, C. 1983. Mycorrhizas and their significance in nodulating nitrogen-fixing plants. Adv. Agron. 36: 1–54.

Berendse, F. 1982. Competition between plant populations with different rooting depths. III. Field experiments. Oecologia 53: 50–55.

Black, J.N. 1958. Competition between plants of different initial seed sizes in swards of subterranean clover (*Trifolium subterraneum* L.) with particular reference to leaf area in the light microclimate. Aust. J. Agric. Res. 9: 299–318.

Black, J.N. 1960. The significance of petiole length, leaf area, and light interception in competition between strains of subterranean clover (*Trifolium subterraneum* L.) grown in swards. Aust. J. Agric. Res. 11: 177–291.

Cavers, P.B. 1983. Seed demography. Can. J. Bot. 61: 3578–3590.

Cideciyan, M.A. and Malloch, A.J.C. 1982. Effects of seed size on the germination, growth and competitive ability of *Rumex crispus* and *Rumex obtusifolius* J. Ecol. 70: 227–232.

Cook, R.E. 1980. Germination and size-dependent mortality in *Viola blanda*. Oecologia 47: 115–117.

Davidson, D.W. and Morton, S.R. 1981. Competition for dispersal in ant-dispersed plants. Science 213: 1259–1261.

Ennik, G.C. and Baan Hofman, T. 1983. Variation in the root mass of ryegrass types and its ecological consequences. Neth. J. Agric. Sci. 31: 325–334.

Fenner, M. 1978. A comparison of the abilities of colonizers and close-turf species to establish from seed in artificial swards. J. Ecol. 66: 953–963.

Fenner, M. 1983. Relationships between seed weight, ash content and seedling growth in twenty-four species of Compositae. New Phytol. 95: 697–706.

Fowler, N. 1984. The role of germination date, spatial arrangement, and neighbour effects in competitive interactions in *Linum*. J. Ecol. 72: 307–318.

Goldberg, D.E. and Werner, P.A. 1983. Equivalence of competitors in plant communities: A null hypothesis and a field experimental approach. Am. J. Bot. 70: 1098–1104.

Grime, J.P. 1973. Competitive exclusion in herbaceous vegetation. Nature 242: 344–347.

Grime, J.P. 1979. Plant strategies and vegetation processes. John Wiley and Sons, Chichester.

Guneyli, E., Burnside, O.C. and Nordquist, P.T. 1969. Influence of seedling characteristics on weed competitive ability of sorghum hybrids and inbred lines. Crop Sci. 9: 713–716.

Harper, J.L. 1977. Population biology of plants. Academic Press, London.

Harper, J.L. and Obeid, M. 1967. Influence of seed size and depth of sowing on the establishment and growth of varieties of fiber and oil seed flax. Crop Science 7: 527–532.

Hartnett, D.C. and Bazzaz, F.A. 1983. Physiological integration among intraclonal ramets in *Solidago canadensis*. Ecology 64: 779–788.

Hutchinson, T.C. 1967. Comparative studies of the ability of species to withstand prolonged periods of darkness. J. Ecol. 55: 291–299.

Kanchan, S. and Jayachandra 1980. Pollen allelopathy: A new phenomenon. New Phytol. 84: 739–746.

Kaufmann, M.L. and McFadden, A.D. 1960. The competitive interaction between barley plants grown from large and small seed. Can. J. Plant Sci. 40: 623–629.

Klemow, K.M. and Raynal, D.J. 1981. Population ecology of *Melilotus alba* in limestone quarry. J. Ecol. 69: 33–44.

Krans, J.V. and Beard, J.B. 1985. Effects of clipping on growth and physiology of 'Merion' Kentucky bluegrass. Crop Sci. 25: 17–20.

Levin, D.A. and Anderson, W.W. 1970. Competition for pollinators between simultaneously flowering species. Am. Nat. 104: 455–467.

Marshall, A.H. and Naylor, R.E.L. 1984. Reasons for poor establishment of direct reseeded grassland. Ann. Appl. Biol. 105: 87–96.

Morishima, H., Oka, H-I. and Chang, W-T. 1961. Directions of differentiation in populations of wild rice, *Oryza perennis* and *O. sativa* F. *spontanea*. Evolution 15: 326–339.
Naylor, R.E. 1980. Effects of seed size and emergence time on subsequent growth of perennial ryegrass. New Phytol. 84: 313–318.
Parrish, J.A.D. and Bazzaz, F.A. 1985. Nutrient content of *Abutilon theophrasti* seeds and the competitive ability of the resulting plants. Oecologia 65: 247–251.
Peters, N.C.B. 1985. Competitive effects of *Avena fatua* L. plants from seed of different weights. Weed Res. 25: 67–77
Rhodes, I. 1968. The growth and development of some grass species under competitive stress: 3. The nature of competitive stress, and characters associated with competitive ability during seedling growth. J. Brit. Grassl. Soc. 23: 330–335.
Rice, E.L. 1979. Allelopathy – An update. Bot Rev. 45: 15–109.
Rice, E.L. and Westoby, M. 1982. Heteroecious rusts as agents of interference competition. Evol. Theor. 6: 43–52.
Ross, M.A. and Harper, J.L. 1972. Occupation of biological space during seedling establishment. J. Ecol. 60: 77–88.
Sari Gorla, M. and Rovida, E. 1980. Competitive ability of maize pollen. Intergametophytic effects. Theor. Appl. Genet. 57: 37–41.
Snaydon, R.W. and Howe, C.D. 1986. Root and shoot competition between established ryegrass and invading grass seedlings. J. Appl. Ecol. 23: 667–674.
Sugiyama, S. and Takahashi, N. 1985. Variation of competitive ability during seedling growth in strains of *Festuca arundinacae* Schreb. J. Japan. Grassl. Sci. 31: 26–33.
Thomas, H. 1984. Effects of drought on growth and competitive ability of perennial ryegrass and white clover. J. Appl. Ecol. 21: 591–602.
Twamley, B.E. 1967. Seed size and seedling vigor in birdsfoot trefoil. Can. J. Plant Sci. 47: 603–609.
Weaver, S.E. 1984. Differential growth and competitive ability of *Amaranthus retroflexus*, *A. powellii* and *A. hybridus*. Can. J. Plant Sci. 64: 715–724.
Werner, P.A. 1975. The effects of plant litter on germination in teasel, *Dipsacus sylvestris* Huds. Amer. Midl. Natur. 94: 470–476.
Wulff, R.D. 1986a. Seed size variation in *Desmodium paniculatum* II. Effects on seedling growth and physiological performance. J. Ecol. 74: 99–114.
Wulff, R.D. 1986b. Seed size variation in *Desmodium paniculatum* III. Effects on reproductive yield and competitive ability. J. Ecol. 74: 115–121.

GERMINATION AND ESTABLISHMENT OF DICOTYLEDONS IN GRASSLANDS

L.J. DE BROECK
Laboratory of Ecology, K.U. Leuven, Kardinaal Mercierlaan 92, B-3030 Louvain, Belgium

Abstract

During a period of five weeks after germination, the establishment of five ruderals and five grassland species in different shade conditions was investigated. The investigated grassland species were *Rumex acetosa, Achillea millefolium, Plantago lanceolata, Daucus carota* and *Trifolium repens*. As ruderals *Stellaria media, Capsella bursa-pastoris, Senecio vulgaris, Spergula arvensis* and *Veronica persica* were chosen. Two weeks after germination, half of the plants were harvested, the rest was harvested five weeks after emergence. Besides dry weight and relative growth rate, a number of morphological characters were measured during the growth period. From differences in growth behavior of ruderals and grassland species insight can be gained into the mechanisms regulating the establishment of dicotyledons in grasslands.

Introduction

Generally, grassland vegetation contains different grass species together with several dicotyledons. It is difficult to identify the factors determining both the species composition and the abundance of these dicotyledons. In literature, a number of factors playing an important role in the establishment of the species are mentioned.

Large seeds (Grime and Jeffrey 1965; Gross 1984; Sheldon 1974; Winn 1985) can be important to survive deep shadow conditions especially in the earliest days of a plant's life; also the morphology of the seedling can be important (Grime and Jeffrey 1965; Pons 1977; Grime 1979; Fenner 1983). Another factor is competition – both for nutrients and for light – and its intensity can increase with increasing density of the grassland (Schenkeveld and Verkaar 1984). Even under strong competition some seedlings are able to survive in shade conditions for a long period, however without showing any growth.

This paper deals with the effect of the light factor, varied in a growthroom from 100% to 0% with five different shade conditions, on the establishment of seedlings. Growth behavior of five grassland species was compared to that of five ruderals. The differences may give us insight into the mechanisms of successful establishment.

Materials and methods

The investigated grassland species are respectively *Rumex acetosa* L., *Achillea millefolium* L., *Plantago lanceolata* L., *Daucus carota* L., *Trifolium repens* L. and the ruderals are *Stellaria media* (L.) Vill., *Capsella bursa-pastoris* (L.) Med., *Senecio vulgaris* L., *Spergula arvensis* L. and *Veronica persica* Poir.

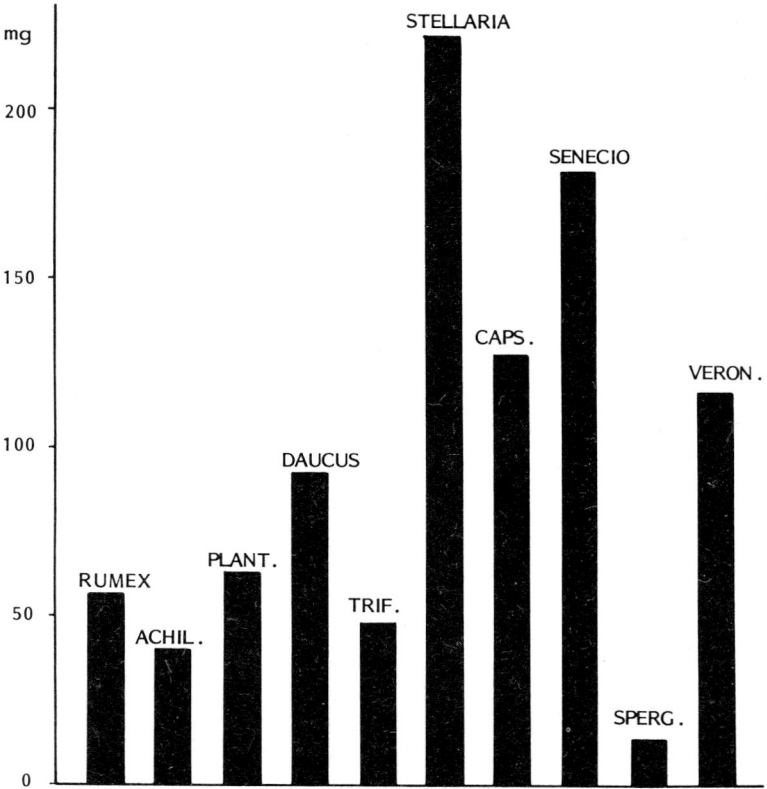

Fig. 1. Mean total dry weight under full light conditions (77 μmol.m^{-2}.s^{-1}) five weeks after germination.

Different levels of light intensity were obtained by using various layers of green shade screen (used normally in horticulture). The screens were stretched over wooden frames. In this way different light intensities were obtained: 0, 1, 2, 3, 4 and 5 layers of shade screen gave 100%, 27%, 17%, 7%, 3% and 0% respectively of the total amount of light (6200 lux = 77 μmol.m^{-2}.s^{-1}) available in the growthroom. The light was provided by Sylvania GRO-LUX tubular lamps. Seeds were collected from plants in their natural habitats.

The seeds were sown on moist filter paper in Petri-dishes, and the germinated seedlings planted out (three or five plants together) in plastic pots (diameter 70 mm) filled with potting compost (ASEF). Because the investigation only concerns the first five weeks of growth, intraspecific competition presumably did not occur.

Ten pots of each species were placed in a growthroom under the six different light conditions with a 12 hours-day at 20°C and a 12 hours-night at 10°C. To keep the soil moist the pots were watered as required. Half of the plants were harvested two weeks after germination and the rest after five weeks. Shoot height, primary root length and number of leaves were recorded. For some of the species, some additional morphological features were determined. Plant parts were dried at 100°C for 24 hours. Since the individual plants under shade conditions were very

Table 1. Mean total dry weight of plants under six different light conditions as a percentage of the dry weight after five weeks at 100% light condition (growth room).

G$_{Total}$	Grassland species						Ruderals			
%	Rumex	Achillea	Plantago	Daucus	Trifolium	Stellaria	Capsella	Senecio	Spergula	Veronica
After two weeks										
100%	4.9	2.9	3.5	3.7	6.2	1.8	1.0	2.7	8.1	2.5
27%	2.5	2.5	2.5	1.5	2.1	0.6	0.3	0.3	3.7	0.4
17%	2.5	2.5	1.4	0.9	1.6	0.5	0.2	0.2	2.4	0.3
7%	2.3	1.8	1.7	1.1	0.9	0.6	0.1	0.2	2.0	0.6
3%	2.3	1.5	1.4	0.7	0.6	0.4	0.1	0.2	1.6	–
0%	–	–	–	0.7	0.6	–	0.3	0.2	2.1	–
After five weeks										
100%	100	100	100	100	100	100	100	100	100	100
27%	11.4	7.6	8.7	10.8	10.9	3.9	3.5	1.9	11.6	4.5
17%	3.7	2.2	4.1	2.9	2.6	0.7	0.7	0.4	5.3	1.0
7%	2.6	1.4	2.5	0.6	1.1	0.4	0.2	0.2	2.2	0.5
3%	2.5	1.4	1.9	0.7	1.4	0.3	0.1	0.1	2.2	+
0%	2.3	+	1.4	0.5	2.1	0.2	0.2	+	1.5	+

small and on the limit of reliable weighing, we took all plants per growth condition together and measured the mean dry weight of roots and shoots.

Mean seed weight and mean seed coat weight of each species were determined, so that the mean embryo weight could be taken into account in determining the relative growth rate for the first two weeks and the subsequent period of three weeks.

Results and discussion

Allocation of dry matter

Figure 1 shows that five weeks after germination the total dry matter production in all grassland species is markedly lower than in the ruderals, with the exception of *Spergula*. *Spergula* stands out, having a very low dry weight.

The difference between full light and 27% of full light conditions is smaller for the grassland species than for the ruderals (Table 1). With the exception of *Spergula*, the ruderals do not grow at all in the 17%, 7%, 3% and 0% treatment during the second growth period. The grassland species still display some growth at the 17% treatment.

Table 2 shows the dry matter allocation to roots under different treatments. The dry matter allocation to roots in ruderals in full light, 27% and 27% of full light is slightly higher than in the grassland species.

Morphological aspects

The morphological responses of the plant may represent important adaptations to shade. Figure 2 shows shoot height and primary root length of all species at two and five weeks after emergence. The grassland species showed elongated petioles (in *Rumex acetosa*) or whole leaves (in *Plantago lanceolata* and *Daucus carota*). The plants remained mechanically firm and self-supporting. The ruderals reacted in two different ways to shade. *Senecio*, *Veronica*, *Spergula* and *Capsella* remained very stunted and showed little or no morphological responses; *Stellaria* showed an extreme elongation of the stem and was unable to mechanically support itself.

The shading provided in this experiment did not increase from the top of the plant to the soil surface. So, the gradient of light occurring in a natural grassland was lacking. That means that in this experiment not shade avoidance (Grime and Jeffrey 1965) but the physiological tolerance to shade (Fenner 1978) is tested.

Figure 3 shows the mean height of the shoots of *Plantago lanceolata* measured at several points in time during the growth period under the six different light conditions. During the first two weeks after germination under all shade conditions the seedlings show a greater extention of their shoots in comparison with those in full light. Between the 16th and the 21st day after germination the juvenile plants in full light start to grow faster and taller than those in the shade. From then on a gradual difference in height correlating with the different shade conditions establishes. The same tendency seems to exist for *Daucus*, *Rumex*, *Achillea*,

Germination and establishment of dicotyledons

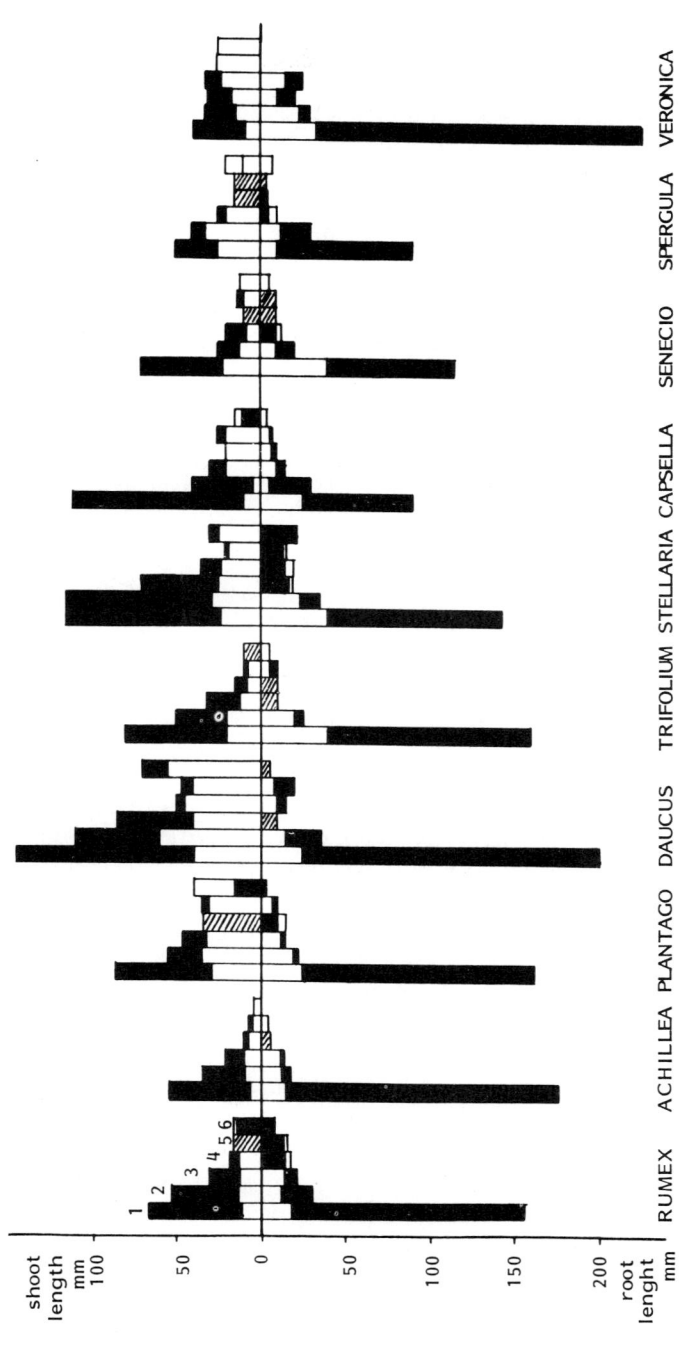

Fig. 2. The shoot and primary root length after two weeks (open columns) and five weeks of growth (closed columns). Hatched columns when length is similar for both periods. The light conditions are 1 = 100%, 2 = 27%, 3 = 17%, 4 = 7%, 5 = 3% and 6 = 0%.

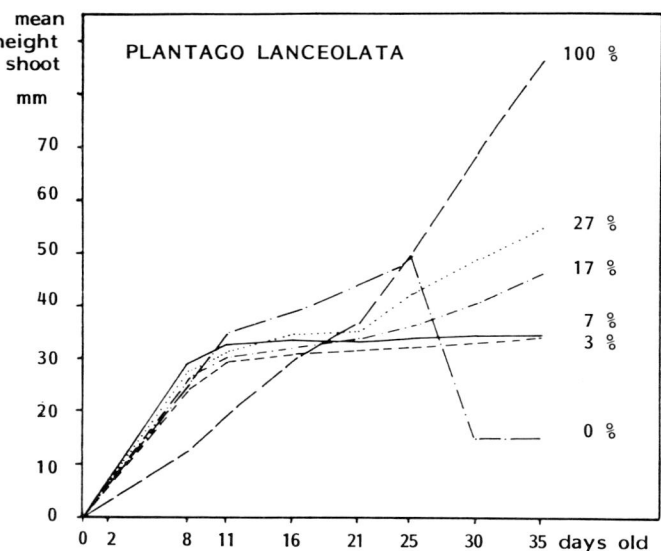

Fig. 3. Mean height of the shoots of *Plantago lanceolata* measured at several points in time during the growth period under six different light intensities.

Capsella and *Veronica*. In contrast, *Trifolium* and *Senecio* are increasingly faster extending their length in full light during the first two weeks than under conditions of shade.

Growth rate and seed size

The amount of seed reserve plays an important role in the establishment of plants in the shade. All our grassland species, with the exception of *Achillea*, have seeds with relatively large food reserves (Table 3). An extreme example is *Plantago lanceolata* where during the first two weeks the relative growth rate is nearly zero which means that the observed 'growth' in fact only represents reallocation of the rather large seed food reserve.

The ruderals tend to have higher relative growth rates than the other species in full light. This is very clear for *Stellaria*, *Capsella* and *Senecio*, already from germination onwards. *Achillea* shows a high relative growth rate during the first two weeks at all light conditions while growth stops completely the next three weeks under shade conditions. *Rumex* grows in the same manner during the first two weeks but during the second period growth does not fully stop. *Daucus*, *Trifolium*, *Spergula* and *Veronica* generally have a lower growth rate during the first two weeks in comparison with the three following weeks and for the lowest light treatments the growth rate is even negative.

Generally, we can conclude that light intensity seems very important for the growth and establishment of dicotyledonous seedlings in grasslands. Within each ecological group some growth characteristics may differ and have an effect on the plant's success but ultimately the whole combination of growth characteristics results in a success or failure in establishment of the seedling in grassland. How-

Table 2. Allocation of dry matter to roots under six different light conditions as a percentage of the total dry matter production.

		Grassland species					Ruderals			
%	Rumex	Achillea	Plantago	Daucus	Trifolium	Stellaria	Capsella	Senecio	Spergula	Veronica
After two weeks										
100%	7.1	12.7	13.6	18.5	18.7	7.7	6.9	11.8	8.2	6.9
27%	14.3	22.7	25.0	6.4	21.0	15.4	20.0	18.0	18.0	20.0
17%	7.1	20.0	22.2	10.0	18.8	16.7	10.0	22.5	33.3	33.3
7%	15.4	28.9	27.3	33.0	14.0	15.4	20.0	23.3	26.7	28.6
3%	15.4	16.7	22.2	18.6	10.0	22.2	30.0	45.0	20.0	–
0%	–	–	–	20.0	26.7	–	12.5	30.0	26.7	–
After five weeks										
100%	16.7	20.3	18.3	17.0	17.1	8.8	7.1	14.8	5.6	18.5
27%	9.2	7.4	9.1	3.4	8.3	4.7	2.9	6.3	4.4	9.6
17%	14.3	14.4	23.1	2.6	10.8	12.5	5.0	11.3	31.4	16.7
7%	13.3	6.7	25.0	16.0	20.0	25.0	10.0	26.7	20.0	11.7
3%	14.3	5.0	33.3	12.9	22.9	28.6	30.0	35.0	40.0	+
0%	15.4	+	22.2	36.0	25.0	25.0	25.0	+	0.0	+

Table 3. Relative growth rate in mg.mg^{-1}.week^{-1} under six different light conditions.

RGR	Grassland species						Ruderals			
mg.mg^{-1} .week	Rumex	Achillea	Plantago	Daucus	Trifolium	Stellaria	Capsella	Senecio	Spergula	Veronica
After two weeks										
100%	0.89	1.17	0.14	0.76	1.02	1.37	1.45	1.85	0.70	0.99
27%	0.55	1.16	-0.02	0.31	0.48	0.82	0.87	0.75	0.31	0.11
17%	0.55	1.09	-0.31	0.07	0.33	0.78	0.53	0.59	0.09	-0.14
7%	0.51	0.94	-0.21	0.17	0.07	0.82	0.31	0.49	0.04	0.28
3%	0.51	0.83	-0.31	-0.06	-0.17	0.64	0.35	0.58	-0.13	–
0%	–	–	–	-0.08	-0.20	–	0.85	0.63	0.02	–
After five weeks										
100%	1.00	1.19	1.12	1.10	0.93	1.35	1.54	1.21	0.84	1.23
27%	0.51	0.34	0.41	0.66	0.54	0.63	0.81	0.62	0.38	0.78
17%	0.14	-0.03	0.35	0.38	0.17	0.10	0.48	0.24	0.27	0.46
7%	0.05	-0.08	0.13	-0.22	0.05	-0.16	0.26	-0.01	0.02	-0.05
3%	0.03	0.00	0.10	0.12	0.30	-0.08	0.03	-0.21	0.12	+
0%	0.20	+	-0.12	-0.09	0.45	–	-0.27	+	0.00	+
Embryo weight	0.47	0.11	1.66	0.74	0.39	0.25	0.07	0.12	0.27	0.40 mg

ever, further studies examining the effects of these characteristics on the capacity of establishment in grassland are needed.

Acknowledgments

I like to thank Prof. Dr. J.A. Van Assche for helpful discussions and comments on earlier versions of this manuscript.

References

Fenner, M. 1978. The susceptibility to shade in seedlings of colonizing and closed turf species. New Phytol 81: 739-744.
Fenner, M. 1983. Seed Ecology. Chapman and Hall, London. 151 pp.
Grime, J.P. and Jeffrey, D.W. 1965. Seedling establishment in vertical gradients of sunlight. J. Ecol. 53: 621-624.
Grime, J.P. 1979. Plant Strategies and Vegetation Processes. John Wiley and Sons, Chichester. 222 pp.
Gross, K.L. 1984. Effects of seed size and growth form on seedling establishment of six monocarpic perennial plants. J. Ecol. 72: 369-387.
Pons, T.L. 1977. An ecophysiological study in the field layer of ash coppice. II. Experiments with Geum urbanum and Cirsium palustre in different light intensities. Acta Bot. Neerl. 26: 29-42.
Sheldon, J.C. 1974. The behaviour of seeds in soil. III. The influence of seed morphology and the behaviour of seedlings on the establishment of plants from surface-lying seeds. J. Ecol. 62: 47-66.
Schenkeveld, A.J.M. and Verkaar, H.J.P. 1984. On the ecology of short-lived forbs in chalk grasslands. Thesis. Utrecht. 180 pp.
Winn, A.A. 1985. Effects of seed size and microsite on seedling emergence of *Prunella vulgaris* in four habitats. J. Ecol. 73: 831-840.

RELATIONSHIPS AMONG SHRUB POPULATION STRUCTURE, SPECIES ASSOCIATIONS, SEEDLING ROOT FORM AND EARLY VOLCANIC SUCCESSION, HAWAII

ROBERT A. WRIGHT* and DIETER MUELLER-DOMBOIS
University of Hawaii at Manoa HA 96822 USA

Abstract

The population structures, species associations, micro-environments and seedling rooting strategies of five shrub species were investigated for 18 months on a 24 years old volcanic cinder deposit, island of Hawaii. The life stage and associations of 500 individuals per species were recorded in belt transects and quadrats located in areas previously surveyed in the 1960's. Beneath-crown precipitation, relative irradiance and soil parameters of a sample of established shrubs and points on open cinder were measured. Root length and biomass were determined on a sample of 20 naturally established seedlings per species for four of the study species.

Population life stage structures varied from invasive to degenerate. The association of seral with pioneer species conformed to predictions of either the facilitation or inhibition model of succession. Moreover, soil parameters, precipitation and relative irradiance changed beneath established shrubs, relative to open cinder.

These micro-environmental changes plus the higher root length:root biomass ratios of pioneer seedlings were proposed as controlling factors in species associations. Population structures were in turn influenced by both time of establishment (the maturity of the population) and species association. Accordingly, the behavior of populations and the trend of succession may be plausibly related to seedling establishment. The dynamics of populations and the successional behavior of species should be viewed as consequences of the natural selection of individuals.

Introduction

Changes in vegetation structure and composition are the outstanding features of terrestrial successions but the driving force behind such macro-scale events is actually the life and death of individual plants. Our objective in this paper is to examine the likelihood that traits of individual plants and their micro-environments mold the associations and population structure of species. We also consider if population structure is a function of the establishment history and maturity of the population. The dependence of structure and dynamics on factors intrinsic to the population would support Mueller-Dombois' (1981) thesis that life cycle dependent phenomena are important in the structuring of vegetation.

The utility of studying succession at complementary levels has been elaborated in studies of particular vegetation types, for example in Mediterranean shrub communities (Debussche *et al.* 1980). While strongly supporting a synthetic approach, we urge a critical and analytical outlook when matching hierarchical levels with ecological problems in light of Rowe's (1961) cautionary exposition on ecological hierarchies and appropriate objects of study. Succession is best understood as a

---

* *Present address*: Department of Forest Science, University of Alberta, Edmonton, Canada T6G 2H1.

population process (Peet and Christensen 1980). However, we ultimately are asking if the path of succession ought to be conceptualized as a partial product of the natural selection of individuals.

To accomplish our objectives we combined ideas and techniques from vegetation science, plant demography and autecology. In the last decade efforts to interlink different levels in the ecological hierarchy (*i.e.*, ecosystem, vegetation, population, organism and gene) have been promoted. White (1985) has persuasively argued for combining the demographic and sociological viewpoints thereby focussing on the birth, life and death of individuals and bringing the process of natural selection to the fore. Several multidisciplinary volumes (*e.g.*, Haeck and Woldendorp 1985; Dirzo and Sarukhan 1984; Karlin and Nevo 1976) have given impetus to efforts at synthesizing phenotype, genotype and population in addressing ecological and evolutionary questions.

A common thread running through these works, and this paper, is an organism-centered approach (see MacMahon *et al.* 1978, 1981). Among botanists, the organism-centered approach has been most eloquently advanced by Harper (1967, 1977, 1982). We believe the most compelling reason for emphasizing the individual organism is its position as the fundamental entity involved in the process of natural selection. We say this, not to deny that the process of natural selection can be considered at other levels (*e.g.*, genes, groups, populations) but to underscore that it is the heritable biological differences among individuals that are necessary to the process of natural selection (see Endler 1986).

An aspect of individual plants that we will be emphasizing is root form. This phenotypic trait is potentially important to the establishment of young plants on droughty substrata. We concentrate on seedling root form in concordance with evidence that events at this stage of the life cycle determine adult pattern and abundance (Grubb 1977; Harper 1977; Gross and Werner 1982).

We will illustrate how several levels of integration can be interrelated using results from our research on primary succession in Hawaii. Our treatment begins with population structure, moves to species associations, then micro-environmental variation, followed by our findings on seedling rooting strategies. We close with a discussion of hierarchical interconnections and natural selection.

Study area

The study site was a 0.4 km² cinderland (the Devastation Area) created by the 1959 eruption of Kilauea Iki on the island of Hawaii. A deposit of cinder, one to many meters thick, killed the pre-eruption montane rain forest. A steep orographic rainfall gradient traverses the study site and fog is an important source of precipitation (Smathers and Mueller-Dombois 1974; Juvik and Ekern 1978). Mean annual precipitation is about 2500 mm at the wet end of the site. The present-day vegetation consists of scattered shrubs (cover ranging from 9–33%, depending on site), separated by open cinder. This site and vegetation were selected for study after Mueller-Dombois (1983a) noted an intriguing pattern of shrub population dieback reminiscent of the cohort senescence and natural canopy dieback phenomena found in the *Metrosideros polymorpha* forests of Hawaii (Mueller-Dombois 1983b, 1986, 1987).

Material and methods

Population structure

Individuals from each of the six most important woody species in the vegetation were described from transects placed in areas sampled in a previous study (Smathers and Mueller-Dombois 1974). The study species included three endemic shrubs (*Dubautia scabra, Rumex giganteus, Vaccinium reticulatum*), two introduced shrubs (*Buddleja asiatica, Rubus penetrans*) and an endemic tree (*Metrosideros polymorpha*). (Nomenclature after St. John 1973). *Dubautia* is a frutescent chamaephyte with a cushion growth form, *Rumex* is a caespitose nanophanerophyte with a well-developed lignotuber and *Vaccinium* is a frutescent caespitose chamaephyte. *Buddleja* is a scapose nanophanerophyte with the form of a small tree and *Rubus* is a caespitose nanophanerophyte with a cane-like growth form. *Metrosideros* is a scapose mesophanerophyte (life form classes after Mueller-Dombois and Ellenberg 1974).

Eleven life stages (Fig. 1) were defined within each species based primarily on developmental (*e.g.*, reproductive maturity) but secondarily on morphometric (*e.g.*, crown height) variables. Note that plants in mature life stages (based on crown foliation) can rejuvenate and enter more vigorous life stages. In contrast, the progression through the non-reproductive life stages is a one-way path defined by irreversible developmental features (with the exception of the SN life stage based on crown foliation).

The rationale for our life stage classification springs from the age state concept of Rabotnov (1969, 1978). Life stage (age state) analysis has been used in the West (*e.g.*, Blackburn and Tueller 1970; Sharitz and McCormick 1973) but has been most highly developed in the Soviet Union (Gatsuk *et al.* 1980).

We used a non-metric multivariate technique known as partial order scalogram analysis (POSA) to formally examine the comparative life stage structure of the six populations. POSA is a special case of multidimensional scalogram analysis (see Zvulun 1978). A general discussion of POSA may be found in Shye (1978) and examples of its use in the behavioral sciences in Levy (1984, 1985).

In the POSA analysis, the proportion of each population in each life stage was compared to the mean proportion for that stage, across the six populations. If the proportion of a species population in a particular life stage, relative to the mean proportion, indicated greater population maturity then the species was assigned a score of two for that stage. If it indicated greater immaturity, it was assigned a score of one. The summed scores for each species population indicate their relative maturity.

Species association

Data were collected by noting whenever an immature plant was dominated by the crown of a mature shrub or tree. Any immature plant not growing directly beneath a dominant plant crown was recorded as associated with open cinder. The model against which we tested for departures from random distribution in space is similar to that used by Auld (1969) and Gross and Werner (1982). Sig-

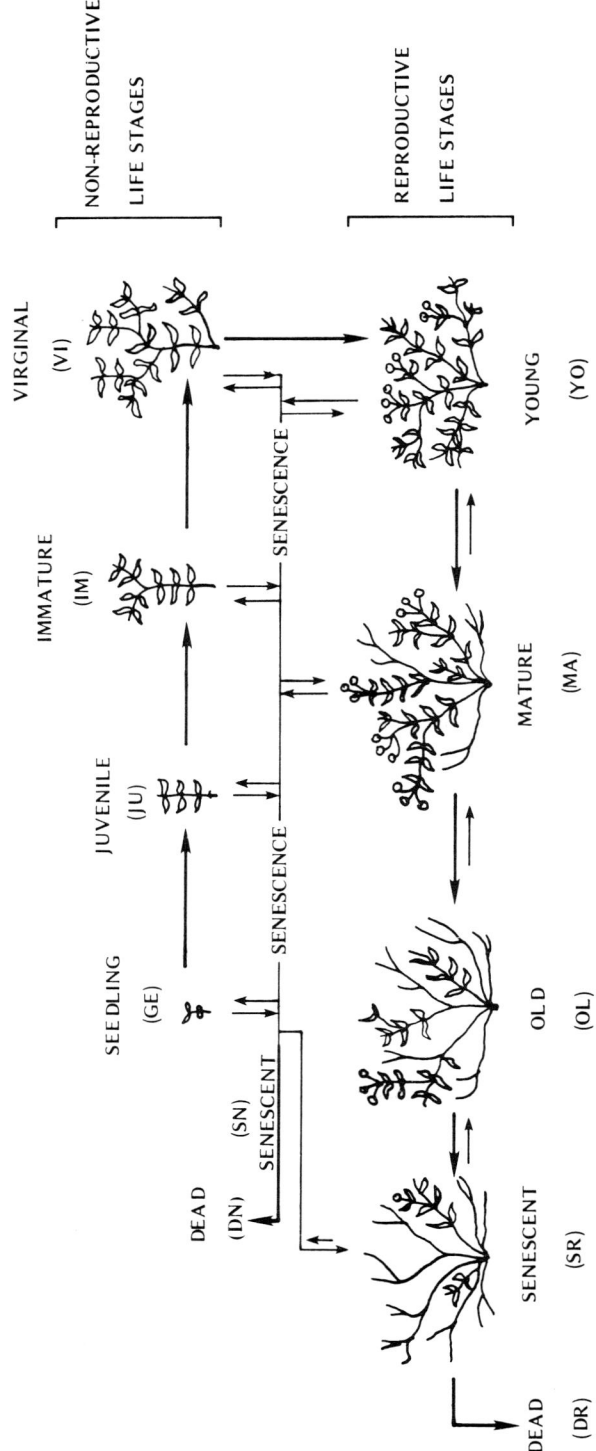

Fig. 1. Life stages and developmental pathways for a hypothetical woody plant. Arrows indicate direction of development. Two-way arrows indicate where transitions may occur in either direction between life stages. No representative stylised drawing could be rendered for the visually variable life stages SN, DN and DR.

nificance of departure from random expectations was evaluated with a χ^2 goodness-of-fit test (Sokal and Rohlf 1981).

Micro-environment

Precipitation (fog and rain), relative irradiance, soil moisture, soil temperature and soil nutrient levels were sampled beneath mature shrub and tree crowns and on open cinder to determine what differences existed between the two major types of micro-environments.

Seedling root form

Approximately 20 seedlings from each of four study species were harvested from their natural habitats. Root and shoot biomass plus maximum root length were measured. Maximum root length was considered an estimator for maximum rooting depth. These data were used to test for differences in rooting strategy among species.

Results

Population structure

Figure 2 shows the population life stage structure of all six study species (hereinafter referred to by genus name only) in July 1983. Three general life stage spectra (*sensu* Rabotnov 1969) were apparent. *Buddleja* exhibited a degenerative spectrum with no seedlings, a low proportion of immature plants and a high proportion of dead and senescing reproductive individuals. Maintenance of this spectrum is unlikely, unless the graduation rate to and from non-reproductive life stages is very high. Unpublished data (Wright 1985) on the population dynamics of *Buddleja* indicated that this graduation rate was not sufficient to maintain the population structure shown in Fig. 2. The lack of seedlings suggests that recruitment may have ceased in this species or is episodic on a cycle longer than one year.

Dubautia and *Vaccinium* both had normal life stage spectra (*i.e.*, spectra in which every life stage is present and the entire life cycle is apparently being completed). In contrast to *Buddleja*, these two species appear to have had lower mortality rates and higher establishment rates in the recent past. This form of life stage spectrum implies population maintenance. Unpublished data in Wright (1985) did indicate that the half-lives (*sensu* Harper 1977) of the *Dubautia* non-reproductive and reproductive sub-populations were, respectively, 2.3 and 14.0 years. These half-lives, plus the failure to establish over the 18 month monitoring period, foretell a gradual decline in numbers if conditions remain constant.

Metrosideros, *Rumex* and *Rubus* all had invasive spectra with high proportions of non-reproductive individuals. We describe this type of life stage structure as invasive in reference to the high proportions of immature plants that give such populations a great potential for expanding their relative cover in the vegetation.

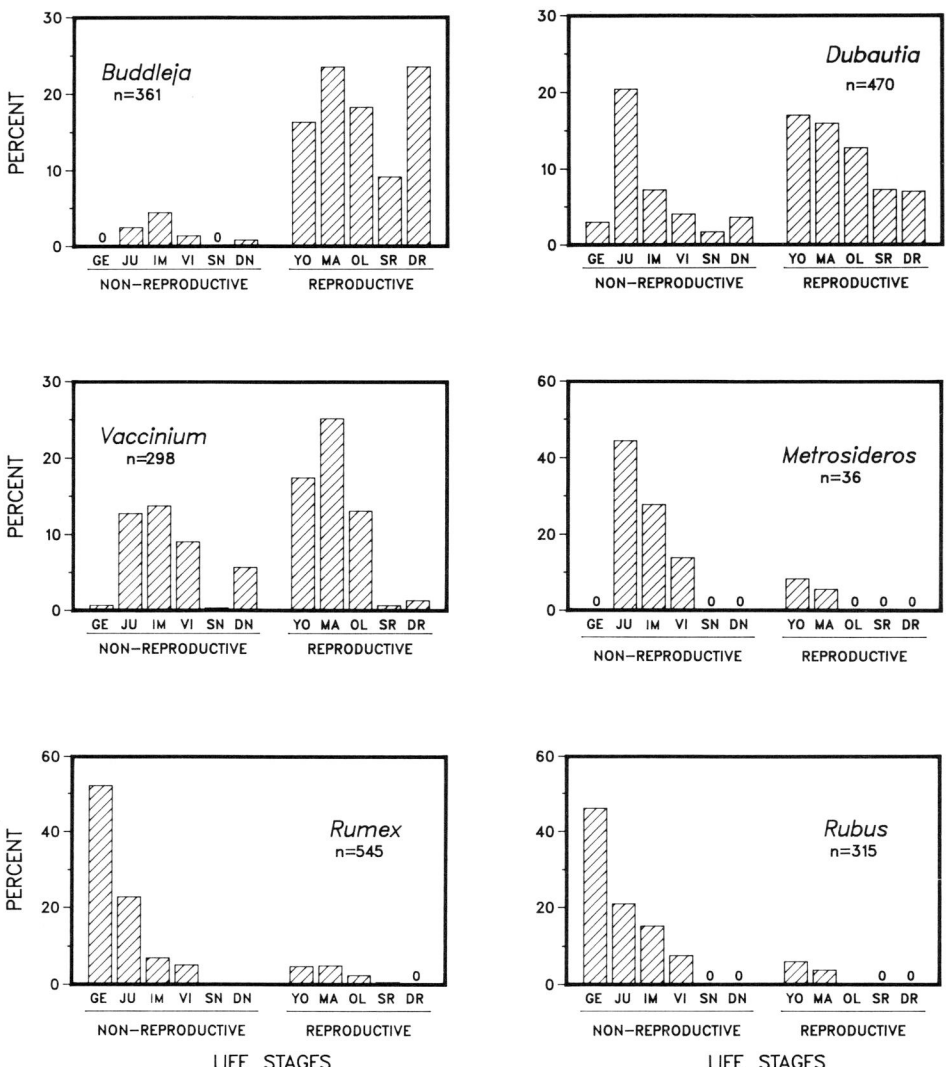

Fig. 2. Overall life stage population structures for the study species. Life stage abbreviations as in Fig. 1.

In our analysis the populations were ordered using a concept of population maturity based on the relative proportions of the life stages in each species. Consider a hypothetical population maturity continuum running from immature, invasive populations of seedlings to mature, degenerate populations of dead adults. All real population structures will match or fall somewhere between these two extremes. The objective was to order our populations along this maturity continuum.

As an example, consider the POSA scores assigned to *Rumex* and *Buddleja* for the seedling life stage (GE) (Table 1). *Rumex*, with the highest seedling proportion (0.52) among the six species was assigned a score of 1, because a high seedling

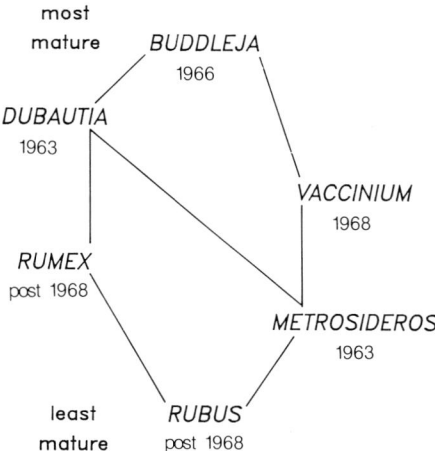

Fig. 3. Relative order of maturity among the species in the Devastation Area as derived from POSA. Species pairs connected by lines are consistently comparable, in terms of relative population maturity, across all 11 life stages.

proportion, relative to the mean proportion, indicated population immaturity along the hypothetical maturity continuum discussed above. Conversely, *Buddleja* was assigned a 2 for its empty seedling stage because an absence of seedlings indicated a degenerate, mature population if, as in this case, a large proportion of the population consisted of mature or senescent plants. Judgements of population maturity were based on the structure of the population, the known history of the population and the concept of a population maturity continuum.

The relative order of maturity among species derived from POSA is shown in Fig. 3. *Buddleja* had the highest total maturity score and appears at the apex of the scalogram. The other species are arrayed below in order of decreasing population maturity. The date associated with each species in Fig. 3 is the first year in which the species attained a frequency of 10% or more (as determined by Smathers 1972). The 10% level was chosen to allow an objective comparison, among species, of times of first significant appearance in the study area vegetation.

We designated *Buddleja*, *Dubautia*, *Vaccinium* and *Metrosideros* as pioneer species, based on their early invasion of the study area. Conversely, we considered *Rubus* and *Rumex* to be seral species based on their late appearance in the Devastation Area. The order of maturity among populations is apparently related to the time of invasion. The pioneers (10% frequency before 1968) had the most mature population structures. The seral species (10% frequency after 1968) had the least mature life stage spectra. *Metrosideros*, a tree, is not readily comparable to the shrubs in terms of population maturity because of its greater longevity and consequently longer population life cycle.

Association analysis

Here we examine the linkage between species association and population structure. This is done within the conceptual framework of the three models of succes-

Table 1. Matrix of life stage maturity scores for the overall study populations in the Devastation Area. Higher score = greater population maturity.

Species	Life stages											Total
	GE	JU	IM	VI	SN	DN	YO	MA	OL	SR	DR	
Bud	2	2	2	2	2	2	2	2	2	2	2	22
Dub	2	2	2	2	1	1	2	2	2	2	2	20
Vac	2	2	1	1	2	1	2	2	2	1	1	17
Rum	1	1	2	2	1	1	1	1	1	1	1	13
Met	2	1	1	1	1	1	1	1	1	1	1	12
Rub	1	1	1	1	1	1	1	1	1	1	1	11

sion proposed by Connell and Slatyer (1977). According to their facilitation model of succession, the pioneers of a site change their immediate environment such that the establishment of their successors is enhanced. In the tolerance model, the environmental changes wrought by the pioneers have little or no effect on the establishment and growth to maturity of later succession species. In the inhibition model, the pioneers prevent establishment of new invaders. In all three models the original pioneers are assumed to change their immediate micro-environment in ways that inhibit their own establishment and growth to maturity.

Connell and Slatyer (1977) contend that the facilitation model is the model most likely to dominate in early primary succession. To test this contention we hypothesized that immature life stages of later seral species (*Rumex* and *Rubus*) would be positively associated with mature life stages of the original pioneers (*Metrosideros, Dubautia, Vaccinium, Buddleja*). To investigate the proposition that all pioneers change their micro-environments in ways that inhibit their own establishment, we hypothesized that immature pioneers would be negatively associated with conspecific adults. Note that this is a partial test of the succession models because only the juvenile phase of the life cycle is considered and not the adult reproductive phase.

Our random model of association (see Methods) predicted that the proportion of immature plants of species x found in a micro-environment would be proportional to the % cover of the micro-environment. For immature life stages of seral species, relative to mature pioneers, failure to reject the hypothesis would constitute support for the tolerance model. Positive deviation would support the facilitation model and negative, the inhibition model. For immature life stages of pioneer species, relative to mature pioneers, negative deviation would support and positive or no deviation would fail to support the three models of succession as described above.

Two distinct species groups were deduced from the association analysis when two broad classes of micro-environments (open cinder and beneath crown) were examined (Table 2). Immature *Buddleja, Dubautia* and *Vaccinium* conformed to random expectation except that immature *Dubautia* were negatively associated with beneath crown micro-environments. In complete contrast, immature *Rumex, Rubus* and *Metrosideros* were negatively associated with cinder and positively associated with beneath crown micro-environments. Apparently establishment of the seral shrubs *Rumex* and *Rubus* and the pioneer tree *Metrosideros* was

Table 2. Summary of associations of immature members of the study species with open cinder and beneath crown micro-environments. + = positive association, − = negative association, none = no significant deviation from the random model, e = expected value of cell < 5.

	Micro-environment	
Species	Cinder	Crown
Bud	none	none
Dub	none	−
Vac	none	none^e
Rum	−	+
Met	−	+^{e}
Rub	−	+

Table 3. Summary of associations of immature members of the study species with three micro-environments. Symbols as in Table 2.

	Micro-environment		
Species	Cinder	Pioneer	Seral
Bud	none	none	none
Dub	none	−	−
Vac	none	+^{e}	none^e
Rum	−	+	+
Met	−	+^{e}	none^e
Rub	−	+	+

facilitated by the presence of mature individuals and inhibited by their absence, relative to open cinder. Other pioneer species were inhibited (*Dubautia*) or unaffected (*Vaccinium* and *Buddleja*) by beneath crown micro-environments, relative to open cinder.

Subdivision of the crown category into pioneer, seral (Table 3) and conspecific, other taxa (Table 4) categories gave a more precise view of species associations. *Rumex* and *Rubus* were positively associated with pioneer crowns, confirming the prediction of our first hypothesis based on the facilitation model. More specifically, *Rumex* (an endemic species) was apparently facilitated by the introduced pioneer, *Buddleja*, but unaffected by one endemic pioneer (*Dubautia*) and inhibited by another endemic pioneer (*Metrosideros*) relative to open cinder (Table 5). Behaving in a different manner, *Rubus* (an introduced species) was apparently facilitated by both *Buddleja* and *Metrosideros* and inhibited by *Dubautia*. Of the pioneer species only *Dubautia* showed inhibition of immature stages by conspecifics leading us to reject our second hypothesis, for *Vaccinium*, *Metrosideros* and *Buddleja*, that mature pioneers would inhibit the establishment of conspecifics. Clearly that prediction of all three succession models is not well-supported by our data.

Micro-environments

Harper (1977) has observed that the abundance of a particular species is governed

Table 4. Summary of associations of immature members of the study species with three micro-environments. Symbols as in Table 2.

	Micro-environment		
Species	Cinder	Conspecific	Other taxa
Bud	none	none	none
Dub	none	$-^e$	$-$
Vac	none	$+^e$	none^e
Rum	$-$	$+$	$+$
Met	$-$	$+^e$	$+^e$
Rub	$-$	$+$	$+$

Table 5. Summary of associations of immature *Rubus* and *Rumex* with six micro-environments. Symbols as in Table 2.

	Micro-environment					
Species	Cinder	Bud	Dub	Rum	Met	Rub
Rum	$-$	$+$	none^e	$+$	$-$	$-$
Rub	$-$	$+$	$-$	$+$	$+$	$+$

by the frequency of micro-environments suitable for seed germination and seedling establishment (provided seed availability is not limiting). Here we examine the micro-environmental changes beneath shrub crowns that might be contributing to the facilitation and inhibition of establishment deduced from the association analysis. Such changes are also of interest because they are assumed to be the initial underlying causes of succession in the facilitation model (Clements 1916; Connell and Slatyer 1977).

Plants may alter the effective precipitation beneath their crowns by intercepting or concentrating fog and rain. Barbour *et al.* (1974) attributed an association between desert herbs and *Larrea* shrubs to the funnelling of rain water down *Larrea* trunks to the beneath crown environment. Interception of fog by snow tussock grass (*Chionochloa rigida*) has also been demonstrated to increase total precipitation, relative to unvegetated ground, in a low alpine area of New Zealand (Mark and Rowley 1976; Mark *et al.* 1980).

During our preliminary reconnaissance we noticed fog drip from the leaves of the shrubs. This led to our investigation of the magnitude of fog interception by plants crowns and artificial interceptors. We found that significant amounts of fog and rain were intercepted by Grunow-type fog interceptors mounted on rain gauges on open cinder. Total precipitation collected in rain gauges placed beneath shrub crowns was also consistently higher than total rain collected on open cinder (Fig. 4). Fog and rain scavenging by shrubs increases effective precipitation to the substratum directly beneath mature shrubs and trees.

Woody plants have been known to improve the nutrient status of soils found directly beneath their crowns (see Lawrence *et al.* 1967; Parker and Muller 1982). We tested for this effect and found significantly higher levels of organic carbon, Kjeldahl nitrogen and available Ca, Mg and K in the 2 mm diameter fraction of the

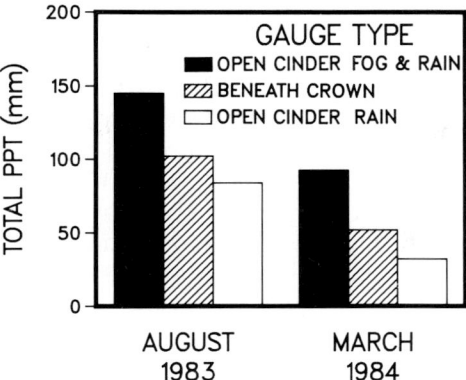

Fig. 4. Comparison of open cinder fog and rain, open cinder rain and beneath crown precipitation at two times. Beneath crown precipitation and open cinder rain not significantly different ($p < .05$).

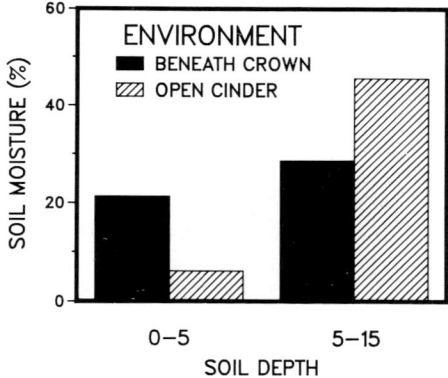

Fig. 5. Comparison of soil moisture beneath *Rumex* shrub crowns and in adjacent open cinder. Both comparisons significant at $p < .01$.

top 5 cm of cinder beneath shrub crowns, compared to open cinder.

Plants also have a dramatic effect on irradiance levels beneath their canopies (Horn 1971; Parker and Muller 1982; Burton and Mueller-Dombois 1984). We found that relative irradiance fell to as low as 0.1% (*Dubautia*), 1.5% (*Buddleja*), 4.0% (*Vaccinium*) and 6.5% (*Rumex*) beneath shrub crowns. Reduced irradiance led to a significantly lower soil surface temperature beneath a mature *Buddleja* shrub (22°C) compared to open cinder (33°C) on a clear day. Turner et al. (1966) found that a similar combination of lower irradiance and soil temperature enhanced survival of Saguaro cactus (*Carnegiea gigantea*) seedlings.

In his work with Sugar pine (*Pinus lambertiana*) seedlings Yeaton (1984) found drought to be an important source of mortality. Moisture stress also caused high mortality in shallow-rooted seedlings of the tree *Sophora chrysophylla* on the island of Hawaii (Scowcroft 1981). With the combination of increased precipitation and soil organic matter and decreased relative irradiance and soil temperature we expected higher soil moisture beneath mature shrub crowns. Our analysis did indicate higher moisture levels beneath crowns than in open cinder in the top 5 cm of soil (Fig. 5).

We also found that soil moisture in the top 5 cm of open cinder fell to levels insufficient for plant maintenance after only two dry, clear days following rainy weather. From January 1983 to July 1984 only four stretches of over 30 days without a dry period of two or more days were recorded in the study area (Wright 1985, unpublished data derived from National Climatic Data Center statistics). Unbroken stretches of wet weather are not common in the Devastation Area and we maintain that this may be a profound barrier to establishment, on open cinder, of species not able to avoid or tolerate low moisture levels in the surface cinder.

The environment of the Devastation Area is one of extremes with high insolation, high substratum temperatures and droughty, nutrient-poor soil. Established shrubs and trees do ameliorate the micro-environment beneath their crowns although in one case (*Dubautia*) the decrease in relative irradiance would prevent the growth of most plant seedlings (Daubenmire 1974) and, specifically, *Metrosideros* seedlings (Friend 1980). The question now becomes: 'Which phenotypic trait might be central to survival in such a harsh environment?'

Seedling rooting strategies

Most shrub populations experience their highest mortality in the early stages of life and exhibit survivorship best described by a power function (*e.g.*, Crisp and Lange 1976; West *et al.* 1979). Consequently, adequate recruitment of seedlings is likely to be a crucial factor in performance of populations and to loom large in the process of succession.

In ecosystems with coarse soils prone to edaphic drought, such as the Devastation Area, selection will favor seedlings with traits that enable avoidance or tolerance of drought. Drought avoidance may be thought of as an ecological (contextual) response and drought tolerance as a physiological (internal) response (Bannister 1976). A species' ability to resist drought will be a combination of both strategies under natural conditions.

Root extension may be considered an integral part of root efficiency (Kramer 1969) and the depth to which roots penetrate one of the most important morphological characteristics of root systems enabling young plants to avoid water stress (Russell 1977). For example, Maruta (1976) found that the inability to obtain sufficient water during short drought periods (nine days) was the major cause of mortality in *Polygonum* seedlings on a volcanic cinder substratum in Japan. Moreover, as noted previously, shallow-rooted *Sophora chrysophylla* seedlings experienced high mortality from water stress in Hawaii (Scowcroft 1981).

With these facts in mind we focussed our attention on rooting strategies that might delimit a species' ability to avoid the droughty conditions frequently encountered in the open cinder habitat (Fig. 5). We do not suggest that other traits such as sclerophylly or adjustments in tissue water potential are unimportant. However, in our opinion, examining rooting strategy was a simple and potentially fruitful method of detecting unambiguous differences between pioneer and seral seedling behavior.

We examined root length and root biomass in seedlings of *Dubautia, Buddleja,*

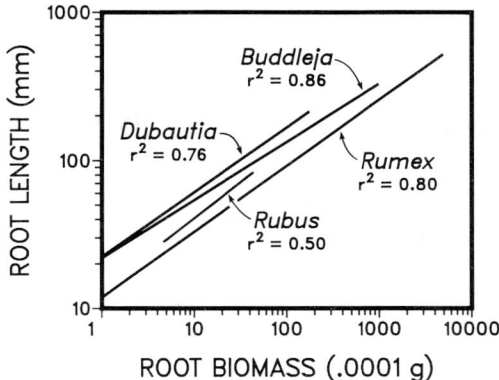

Fig. 6. Log root length vs. log root biomass regression lines for seedlings and juveniles of four shrub species. Slopes are homogeneous but y-intercepts are significantly different ($p < .05$).

Rubus and *Rumex* to document any specific differences that might account for the association pattern of immature plants in the study area. Our data on 20 naturally occurring seedlings of each species revealed no trend in root: shoot biomass ratios helpful to our understanding of the observed associations. However, examination of the relationship between root length and root biomass was illuminating (Fig. 6). Covariance analysis of the root data showed that the early pioneers of open cinder (*Dubautia* and *Buddleja*) had significantly longer maximum root lengths, for a particular root biomass, than the seral invaders of beneath crown micro-environments (*Rumex* and *Rubus*). (Tables 6 and 7). We suggest this trait enables the pioneer shrubs to establish on drought-prone open cinder and restricts the seral shrubs to establishment in moister beneath crown micro-environments.

Discussion and conclusions

We consider the study populations to be dynamic entities temporarily occupying regions on a population life cycle or maturity continuum. *Buddleja* is at the degenerate, over-mature end of the continuum and *Rubus* and *Metrosideros* are at the invasive, immature end. The normal, mature populations of *Dubautia* and *Vaccinium* fall between these two extremes. Our analysis showed that the maturity of the populations was related to their time of establishment in the study area, the more mature populations being the first to establish.

Our concept of a population maturity continuum is similar to Rabotnov's (1978), 'great cycle of populations'. Taken in aggregation over time the population structural changes we have observed are translated into change in vegetation structure. It is this aggregated process of population maturation that gives rise to the patterns and phases in vegetation first elaborated by Watt (1947). We suggest that factors intrinsic to the population, such as life stage specific mortality rates, are the forces that drive populations along our proposed maturity continuum.

We also wish to make explicit the connection between population maturity and species associations. The dichotomy in associations among the six species (Table 2)

Table 6. Comparison of least squares adjusted means for log root length among four study species. SEM = standard error of the mean.

Species	Log root length LS mean	SEM	I	\multicolumn{4}{c}{Prob > |T| HO:LS mean (I) = LS mean (J)}
				1	2	3	4
Budd	1.978	0.0324	1	–	.023	.016	.146
Rub	1.875	0.0336	2	.023	–	.377	.0004
Rum	1.823	0.0435	3	.016	.377	–	.0001
Dub	2.042	0.030	4	.146	.0004	.0001	–

Table 7. Regression equations of log root length (mm) on log root biomass (0.0001 g) with predicted values of root length at biomass = 0.004 g.

Species	Regression equation	Predicted length (mm) at biomass = .004 g
Dubautia	log y = 1.35 + .44 log x	113.5
Buddleja	log y = 1.34 + .39 log x	92.2
Rubus	log y = 1.13 + .48 log x	79.2
Rumex	log y = 1.08 + .45 log x	63.2

parallels the order in population maturity described in the POSA analysis (Fig. 3). The three most mature populations were not significantly associated with either the generalized beneath crown or open cinder micro-environments. The three least mature populations were all positively associated with beneath crown micro-environments and negatively associated with open cinder. We conclude that the relative population maturity of species is related to their spatial and successional associations with other plant species as well as to their time of establishment.

As the total cover of beneath crown micro-environments grows, the populations of species associated with such environments is likely to expand. Conversely, the populations of species associated with open cinder will necessarily shrink. Thus the population structure of plant species at this stage of succession may be plausibly linked with the changes in relative importance of micro-environments that act as safe sites for establishment.

Results of our association analysis for *Dubautia* are similar to those of Gross and Werner (1982) who found that pioneer establishment was inhibited by any sort of ground cover. In contrast, facilatory relationships of immature *Rubus*, *Rumex* and *Metrosideros* with certain beneath crown micro-environments resembles that found between desert shrubs and annuals (Went 1942), desert shrubs and saguaro seedlings (Steenbergh and Lowe 1969) and *Larrea* shrubs and *Opuntia* seedlings (Yeaton 1978).

Evidently both the inhibition and facilitation models are valid descriptors of successional relationships between species in the Devastation Area. However, we concur with Turner (1983) that associations between species need to be examined individually to determine if they conform to one of the three models of Connell and Slatyer (1977). In addition, we maintain that the models need to be elaborated to account for the diversity of associational and successional relationships in real communities. General answers to specific questions about how Devastation Area shrubs and trees behave eluded us and may be generally unattainable (see Bartholomew 1986).

Delving into the nature of micro-site variation in the Devastation Area we found that established shrubs and trees profoundly changed the micro-environment beneath their crowns. The interception of fog and rain by leaves and branches, coupled with higher soil organic matter and lower irradiances and soil surface temperatures, produced relatively high soil moisture levels beneath shrub crowns. These altered conditions are facilitating the establishment of receptive plant species. Conversely, the low relative irradiance beneath *Dubautia* crowns may account for the inability of seedlings to establish under shrubs of this species. We agree with Shmida and Whittaker (1981) that biological modification of micro-sites is probably a crucial determinant of the population dynamics of individual species

Our analysis of seedling root form indicated that pioneer species have a growth strategy which enables them to send roots down to deeper cinder more economically than seral species. Here they encounter higher, and presumably more reliable, water supplies (Fig. 5). For relatively deep-rooted pioneer seedlings the principal safe site for establishment appears to be open cinder. For relatively shallow-rooted seral seedlings the principal safe site appears to be the ameliorated micro-environment beneath mature shrubs and trees.

Our research in the Devastation Area leads us to the conclusion that the path of succession will be charted by the population dynamics of the participating species which in turn hinge on species associations dependent on the performance of individual plants in a variety of micro-environments. We also propose that vegetation and individual form an interactive system with the selection of individuals ultimately producing vegetation dynamics and the structure of vegetation influencing the selection of individuals. Could succession and population dynamics also be viewed as products of the natural selection process centered around individuals? Addressing this question requires adoption of a population genetic perspective but the resultant wedding of ecological and evolutionary thought could lead to a clearer understanding of the coevolutionary relationship between a vegetation and its constituent species.

Acknowledgments

R.A. Wright would like to thank J.S. Rowe and B. Weichel for their critical comments on an initial transcript of our paper. This paper is based on the thesis research of R.A. Wright which was partially supported by a grant-in-aid from the Pacific Tropical Botanical Garden, Hawaii.

References

Auld, B.A. 1969. The distribution of *Eupatorium adenophorum* Spreng. on the far north coast of New South Wales. J. Proc. R. Soc. of N. S. Wales 102: 159–161.

Bannister, P. 1976. Introduction to physiological plant ecology. John Wiley & Sons, New York, USA.
Barbour, M.G., Diaz, D.V. and Breidenbach, R.W. 1974. Contributions to the biology of *Larrea* species. Ecology 55: 1199-1215.
Bartholomew, G.A. 1986. The role of natural history in contemporary biology. BioScience 36: 324-329.
Blackburn, W.H. and Tueller, P.T. 1970. Pinyon and juniper invasion in black sagebrush communities in east-central Nevada. Ecology 51: 841-848.
Burton, P.J. and Mueller-Dombois, D. 1984. Response of *Metrosideros polymorpha* seedlings to experimental canopy opening. Ecology 65: 779-791.
Clements, F.E. 1916. Plant succession: an analysis of the development of vegetation. Carnegie Institution of Washington, Publication 242, Washington, D.C., USA.
Connell, J.H. and Slatyer, R.O. 1977. Mechanisms of succession in natural communities and their role in community stability and organization. Am. Nat. 111: 1119-1144.
Crisp, M.D. and Lange, R.T. 1976. Age structure, distribution and survival under grazing of the arid-zone shrub *Acacia burkittii*. Oikos 27: 86-92.
Debussche, M., Escarré, J. and Lepart, J. 1980. Changes in Mediterranean shrub communities with *Cytisus purgans* and *Genista scorpius*. Vegetatio 43: 73-82.
Daubenmire, R.F. 1974. Plants and environment: a textbook of plant autecology, 3rd Edition. John Wiley and Sons, London, England.
Dirzo, R. and Sarukhan, J. 1984. Perspectives on plant population ecology. Sinauer Associates, Sunderland, USA.
Endler, J.A. 1986. Natural selection in the wild. Princeton University Press, Princeton, USA.
Friend, D.J. 1980. Effect of different photon flux densities (PAR) on seedling growth and morphology of *Metrosideros collina* (*Forst*). Gray. Pac. Sci. 34: 93-100.
Gatsuk, L.E., Smirnova, O.V., Vorontzova, L.I., Zaugolnova, L.B. and Zhukova, L.A. 1980. Age states of plants of various growth forms: a review. J. Ecol. 68: 675-696.
Gross, K.L. and Werner, P.A. 1982. Colonizing abilities of 'biennial' plant species in relation to ground cover: implications for their distributions in a successional sere. Ecology 63: 921-931.
Grubb, P.J. 1977. The maintenance of species richness in plant communities: the importance of regeneration niche. Biol. Rev. 52: 107-145.
Haeck, J. and Woldendorp, J.W. (eds) 1985. Structure and functioning of plant populations 2: phenotypic and genotypic variation in plant populations. North-Holland Publishing, Amsterdam, the Netherlands.
Harper, J.L. 1967. A Darwinian approach to plant ecology. J. Ecol. 55: 247-270.
Harper, J.L. 1977. Plant population biology. Academic Press, London, England.
Harper, J.L. 1982. After description. In: E.I. Newman, The Plant Community as a Working Mechanism, pp. 11-25. Blackwell Scientific Publications, Oxford, England.
Horn, H.S. 1971. The adaptive geometry of trees. Princeton University Press, Princeton, USA.
Juvik, J.O. and Ekern, P.C. 1978. A climatology of mountain fog on Mauna Loa, Hawaii island. Water Resources Research Center, Technical Report No. 118. University of Hawaii, Honolulu, USA.
Karlin, S. and Nevo, E. 1976. Population genetics and ecology. Academic Press, New York, USA.
Kramer, P.J. 1969. Plant and soil water relationships. McGraw-Hill, New York, USA.
Lawrence, D.B., Schoenike, R.E., Quispel, A. and Bonds, G. 1967. The role of *Dryas drummondii* in vegetation development following ice recession at Glacier Bay, Alaska, with special reference to its nitrogen fixation by root nodules. J. Ecol. 55: 793-813.
Levy, S. 1984. Partial orders of Israeli settlements by adjustive behaviors. Israel Social Science Research 2: 44-65.
Levy, S. 1985. Partial order analysis of crime indicators. Social Indicators Research 16: 195-199.
MacMahon, J.A., Phillips, D.L., Robinson, J.V. and Schimpf, D.J. 1978. Levels of biological organization: an organism-centered approach. BioScience 28: 700-704.
MacMahon, J.A., Schimpf, D.J., Andersen, D.C., Smith, K.G. and Bayn, R.L. 1981. An organism-centered approach to some community and ecosystem concepts. J. Theor. Biol. 88: 287-307.
Mark, A.F. and Rowley, J. 1976. Water yield of low-alpine snow tussock grassland in central Otago. J. Hydrol. (N.Z.) 15: 59-79.
Mark, A.F., Rowley, J. and Holdsworth, D.K. 1980. Water yield from high altitude snow tussock grassland in central Otago. Tussock Grasslands and Mountain Lands Institute Review 38: 21-33.

Maruta, E. 1976. Seedling establishment of *Polygonum cuspidatum* on Mt. Fuji. Jap. J. Ecol. 26: 101–105.
Mueller-Dombois, D. 1981. Vegetation dynamics in a coastal grassland of Hawaii. Vegetatio 46: 131–140.
Mueller-Dombois, D. 1983. Population death in Hawaiian plant communities: a causal theory and its successional significance. Tuexenia 3: 117–130.
Mueller-Dombois, D. 1983. Canopy dieback and successional processes in Pacific forests. Pac. Sci. 37: 317–325.
Mueller-Dombois, D. 1986. Perspectives for an etiology of stand-level dieback. Ann. Rev. Ecol. Syst. 17: 221–243.
Mueller-Dombois, D. 1987. Natural dieback in forests. BioScience (in press).
Mueller-Dombois, D. and Ellenberg, H. 1974. Aims and methods of vegetation ecology. John Wiley and Sons, New York, USA.
Parker, V.T. and Muller, C.H. 1982. Vegetational and environmental changes beneath isolated live oak trees (*Quercus agrifolia*) in a California annual grassland. Am. Midl. Nat. 107: 69–81.
Peet, R.K. and Christensen, N.L. 1980. Succession: a population process. Vegetatio 43: 131–140.
Rabotnov, T.A. 1969. On coenopopulations of perennial herbaceous plants in natural coenoses. Vegetatio 19: 87–95.
Rabotnov, T.A. 1978. On coenopopulations of plants reproducing by seed. In: Freysen, A.H.J. and Woldendorp, J.W. (eds), Structure and functioning of plant populations, pp. 1–26. North-Holland, Amsterdam, the Netherlands.
Rowe, J.S. 1961. The level of integration concept and ecology. Ecology 42: 420–427.
Russell, R.S. 1977. Plant root systems: their function and interaction with the soil. McGraw-Hill, New York, USA.
St. John, H. 1973. List and summary of the flowering plants in the Hawaiian islands. Pacific Tropical Botanical Garden, Mem. No. 1, Lawai, Hawaii. 519 pp.
Scowcroft, P.G. 1981. Regeneration of mamane: effects of seedcoat treatment and sowing depth. For. Sci. 27: 771–779.
Sharitz, R.R. and McCormick, J.F. 1973. Population dynamics of two competing annual plant species. Ecology 54: 723–740.
Shmida, A. and Whittaker, R.H. 1981. Pattern and biological microsite effects in two shrub communities, southern California. Ecology 62: 234–251.
Shye, S. 1978. Partial order scalogram analysis. In: S. Shye (ed.), Theory construction and data analysis in the behavioral sciences. Jossey-Bass, San Francisco, USA.
Smathers, G.A. 1972. Invasion, early succesion and recovery of vegetation on the 1959 Kilauea volcanic surfaces, Hawaii Volcanoes National Park, Hawaii. PhD. dissertation, University of Hawaii, Honolulu, USA.
Smathers, G. and Mueller-Dombois, D. 1974. Invasion and recovery of vegetation after a volcanic eruption in Hawaii. National Park Service Scientific Monograph Series, No. 5.
Sokal, R.R. and Rohlf, F.J. 1981. Biometry, 2nd edition. W.H. Freeman, San Francisco. USA.
Steenbergh, W.F. and Lowe, C.H. 1969. Critical factors during the first years of life of the saguaro (*Cereus giganteus*) at Saguaro National Monument, Arizona. Ecology 50: 825–834.
Turner, R.M., Alcorn, S.M., Olin, G. and Booth, J.A. 1966. The influence of shade, soil and water on saguaro seedling establishment. Bot. Gaz. 127: 95–102.
Turner, T. 1983. Facilitation as a succesional mechanism in a rocky intertidal community. Am. Nat. 121: 729–738.
Watt, A.S. 1947. Pattern and process in the plant community. J. Ecol. 35: 1–22.
Went, F.W. 1942. The dependence of certain annual plants on shrubs in southern California deserts. Bull. Torrey Bot. Club 69: 100–114.
West, N.E., Rea, K.H. and Harniss, R.O. 1979. Plant demographic studies in sagebrush-grass communities of southeastern Idaho. Ecology 60: 376–388.
White, J. 1985. The population structure of vegetation. In: J. White (ed.), The Population Structure of Vegetation, pp. 1–14. Handbook of vegetation science, part III. Dr W. Junk Publishers, Dordrecht, the Netherlands.
Wright, R.A. 1985. Shrub population dynamics and succession on volcanic cinder in Hawaii. M.S. Thesis, University of Hawaii at Manoa, Honolulu, USA.
Yeaton, R.I. 1978. A cyclical relationship between *Larrea tridentata* and *Opuntia leptocaulis* in the northern Chihuahuan desert. J. Ecol. 66: 651–656.

Yeaton, R.I. 1984. Aspects of the population biology of sugar pine (*Pinus lambertiana* Dougl.) on an elevational gradient in the Sierra Nevada of central California. Am. Midl. Nat. 111: 126–137.
Zvulun, E. 1978. Multidimensional scalogram analysis: the method and its application. In: S. Shye (ed.), Theory Construction and Data Analysis in the Behavioral Sciences, pp. 237–264. Jossey-Bass, San Francisco, USA.

VERTICAL STRUCTURE OF *SPARTINA MARITIMA* AND *SPARTINA DENSIFLORA* IN MEDITERRANEAN MARSHES

M.E. FIGUEROA and E.M. CASTELLANOS
Dept. de Ecologia, Univ. de Sevilla, Aptdo. 1095, Sevilla, Spain

Abstract

The vertical structure, above- and below-ground biomass of *Spartina maritima* and *Spartina densiflora* in South Western Spanish marshes is described. The mediterranean climate in this area, modified by oceanic influences, induces a nearly continuous year-round growth, *Spartina maritima* and *S. densiflora* are pioneer plants in sandy and muddy soils, appearing primarily in monospecific stands, although in recently formed mud flats there are some mixed communities. The average leaf area index (LAI) in *S. maritima* is 1.7. *S. densiflora*, more abundant than *S. maritima* in these marshes, shows a LAI of 11.0 in mature stands. Above-ground biomass (dry weight) in *S. maritima* is $300-2000$ g.m^{-2} during the summer peak; in *S. densiflora* it is $3000-7000$ g.m^{-2}. Below-ground biomass is 2000 g.m^{-2} and 12000 g.m^{-2} in *S. maritima* and *S. densiflora*, respectively. Light extinction curves in the two species are quite different. *S. densiflora* is an American neophyte, strongly competitive in relation to other species, replacing them completely when given suitable conditions. With its vertical structure it competes very effectively for space and light and once gaining a hold it practically stops succession.

Introduction

Different species of the genus *Spartina* dominate salt marshes on the Atlantic coast of North America and Europe. In Spanish salt marshes *S. maritima* (Curtis) Fernald, and *S. densiflora* Brong. occur.

Spartina densiflora is a competitive American neophyte and displaces *S. maritima* in tidal mud flats. This situation could lead to conservation problems with respect to *S. maritima*, and other associated species such as *Sarcocornia perennis* (Miller) A.J. Scott, *Halimione portulacoides* (L.) Aellen, and *Limonium angustifolium* (Tausch) Degen.

The structural design of a plant is the phenotypic expression of genetic traits that are subject to selection (Smith 1984). In this work the leaf stratification has been studied along with leaf area index and light penetration in stands of *S. maritima* and *S. densiflora* in SW. Spain. The results have been interpreted in terms of marsh succession.

Methods

Our study was done in tidal marshes of the Odiel River, Gulf of Cadiz (SW, Spain). The tidal range is $1-3$ m. The climate is Mediterranean, modified by oceanic influences.

Spartina maritima is a caespitose and rhizomatous plant with flat leaves and stems of $10-90$ cm. *S. densiflora* is very robust, rhizomatose, densely caespitose with convolute leaves and stems of $50-200$ cm.

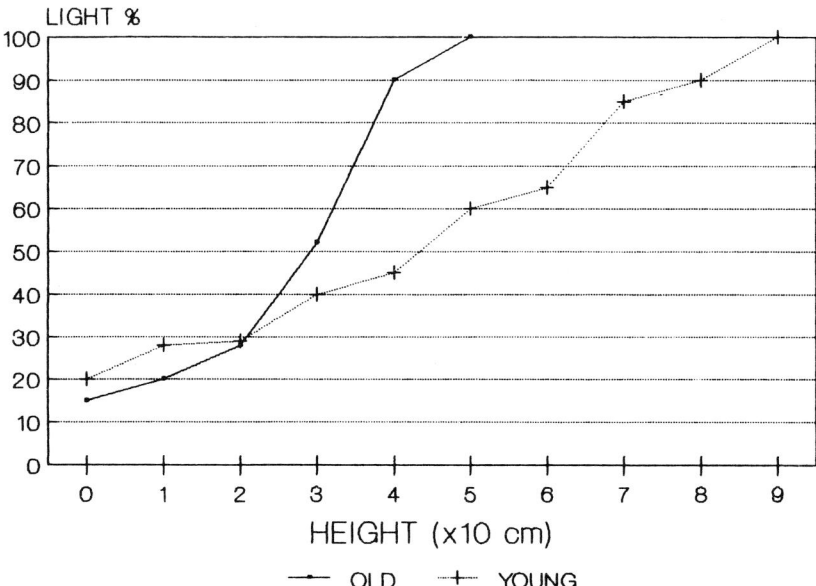

Fig. 1. Light extinction in young (90 cm) and old (50 cm) stands of *Spartina maritima*.

The vertical distribution of leaves, LAI (one-sided), tiller density, canopy depth (thickness of layer with leaves), and leaf number of each plant was measured in monospecific *S. maritima* stands of different ages (less than 10 years old and over 40 years old). LAI (live and dead) was measured in *S. densiflora* in a monospecific old stand (over 40 years old). Such old stands are common. Light penetration in canopies was measured using a quantum sensor (Li-Cor 190) for Photosynthetically Active Radiation. It was measured 10 times at each height at every 10 cm from the ground in *S. maritima* (young and old) and *S. densiflora* (young, less than four years; old, over 40 years) stands at noon on the same day during spring time.

Results and discussion

Spartina maritima and *S. densiflora* show a nearly continuous growth during the year, only interrupted for a short time in December and January. This is due to the favorable year round climate in this area. *Spartina densiflora* is more abundant, having a biomass (dry weight) of 3000–7000 g.m^{-2}, above-ground and 7000–12000 g.m^{-2} below-ground. *S. maritima* shows a biomass of 300–2000 g.m^{-2} above-ground, and 500–3000 g.m^{-2} below-ground.

Spartina maritima exhibits a unimodal distribution of leaves (monolayer) with LAI and canopy depth changing from early successional sites to mature stands (Table 1). The net above-ground production in young and old sites is ca. 1000 and 300 g.m^{-2}.yr^{-1} (dry weight), respectively (Figueroa 1987).

Light extinction in the canopy partly depends on the pattern of leaf arrangement. *S. maritima* canopies are higher in young stands than in older ones (Fig. 1). This partly explains why late-successional stands of *S. maritima* only occur in

Table 1. Structural characteristics of *Spartina maritima* stands.

	Max. height (cm)	Canopy depth (cm)	LAI m².m⁻²	Tiller density (m⁻²)	No. of leaves per tiller
Young stand (10 yrs)	90	27	1.9	1200	5.3
Old stand (over 40 yrs)	50	12	1.5	2400	4.0

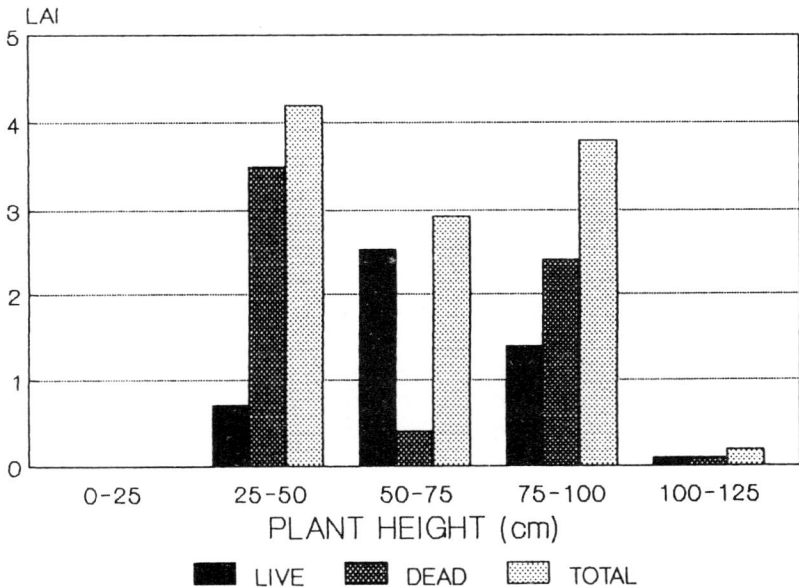

Fig. 2. Live, dead and total leaf area index in *Spartina densiflora* at different levels from the ground.

marginal habitats such as salt pans and eroded creek margins. In more favorable habitats *S. maritima* is outcompeted by *Sarcocornia perennis* and *Halimione portulacoides*.

Spartina densiflora grows in very dense stands with LAI values of 11.0 (live and dead leaves on the plant taken together). Figure 2 shows LAI (live, dead and total) in a stand of over 40 years old in layers from ground level to 125 cm high. *S. densiflora* does not shed dead leaves, thus showing a bimodal leaf distribution (essentially two layers of leaves) in tall plants (over 120 cm).

The light extinction inside young and old stand canopies is shown in Fig. 3. Light extinction is greater in *S. densiflora* than in *S. maritima* canopies. This, together with a higher net primary production and a very high standing crop because dead leaves are not shed in *S. densiflora*, promotes the replacement of *S. maritima*. When *S. densiflora* colonizes fresh mud no other plants succeed in entering the area because the species completely occupies the ground. *S. densiflora* can enter an area with *S. maritima* and can stop the latter's spread. The strong

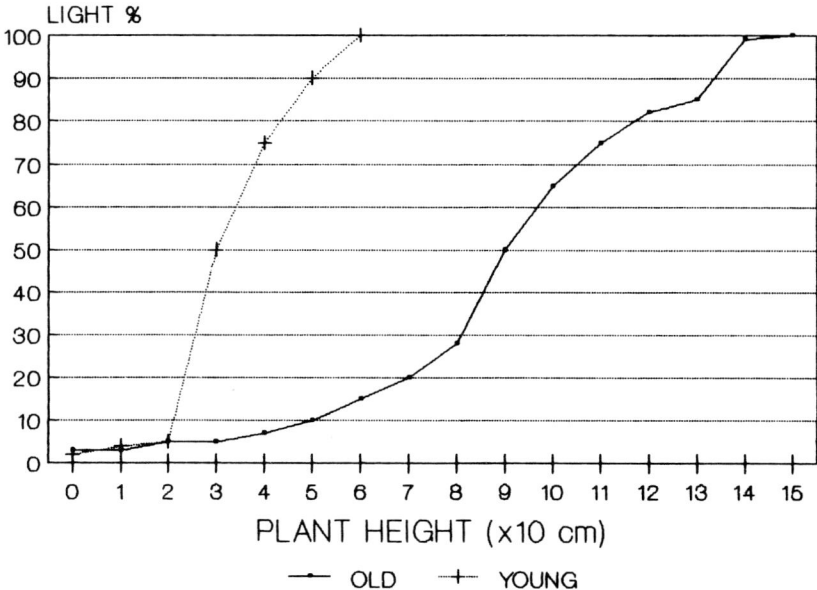

Fig. 3. Light extinction in young (60 cm) and old (150 cm) stands of *Spartina densiflora*.

light extinction in the canopy, together with its occupation of space prevents the establishment of new plant species in mature *S. densiflora* stands and impedes succession.

Acknowledgments

A grant from the Junta de Andalucía is greatly appreciated. Useful suggestions from Dr. García Novo and Dra. Luque are also acknowledged. We thank Consolación Heredia and Carmelo Escot for technical assistance.

References

Figueroa, E. 1987. Ecologia de las Marismas del Odiel y del Tinto. Bases Cientificas para la proteccion de los Humedales Españoles. Real Academia de Ciencias Exactas, Fisicas y Naturales, Madrid, pp. 269–282.

Smith, B.H. 1984. The optimal design of a herbaceous body. Am. Nat. 123: 197–211.

VARIATION IN BIOMASS AND ARCHITECTURE OF PLANTS DUE TO SMALL-SCALE ENVIRONMENTAL HETEROGENEITY

L.A.P. LOTZ* and H. OLFF
Institute for Ecological Research, Weevers' Duin, Duinzoom 20a, 3233 EG Oostvoorne, the Netherlands

Abstract

The vegetation on a former beach plain showed a marked differentiation in structure (plant biomass and height), which was related to spatial variation in nutrient supply and water saturation of the soil. The environmental impact on the biomass and architecture of individual plants was studied using *Plantago major* L. ssp. *pleiosperma* Pilger. Inbred lines, reared from plants sampled all over the beach-plain area, were grown in a greenhouse with nutrient supply and degree of water saturation as experimental factors. Variances in biomass of the shoot, leaf-area ratio, number of leaves, and leaf morphology were mainly determined by the experimental factors and thus based on phenotypic plasticity. There were, however, differences between plants from different origins (specific microhabitats within the study site) in shoot biomass and the number of leaves per plant, and in the level of phenotypic plasticity in these characteristics. Inbred lines originating from low-lying patches where the soil was deficient in nutrients reached a higher shoot biomass under similar stress conditions in the greenhouse, when compared to inbred lines from other microhabitats. Besides this higher biomass these plants also had a higher proportion of flowering plants. In some treatments plants from these nutrient-deficient soils had more leaves than plants from the other subsites. Variances of other characteristics of plant architecture, shoot-root ratio, leaf-area ratio and leaf morphology, were not explained by the origin of the inbred lines.

It was suggested that, in the low-lying patches of this beach plain, selection in *P. major* ssp. *pleiosperma* had resulted in a relatively higher shoot biomass, possibly due to a more efficient use of nutrients. This genetic differentiation in biomass accumulation due to small-scale environmental heterogeneity is discussed in terms of environmental constraints, vegetation structure, mosaic environment, gene flow, and adaptation.

Introduction

Within-population variability in plant size is an item of interest in population biology (*e.g.*, Van Andel *et al.* 1984; Weiner and Solbrig 1984; Weiner 1985). Differences in plant size (or growth rate) within populations may be caused by age differences, genetic variation, heterogeneity of resources, competition, or the effects of herbivores, parasites or pathogens (Weiner and Solbrig 1984). Little is known, however, about the relative importance of each of these factors (Weiner 1985). In order to understand the evolutionary significance of within-population variability in plant size the relative importance of genetic differentiation and the impact of environmental factors (phenotypic plasticity) on the establishment of these distributions should be assessed. In this respect, a pioneer study on the relationship between genetic variation and variability in plant size was made by Gottlieb (1977), who found a genetic similarity between the opposite extremes of the size

---

* *Present address*: Centre for Agrobiological Research, P.O. Box 14, 6700 AA Wageningen, The Netherlands.

distribution within an annual plant population. Therefore, he concluded that the observed differences in plant size were due to phenotypic plasticity and were unlikely to be the result of evolutionary changes.

As in plant size variability in plant architecture may also occur in populations. This morphological variation may also have evolutionary significance. For example, by altering the growth form a plant may assimilate or utilize resources more efficiently under specific conditions than neighboring plants (see for a recent review Givnish 1986). Differences in plant architecture may be induced, too, by the possible causes of differences in plant size. Therefore, relating genetic and environmentally induced variation of both plant size and architecture seems interesting.

The genetic structure of plant populations is influenced e.g., by selective forces, rates of gene flow, and random factors. When coupled with restricted gene flow, spatial variation in selective forces may stimulate local differentiation. If, on the contrary, the scale of spatial environmental heterogeneity is small compared to the mean distance of gene flow (i.e., the progeny of an individual plant experiences two or more possible habitats) there may be a selection for high levels of phenotypic plasticity of specific traits (Bradshaw 1965; Via and Lande 1985; Schlichting 1986). In demographic models of life-history evolution Caswell (1983) distinguished variables forming part of a response system and 'essential' variables, which must be kept within a certain range for continued survival. He stated that plasticity is adaptive to the extent that it contributes to the homoeostasis of these fitness-related variables.

The present study is part of a multidisciplinary research project on variability in life-history characteristics within a population of *Plantago major* L. ssp. *pleiosperma* Pilger on a former beach plain (Lotz and Olff, in press). Plants of this taxon are short-lived perennials with a rosette-growth form. At the study site the environment varies spatially in nutrient availability and water saturation of the soil, and in vegetation structure (Troelstra *et al.*, in press). This mosaic environment consists of a low-lying area in which there are patches of two size classes (diameters 0.5–1.5 and 20–40 m, respectively). At each microsite plants of *P. major* ssp. *pleiosperma* grow and reproduce. For that reason, variation in plant characteristics may be related to the scale of environmental heterogeneity.

The present paper is concerned with variability in plant size and architecture of *P. major* ssp. *pleiosperma* on this beach plain. An attempt is made to answer the following questions:

1. What is the relative contribution of genetic differentiation and phenotypic plasticity to the variability in plant size and architecture within this population?
2. What is the relationship between the scale of environmental variation and the level of phenotypic plasticity in plant size and architecture?
3. If, in this population, plant size is strongly correlated with fitness (*i.e.*, essential variable *c.f.* Caswell 1983) can it be demonstrated that variation of an other plant characteristics may contribute to a relative homoeostasis of the optimal plant biomass?

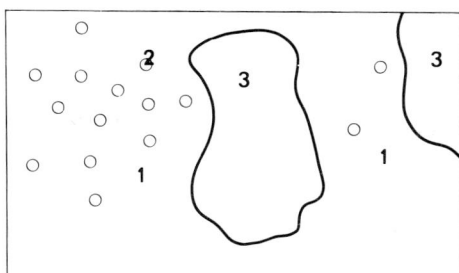

Fig. 1. Sketch of the mosaic environment of the Oostvoornse Meer site. Three subsites are distinguished: Subsite 1, basic area with very open grassland; subsite 2, patches (0.5–1.5 m in diameter) with a high plant cover of mainly grasses; subsite 3, patches (20–40 m in diameter) with shrubs of *Hippophaë rhamnoides* L.

In order to answer these questions inbred lines obtained from plants of *P. major* ssp. *pleiosperma*, sampled from three microsites on the beach plain, were studied in a greenhouse experiment with nutrient supply and degree of water saturation of the substrate as experimental factors.

Several authors (*e.g.*, Antonovics and Primack 1982; Watkinson and Gibson 1985) stressed the importance of evaluating adaptive values of life-history traits in the natural habitat over the whole life cycle. Together with this greenhouse experiment, in which the impact of two specific environmental factors on plants has been studied, long-term investigations on variation in plant biomass and architecture, selection differentials and gene flow in the natural habitat itself are currently in progress.

Materials and methods

Study site

The population of *P. major* L. ssp. *pleiosperma* Pilger that was studied is located near Oostvoornse Meer, a brackish lake on the southwest coast of the Netherlands. The Oostvoornse Meer site is a former beach plain, embanked in 1966. Because of differences in micro-relief, nutrient availability (for a detailed description see Troelstra *et al.*, in press), and vegetation structure three subsites, forming a mosaic environment, can be distinguished (Fig. 1, Table 1). Subsite 1 is a low-lying area. The soil is low in nutrients and highly water-saturated. In winter and occasionally in summer, this area is flooded after heavy rain. The vegetation is rich in species. Cover by higher plants is relatively low (30%). At this microsite aboveground biomass and leaf size of plants of *P. major* ssp. *pleiosperma* are significantly smaller than at the other microsites. Seed production by plants of this taxon is also very low, 5.4 (S.D. = 4.2, n = 10) mg per plant. Subsite 2 consists of small elevations (0.07–0.015 m high), which are mainly covered with a short dense grassy vegetation. At these patches the soil is relatively high in nutrients and or-

Table 1. Estimates of characteristics of the habitat (A) and of plants of P. major ssp. pleiosperma (B) at the study site Oostvoornse Meer. Three subsites were distinguished (See Fig. 1). Data from Troelstra et al. (in press) and Lotz and Olff (in press). Significant differences ($p < 0.05$) between subsites are indicated by different letters.

	Subsite 1	Subsite 2	Subsite 3
A. Soil (0–10 cm, $n = 3$)			
Organic matter (%)	1.3a	2.7b	1.4a
Total N (g·m^{-2})	66.4a	117.6b	68.7a
N-NO$_3^-$ after 20 weeks incubation at 30°C			
(g·m^{-2})	7.8a	17.1c	11.9b
(% total N)	11.7a	14.5ab	17.3b
Standing crop herbs and grasses (g·m^{-2}, $n = 3$)			
0–5 cm	14.4a	94.4b	198.4b
> 5 cm	0	0.8	24.0
B. Plant characteristics ($n = 10$)			
Number of leaves	4.7	5.9	5.0
Length largest leaf (mm)	16a	40	99c
Width largest leaf (mm)	6a	13	23c
Above-ground biomass (g)	0.018a	0.123b	0.261b

ganic matter and *P. major* ssp. *pleiosperma* plants have a higher biomass and produce more seeds than at subsite 1, 76.5 (S.D. = 63.9, $n = 10$) mg per plant. Subsite 3 is covered by *Hippophaë rhamnoides* L. Between these shrubs the standing crop of the herb layer is relatively high. The *P. major* ssp. *pleiosperma* plants have a relatively high biomass. The leaf size is larger than at the other microsites. Seed production is also relatively high, 162.8 (S.D. = 192.3, $n = 10$) mg per plant. The amount of total nitrogen in the soil at this subsite is relatively low. Since relative mineralization rates were high compared to samples from the other subsites (Table 1), we assume that the soil of subsite 3, in common with the soil of the patches of subsite 2, has a higher nitrogen availability than the soil of subsite 1.

Greenhouse experiment

On April 25, 1985 seeds of six inbred lines from each subsite were sown in containers with moist dune sand in a greenhouse (20°C). At least 95% of the seeds of each inbred line germinated within three days. Twenty days after sowing, seedlings were transplanted into pots with silver-sand (contents 0.42 l, one plant per pot). At the start of the experiment three different nutrient levels and two levels of water saturation were created, together forming a full factorial treatment combination. Nutrient levels were made by adding different quantities of a solution twice as strong as prescribed by Steiner (1968): the Low treatment 2.6 ml, the Intermediate treatment 38 ml, and the High treatment 75 ml. Previous experiments (Lotz and Blom 1986) justified the expectation that the biomass accumula-

Table 2. Effects of nutrient supply and level of water saturation on shoot biomass, some morphological characteristics, and seed production of inbred lines of *Plantago major* ssp. *pleiosperma* in the greenhouse. Inbred lines originated from three origins (subsite 1, 2, and 3) within a beach-plain population. Effects were tested in a nested analysis of variance. Data were summarized from Lotz and Olff (in press). In the analysis of variance of number of seeds per plant data from the low nutrient supply were excluded because of some empty cells. Abbreviations: Nut = nutrient supply; Wat = level of water saturation; Line = inbred line; Orig = origin. Levels of significance: ns = not significant; * $p < 0.05$; ** $p < 0.01$; *** $p < 0.001$.

Source of variation	Dependent variables				
	Dry weight shoot[1]	Shoot-root ratio	Leaf-area ratio	Number of leaves	Petiole ratio (dry weight petioles/ dry weight shoot)
Nut	***	***	***	***	***
Wat	ns	***	ns	***	***
Nut × Wat	***	*	*	***	ns
Lin	***	***	***	*	***
Nut × Line	***	***	***	ns	**
Wat × Line	***	***	ns	ns	ns
Nut × Wat × Line	***	***	ns	ns	*
Orig	*	ns	ns	**	ns
Nut × Orig	*	ns	ns	ns	ns
Wat × Orig	ns	ns	ns	ns	ns
Nut × Wat × Orig	ns	ns	ns	*	ns

Source of variation	Dependent variables			
	Leaf length	Leaf width	Leaf length/width	Number of seeds per plant[1]
Nut	***	***	*	***
Wat	***	***	***	**
Nut × Wat	ns	***	ns	ns
Line	***	***	***	***
Nut × Line	***	***	ns	ns
Wat × Line	***	**	ns	*
Nut × Wat × Line	***	*	**	ns
Orig	ns	ns	ns	ns
Nut × Orig	ns	ns	ns	ns
Wat × Orig	ns	ns	ns	ns
Nut × Wat × Orig	ns	ns	ns	ns

[1] Analysis of variance was performed after log-transformation.

tion at these nutrient levels would be within the same range as that found at the study site. Micro-nutrients were supplied to each plant by adding 75 ml solution containing micro-nutrients with concentrations twice as high as prescribed by Smakman and Hofstra (1982). The high level of water saturation (treatment 'Waterlogged') was created by placing pots with silver-sand in low containers with water (the water level was kept permanently at 5 cm above the container bottoms). The substrate of the pots was fully saturated by water which entered the pots through holes in the bottom. In the pots with low water saturation (treatment 'Drained') soil moisture was kept at 20% by weight. The number of replicates per inbred line was five. Pots were placed in a greenhouse (20°C). In order to maintain differences in treatment level during the experiment, uptake of minerals by the plants was compensated for as described by Lotz and Blom (1986).

All plants were harvested on day 88 after sowing. The following characteristics were measured for each plant: number of leaves, length and width of the largest leaf, fresh and dry weight of leaf blades, petioles and roots, leaf area, and number of seeds. Shoot-root ratio was computed as fresh weight of leaves per fresh weight of roots.

Data were analyzed in a nested analysis of variance (SPSS-subprogram MANOVA, Nie and Hull 1981) with independent variables: supply of nutrients, level of water saturation, origin, and inbred line nested within origin. Comparisons of means were made by performing the Student-Newman-Keuls multiple range test. If treatment x origin effects were significant the difference in response to treatment (*i.e.*, difference in level of phenotypic plasticity) was tested in a trend analysis (see Lotz and Blom 1986). Therefore, linear regression coefficients were computed for each inbred line ($d.f.$ = 2). In a one-way analysis of variance differences in these regression coefficients between origins were tested against the mean squares due to the inbred line within an origin and nutrient supply ($d.f.$ = 10). Differences in frequencies of flowering plants at day 88 after sowing were tested by fitting a log-linear model (BMDP-program PLR, Dixon 1983).

Results

The variance of the dry weight of the shoot at the end of the experiment was explained by nutrient supply, the interaction of nutrient supply and level of water saturation, and interactions of inbred lines with nutrient and water treatments (Table 2). Besides, significant effects existed due to origin and the interaction of nutrient supply with origin (the relative contribution of these added variance components to the total sum of variance was 0.4%, each, whereas 93.0% of the total variance was explained by both the experimental treatments). A significant effect due to origin represents genetic differentiation in shoot biomass between inbred lines of the different subsites. A significant effect due to the interaction of nutrient supply with origin indicates genetic variation for phenotypic plasticity between inbred lines of different origins. In the low nutrient level plants originating from subsite 1 had a higher shoot biomass than plants from subsite 3 (Fig. 2, $p < 0.05$). In the treatment combination 'low nutrient supply and drained' this contrast was significant, too, for plants originating from subsite 1 and 2 ($p < 0.05$). Linear regression coefficients (means over six inbred lines) between shoot biomass and level of nutrient supply were for plants from subsite 1, 2 and 3: 0.80, 0.89 and 0.91, respectively. Irrespective of the level of water saturation, plants from subsite 1 demonstrated less phenotypic plasticity and, therefore, relatively more homoeostasis in shoot biomass than plants from subsites 2 and 3, whereas differences due to nutrient supply in response of shoot biomass between plants

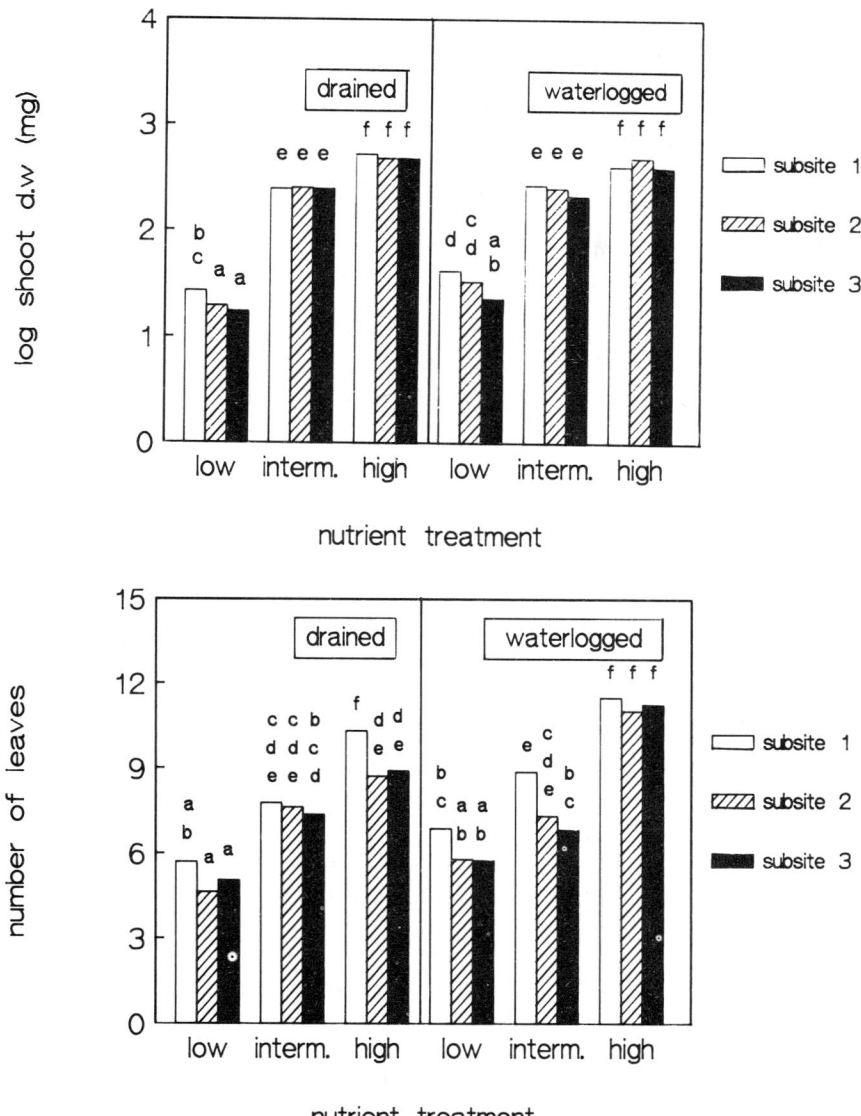

Fig. 2. The effect of nutrient supply and level of water saturation on shoot biomass (A) and number of leaves per plant (B) of *Plantago major* ssp. *pleiosperma*. Plants originated from three subsites within a beach plain. Means are given over six inbred lines with five replicas each. Significant differences between origin, tested against the error due to inbred line, are indicated by different letters ($p < 0.05$).

from subsites 2 and 3 were not significant (tests of contrasts between origins in linear regression coefficients: $p < 0.05$ and $p < 0.05$, respectively).

In the low-nutrient treatment differences in shoot biomass between origins were coupled to different proportions of flowering plants ($p < 0.05$). On day 88 after sowing, in the treatment combination 'low nutrient supply and drained' the percentage plants from subsite 1, 2, and 3 that flowered were: 37, 20, and 30, respectively. In the treatment combination 'low nutrient supply and waterlogged' these percentages were 83, 53 and 40. In the other treatments, however, all plants of each subsite flowered at the end of the experiment. At the high and intermediate nutrient treatments seed production for plants from different origins was not significantly different (Table 2).

The significant effect of the origin on the biomass of the shoot was accompanied by an origin effect on the number of leaves per plant (Table 2). The relatively high level of homoeostasis in shoot biomass of plants from subsite 1 was, at least partially, related to less variation in the production of individual leaves. This covariate effect, however, was not strong ($r^2 = 0.28$). Besides, *a posteriori* contrasts between origins in number of leaves per plant were only significant in the treatment combinations 'high nutrient supply and drained' and 'intermediate nutrient supply and waterlogged'. In these treatment combinations plants from subsite 1 had more leaves (Fig. 2).

For shoot-root ratio, leaf-area ratio, petiole ratio (dry weight of petioles / dry weight of the shoot), leaf length, leaf width and the ratio leaf length / leaf width treatment effects were significant (Table 2). For these characteristics no significant effects due to origin were present. In contrast to this, the results demonstrate for all these characteristics significant effects of inbred line, in part dependent on treatments. These last-mentioned effects (with respect to the shoot biomass they jointly explained 3.3% of the total sum of variance) represent genetic variation which is not related to the distinct subsites.

Discussion

The main objective of the present paper is to assess the relative contribution of genetic differentiation and phenotypic plasticity to the establishment of plant-size variability of *P. major* ssp. *pleiosperma* in a relatively young (20 years) mosaic environment. In order to investigate the ecological and evolutionary significance of phenotypic plasticity, several authors (e.g., Schlichting 1986) emphasized the importance of applying experimental factors of which the variability is relevant to the life history of an individual plant. Besides, responses of that specific plant should also be studied within the ranges of environmental factors, that are found in the natural habitat. In the present study the impact of two soil characteristics, nutrient supply and level of water saturation, was studied using inbred lines. The selection of these experimental factors and the ranges used, both based on earlier investigations on soil characteristics of the habitat and demography (Troelstra *et al.*, in prep., unpubl. results), were expected to fulfill the above-mentioned conditions.

In the greenhouse experiment variances in biomass of the shoot, and moreover, in shoot-root ratio, leaf-area ratio, and several other morphological characteristics of the shoot, were mainly determined by the experimental factors nutrient supply and degree of water saturation and thus based on phenotypic plasticity. Less than 1% of the total sum of variance of shoot biomass was explained by the subsite the inbred lines originated from. Inbred lines originating from low subsites with a nutrient-deficient soil (subsite 1) had a higher biomass of the shoot and the root (data of root biomass have not been presented, but can be computed from the shoot-root ratios) in the treatment combination 'low nutrient supply and high level of water saturation' than plants from other subsites. Although these differences were relatively small (Fig. 2), this higher biomass was coupled with a higher proportion of flowering plants. Therefore it is suggested that flowering of *P. major* ssp. *pleiosperma* is determined by a developmental size threshold, which is probably reached by a greater number of plants from subsite 1 than from the other subsites. At the intermediate and high nutrient level all the plants from the three subsites flowered. In these treatments seed production of plants from different subsites was not significantly different. The relatively high level of homoeostasis of shoot biomass in plants from subsite 1 was not related to a higher level of phenotypic plasticity in variables of plant architecture as shoot-root ratio or leaf-area ratio. From estimates of the total uptake of nutrients (N, P, and K), however, it was concluded that inbred lines of the three subsites differed in level of phenotypic plasticity in nutrient use (Lotz and Olff, unpublished results). Plants from subsite 1 might have a higher biomass at the low-nutrient supply because of a higher efficiency of this nutrient use (a variable thus forming part of a response system *cf.* Caswell 1983) in these nutrient-deficient conditions than plants from subsite 2 or 3.

No differences in leaf morphology were found between plants from different subsites. The numbers of leaves per plant from different origins were, however, significantly different. The relatively high level of homoeostasis in shoot biomass was coupled with relatively little variation in number of leaves per plant. For that reason, at the low nutrient treatment, plants from subsite 1 should have more leaves than plants from the other subsites. A higher number of smaller leaves, which is related in the natural habitat to a more prostrate growth form (unpublished results) might have adaptive value in the very open rabbit grazed vegetation of subsite 1. However, *a posteriori* contrasts (Fig. 2) do not support this suggestion, probably due to relatively large variances or small sample sizes. In only two treatment combinations, 'high nutrient supply and drained' and 'intermediate nutrient supply and waterlogged', plants from subsite 1 had more leaves than plants from other subsites. Moreover, the effect of origin on the length and width of the largest leaf was not significant (Table 2).

The results of the present study on inbred lines demonstrate genetic variation in biomass accumulation (and in level of phenotypic plasticity in this characteristic), which has been related to a pattern of environmental variation within a mosaic environment. The relationship between the demonstrated genetic differentiation and the scale of environmental heterogeneity in the natural habitat depended on the treatment. In the treatment combination 'low nutrient supply and high level of water saturation' shoot dry weights of plants from subsites 1 and 2 were significantly different from those of plants from subsite 3; this variation corresponds with a scale of environmental heterogeneity at the beach plain of about 10–40 m. In the treatment combination 'low nutrient supply and low level of water saturation' dry weights of the shoot were even significantly different between plants from subsite 1 and subsite 2. Both subsites form part of a small-scale heterogeneous environment that varies over distances of about 0.5–3.0 m. Lotz and Spoormakers (1988) also demonstrated differences in phenotypic plasticity in seed-yield components between inbred lines from the three subsites. Their results as well as the present results suggest that the level of gene flow within the beach-plain population is relatively low compared to the spatial variation in selection (*cf.* Bradshaw 1965; Via and Lande 1985; Slatkin 1985). *P. major* is self-compatible and wind pollinated with a high self-fertilization rate (Van Dijk and Van Delden 1981). Van Dijk (1985) estimated the mean gene transport per generation within a population of *P. major* ssp. *major* to be only 0.11 – 0.35 m. It might be hypothesized that within the present beach-plain population similar levels of gene flow occur and that these levels are low enough to establish a genetic structure within the twenty years this area had been colonized by higher plants. To test this hypothesis gene flow and relative selection pressures are currently being studied at the Oostvoornse Meer site.

Acknowledgments

We thank J.M.M. van Damme and J.W. Woldendorp for helpful comments on the manuscript. These investigations were supported by the Foundation for Fundamental Biological Research (BION), which is subsidized by the Netherlands Organization of Pure Research (ZWO), Grassland Species Research Group Publ. No. 140.

References

Antonovics, J. and Primack, R.B. 1982. Experimental ecological genetics in *Plantago* VI. The demography of seedling transplants of *P. lanceolata*. J. Ecol. 70: 55–75.
Bradshaw, A.D. 1965. Evolutionary significance of phenotypic plasticity in plants. Adv. Genet. 13: 115–155.
Caswell, H. 1983. Phenotypic plasticity in life-history traits: demographic effects and evolutionary consequences. Amer. Zool. 23: 35–46.
Dixon, W.J. 1985. BMDP Statistical Software. University of California Press. Berkeley.
Givnish, T.J. (ed.) 1986. On the economy of plant form and functioning. Proceedings of the Sixth Maria Moors Cabot Symposium 'Evolutionary Constraints on Primary Productivity: Adaptivity patterns of Energy Capture in Plants'. Harvest Forest, August 1983. Cambridge University Press, Cambridge.
Gottlieb, L.D. 1977. Genotypic similarity of large and small individuals in a natural population of the annual plant *Stebanomeria exigua* ssp. *coronaria* (Compositae). J. Ecol. 65: 127–134.
Lotz, L.A.P. and Blom C.W.P.M. 1986. Plasticity in life-history traits of *Plantago major* L. ssp. *pleiosperma* Pilger. Oecologia 69: 25–30.
Lotz, L.A.P. and Olff, H. (in prep.) Differentiation in life-history traits within a population of *Plantago major* L. ssp. *pleiosperma* Pilger. II. The effect of nutrient supply and water saturation of the soil.
Lotz, L.A.P. and Spoormakers, L.D.H. (1988). Differentiation in reproductive characteristics within a population of *Plantago major* L. ssp. *pleiosperma* Pilger. Oecologia Plant 9: 11–18.
Nie, N.H. and Hull, C.H. 1981. SPSS Update 7–9: New procedures and facilities for releases. McGraw-Hill, New York.
Schlichting, C.D. 1986. The evolution of phenotypic plasticity in plants. Ann. Rev. Ecol. Syst. 17: 677–693.
Slatkin, M. 1985. Gene flow in natural populations. Ann. Rev. Ecol. Syst. 16: 393–340.
Smakman, G. and Hofstra, J.J. 1982. Energy metabolism of *Plantago lancolata*, as affected by change in root temperature. Physiol. Plant. 56: 33–37.
Steiner, A.A. 1968. Soilless culture. Proc. 6th Coll. Int. Potash Inst., pp. 324–342, Florence.

Troelstra, S.R., Lotz, L.A.P., Wagenaar, R. and Sluimer, L. (in prep.). Temporal and spatial variability of soil properties in a former beach plain.

Van Andel, J., Nelissen, H.J.M., Wattel, E., Van Valen, T.A. and Wassenaar, A.T. 1984. Theil's inequality index applied to quantify population variation of plants with regard to dry matter allocation. Acta Bot. Neerl. 33: 161–175.

Van Dijk, H. 1985. Genetic variability in *Plantago* species in relation to their ecology. Thesis. University of Groningen, the Netherlands.

Van Dijk, H. and Van Delden, W. 1981. Genetic variability in *Plantago* species in relation to their ecology. I. Genetic analysis of the allozyme variation in *P. major* subspecies. Theor. Appl. Genet. 60: 285–290.

Via, S. and Lande, R. 1985. Genotype-environment interactions and the evolution of phenotypic plasticity. Evolution 39: 505–512.

Watkinson, A.R. and Gibson, C.C. 1985. Life-history variation and the demography of plant populations. In: J. Haeck and J.W. Woldendorp (eds), Structure and Functioning of Plant Populations/ 2. Phenotypic Plasticity and Genotypic Variation in Plant Populations. North-Holland Publishing Company, Amsterdam, pp. 105–113.

Weiner, J. 1985. Size hierarchies in experimental populations of annual plants. Ecology 66: 743–752.

Weiner, J. and Solbrig, O.T. 1984. The meaning and measurement of size hierarchies in plant populations. Oecologia (Berlin) 61: 334–336.

ASPECTS OF THE STRUCTURE OF CLONAL PERENNIAL HERBS

MICHAEL J. HUTCHINGS and ANDREW J. SLADE
School of Biological Sciences, University of Sussex, Falmer, Brighton, Sussex, U.K., BN1 9QG

Abstract

Aspects of the spatial structure and performance of annually-replaced populations of ramets in clonal herbs are discussed, with particular reference to *Glechoma hederacea*. Structural plasticity, which results in different patterns of ramet placement in clones growing under different conditions, can be interpreted as a property which allows them to forage efficiently for essential resources. The expression of plasticity in foraging behaviour under different biotic and abiotic conditions is illustrated using results from greenhouse experiments. Morphological alterations in clone architecture under different growing conditions result in a high intensity of occupation of favourable sites, through intensive foraging, and in a low intensity of occupation of unfavourable sites by means of extensive foraging.

In clonal species which display physiological integration between their constituent ramets, the habitat is occupied efficiently by a network of large ramets. When physiological integration is disrupted, a greater density of smaller ramets is produced, and this may be accompanied by density-dependent ramet mortality. The effects of disrupting integration between ramets are illustrated using results from manipulative experiments in which ramets are isolated from each other by severing. The structure and population biology of ramet populations of clonal perennial herbs differ between field sites and between hermaphrodite and male sterile clones in the gynodioecious *Glechoma hederacea*. Different subsets of ramets (fertile, sterile) within clones differ in their sizes and their survivorship.

Introduction

Clonal plants are important components of a range of communities at many latitudes and altitudes. Some are among the most successful species in terms of their geographical distribution and colonizing ability (Silander 1985), many form extensive monoclonal stands for considerable periods, resisting invasion by other species (Hutchings and Bradbury 1986), and several are included among those with the longest recorded life spans for plants (Cook 1983). However, despite their ecological importance, clonal plant species had been little studied until the last decade, during which there has been a marked increase of interest in their ecology and evolution (Jackson *et al.* 1985; Harper *et al.* 1986).

The fact that so little data had been collected on clonal plants until recently is surprising, given their major contribution to the structure of natural vegetation and the ease with which highly controlled experiments can be performed on many species under greenhouse conditions. However, sufficient data are now available to enable us to begin to understand some aspects of the structure of clonal herbs. We confine the following discussion to clonal herbs with short-lived ramet populations which are replaced annually. There are four major structural components of plant populations, namely performance, spacing, age and genetic structure (Hutchings 1986). The genetic structure of clonal populations has been extensively reviewed by Silander (1985). In this paper we focus upon two of the three remaining components: (i) the spatial structure of clonal herbs as it is affected by plastic architectural responses to abiotic growing conditions and by the physiologically integrated behaviour of connected ramets, and (ii) the performance and demography of ramet populations. Consideration of the age structure of clonal herbs would necessitate considerable expansion of this paper to include information on clones with longer-lived ramet populations. The data upon which this paper is based include field studies of the population biology of a range of clonal herbs and controlled experiments performed under greenhouse conditions. However, we draw heavily upon recent research on *Glechoma hederacea* L. for which a large amount of relevant information is now available.

Clone structure and differential foraging in heterogeneous environments

Genetically identical ramets of *Glechoma hederacea* clones can easily be propagated in the greenhouse. Growth of these ramets eliminates much of the plant-to-plant variation usually observed even under standard experimental conditions, and therefore the morphological plasticity shown by clones under different growing conditions can be clearly analyzed. *G. hederacea* clones consist of ramets developed at nodes along stolons (Slade and Hutchings 1987a). Secondary stolons may develop from axillary buds situated at the nodes, although control of their growth by apical dominance ensures that this only occurs a few nodes behind the stolon tip. Each ramet has a determinate structure, and consists of two erect petioles each bearing a single horizontal reniform to cordate leaf. For experimental purposes each ramet can be rooted in a separate pot, without severing its connections with other ramets, and each ramet can thus be treated independently with any specified combination of growing conditions (Slade and Hutchings 1987a–c).

Although this general description of clone architecture is accurate, experimental studies of *G. hederacea* clones grown under different soil nutrient and incident light levels demonstrate marked plasticity in many morphological characteristics (Table 1). Among these are significant reductions in the frequency of stolon branching and significant increases in mean stolon internode lengths when availability of either nutrients or light decreases (Slade and Hutchings 1987a–c). Branching frequency and internode length are two of the three major characteristics which define the architecture, and determine the spatial arrangement of ramets in stoloniferous and rhizomatous species. The third characteristic is branching angle. Although our experiments were not designed for rigorous investigation of branching angle in *G. hederacea*, the results agree with the only other published data about this variable (Sutherland, in press) in demonstrating no systematic change as conditions of growth were altered. In addition to these changes in branching frequency and internode lengths, the proportional allocation of dry matter to stolons increases under both low nutrient and low light supply, and the dry weight per unit length of stolons decreases, particularly under low light. A significantly greater proportion of plant biomass is devoted to roots under low

Aspects of the structure of clonal perennial herbs 123

Table 1. Summary table of responses by *Glechoma hederacea* to availability of light and soil nutrients (for detailed description of the treatments applied see text and Slade and Hutchings 1987c).

Characteristics measured	Treatment applied		
	High light High nutrients	High light Low nutrients	Low light High nutrients
Number of ramets per clone	————————————Decreases————————————>		
Number of stolon branches per clone	————————————Decreases————————————>		
Internode lengths	————————————Increases————————————>		
Total clone dry weight	————————————Decreases————————————>		
Dry weight per unit length of stolons	High	Intermediate	Low
Dry weight of leaves per clone	High	Low	Very low
Dry weight of roots per clone	<—— No difference ——>		Low
Dry weight of petioles per clone	High	Low	Very low
Dry weight of stolons per clone	<—— No difference ——>		Low
Proportional allocation of weight to leaves	High	Low	High
Proportional allocation of weight to petioles	Intermediate	Low	High
Proportional allocation of weight to stolons	Low	High	High
Proportional allocation of weight to roots	Intermediate	High	Low
Leaf area per clone	————————————Decreases————————————>		
Specific leaf area	High	Low	High
Chlorophyll content of leaves	High	Low	Not measured
Growth form	Phalanx ————————————————————> Guerilla		

nutrient levels, whereas under low light intensity, clones allocate a significantly greater proportion of their dry weight to petioles (Slade and Hutchings 1987a, b).

These results from plants grown under greenhouse conditions are substantiated by the few data which can be gleaned from populations of clonal species growing under field conditions (Lovett Doust 1981a, b; Harper 1983; Hutchings and Slade 1988) and from other experimental investigations (Ginzo and Lovell 1973). Clone morphology displays changes which promote spread in two dimensions and close packing of ramets under favourable growing conditions, and unidirectional spread and low densities of ramets under poor conditions. In terms of the descriptions of clone growth form as phalanx or guerilla (Clegg 1978), the results show that clone morphology alters in the direction of being more guerilla when resources are less abundant. Most resource accumulation by clonal herbs is carried out by the ramets, which photosynthesize and take up water and nutrients (although some photosynthesis is also carried out by stolons). In terms of foraging behaviour, where foraging is defined as the process whereby an organism searches or ramifies within its habitat in the activity of acquiring essential resources (Slade and Hutchings 1987a), clones change from foraging intensively (packing ramets at a high density) in areas of abundant resources, to foraging extensively (with a lower density of ramets) when resources are scarcer.

Integration between clonal ramets

Plasticity in clone architecture as described above can be interpreted as beneficial to the clone because it enables resource acquiring activities to be concentrated in the areas of habitat from which the rewards are likely to be greatest. In order for efficient foraging behaviour to be achieved by a clonal plant with ramets which sample different parts of a heterogeneous environment, there must be a strong element of integrated behaviour between the ramets. Investigations have shown a potential for physiological integration between connected ramets, through translocation between parts of the plant, of both photoassimilates and soluble nutrients, in some cases over long distances (see Pitelka and Ashmun 1985). Bidirectional translocation may occur, but generally acropetal translocation substantially exceeds basipetal translocation in undisturbed clones (Guttridge 1959; Norton and Wittwer 1962; Hoshino 1972; Grindey 1975; Ginzo and Lovell 1973a, b; Newell 1982; Noble and Marshall 1983; Alpert and Mooney 1986; Slade and Hutchings 1987c).

Integration between ramets can only occur while ramets remain physically connected. Thus, those clonal species in which connections between established ramets persist for long periods are more likely to produce a structure which will forage efficiently than those species in which connections decay soon after ramets have established. Similarly, any activity which breaks inter-ramet connections prematurely will disrupt integration and prevent the development of a structure which can forage efficiently. It may also result in the death of small fragments of the clone which are suffering strong competition and incapable of acquiring sufficient resources for survival.

In the absence of disturbance, density-dependent ramet mortality seems to be uncommon in clonal species which maintain inter-ramet connections for the whole of a growing season or longer (Hutchings 1979). This is perhaps particularly important in species with a phalanx structure. One of the defining characteristics of phalanx clones is that, in contrast to guerilla clones, their ramets experience contacts mainly with other ramets of the same clone, rather than with ramets of other clones or with other species. A phalanx clone which forages efficiently consists of mature ramets with a size and density giving a biomass per unit area just below that at which density-dependent mortality will occur, the resources supplied by the habitat will be optimally utilized and the likelihood of being supplanted by another species will be very low (Hutchings 1979). Correlative inhibition of bud development prevents the simultaneous growth of many of the bud initials in the intact clone, the result being production of a population of ramets with a density and total biomass which can be supported by the habitat with no density-dependent mortality. As discussed below, the lack of intense competition in these ramet populations is reflected in their size frequency histograms, which show little of the tendency to develop very marked skewness seen in populations of genetically distinct plants under severe competition (Hutchings and Barkham 1976).

The moderating effects of correlative inhibition upon ramet development and of physiological integration on competition between ramets within clones and on the spatial structure of clones, have been investigated in experiments in which inter-ramet connections were severed prematurely. In studies on *Solidago cana-*

Aspects of the structure of clonal perennial herbs

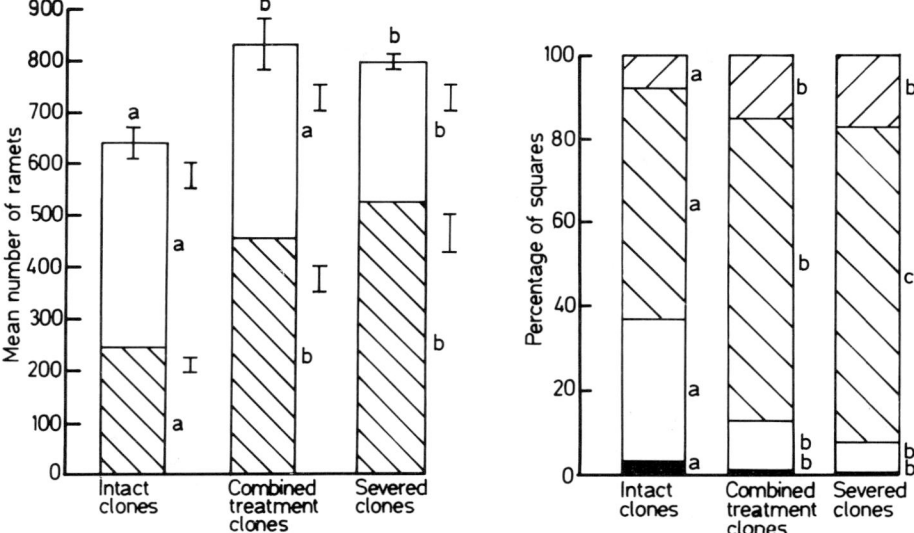

Fig. 1. Mean total number (±S.E.) of ramets produced by pairs of clones of *Glechoma hederacea*. ▨ = rooted ramets, □ = unrooted ramets. Standard error bars for the number of rooted and unrooted ramets are shown at the side of each treatment. Significant differences (at least $p < 0.05$) between the number of ramets produced in each treatment are indicated by different letters.

Fig. 2. Percentage of 2 × 2 cm squares occupied by leaf tissue of *Glechoma hederacea*. ■ = squares with no leaf tissue present; □ = squares with from 1 to 50% leaf cover; ▨ = squares with 51 to 95% leaf cover; ▧ = squares with > 95% leaf cover. Significant differences (at least $p < 0.05$) between the percentage of squares with different classes of cover in the three treatments are indicated by different letters.

densis L., the date on which ramets were severed had significant impacts upon the probability of survival and flowering, both of which declined the earlier that severing took place in the growing season (Hartnett and Bazzaz 1983). In experiments on *Glechoma hederacea*, three types of treatments were established in 56 × 56 cm boxes (Slade and Hutchings 1987e). Two ramets were planted 40 cm apart in each treatment, at opposite corners of the boxes, and clones were allowed to develop from each ramet for a period of 10 weeks. The treatments were as follows: (i) all stolon connections between ramets were severed as soon as new ramets had rooted, (ii) all stolon connections between ramets of one clone were severed as soon as they had rooted, whereas all stolon connections on the other clone were allowed to remain intact (combined treatment), (iii) all stolon connections of both clones were allowed to remain intact. Treatments were replicated four times. By the end of the growing period, clones had produced rooted ramets within the boxes and large numbers of unrooted ramets on stolons which had grown beyond the edges of the boxes. Coverage of the box surfaces by the leaves of the ramets was estimated by allocating squares of a 2 × 2 cm grid (784 squares per box) placed over the box into one of four different cover categories: (i) no leaf cover, (ii) 1–50% leaf cover, (iii) 51–95% leaf cover, (iv) > 95% leaf cover. The above ground parts of the clones were then harvested, keeping the parts of clones which had grown beyond the box edges separate from the parts within the boxes.

Table 2. Mean densities and dry weights (± S.E.) of experimental clones of Glechoma hederacea.

	Intact clones	Combined treatments clones	Severed clones
Mean density of rooted ramets (no. per 10 × 10 cm)	8.1 ± 0.7a	15.0 ± 0.86b	17.4 ± 0.8b
Mean dry weight per rooted ramet (g)	0.117 ± 0.009a	0.089 ± 0.004b	0.080 ± 0.004b
Mean dry weight per unrooted ramet (g)	0.136 ± 0.012a	0.093 ± 0.002b	0.085 ± 0.007b
Total dry weight of rooted ramets (g)	28.77 ± 3.54a	40.47 ± 3.66b	41.92 ± 3.05b
Total dry weight of unrooted ramets (g)	53.50 ± 3.90a	35.34 ± 3.31b	23.56 ± 4.80b
Percentage of biomass outside boxes	65.0	46.6	36.0
Total dry weight of clone (g)	82.27 ± 5.80a	75.81 ± 6.15a	65.48 ± 3.12b

Significant differences (at least $p < 0.05$) between mean values for the different treatments are indicated by different letters.

bers of rooted and unrooted ramets were counted, and the total dry weights of the rooted and unrooted ramets were recorded separately.

In treatments in which the stolons of both clones were severed, significantly more ramets ($p < 0.001$) rooted within the boxes than in the treatments which were left intact (Fig. 1). Conversely, greater lateral spread of the intact clones resulted in a significantly ($p < 0.05$) greater number of unrooted ramets beyond the edges of the boxes. In both respects the combined treatment produced an intermediate number of ramets. The spatial arrangement of rooted ramets in the severed treatment resulted in more complete ground cover and more leaf overlap between adjacent ramets (Fig. 2), and the mean dry weight of both rooted and unrooted ramets was significantly lower than in the intact clones. The total dry weights of intact clones, and of their unrooted ramets, were significantly greater than those of severed ramets, but the total weight of rooted ramets was significantly greater in the severed clones (Table 2).

These results demonstrate a marked increase in the intensity with which territory is occupied by the ramet progeny of severed clones. Although the average size of their ramets declined, the potential for inter-ramet competition increased in these clones because more ramets and a greater biomass were produced per unit area. However, the potential for colonization of new territory was markedly reduced in the severed clones. This is shown by the far smaller biomass and number of ramets, and lower proportion of total clone biomass which is produced beyond the box edges (Fig. 1, Table 2), and it is probable that under natural conditions the smaller size of the ramets produced after severing would render them more susceptible to competition from neighbouring vegetation. If the experiment had continued for longer, it is likely that increased mortality of rooted ramets would have been observed in these severed clones together with a lower rate of establishment of new ramets.

The observed sizes and densities of ramets in this experiment might have important consequences for the long-term persistence of ramets in the population. The maximum sustainable mean plant weight which will not result in density dependent mortality at different plant densities can be calculated from the linear $-3/2$ power law equation (Yoda et al. 1963; Westoby 1984). Any plant population in which the accumulation of biomass results in transgression above this line will suffer density dependent mortality until its combination of mean plant weight and density comes to lie beneath this limiting line once again. Although undisturbed ramet populations of clonal herbs do not appear to thin along this line, they do appear, just like populations of genets, to be subject to the limitations which it sets to biomass accumulation at different ramet densities (Hutchings 1979). Most documented examples of the $-3/2$ power law suggest that in the equation $\log w = \log K - 3/2 \log d$ (where w = mean plant weight, d = plant density and K is a constant), the value of log K at a density of one plant per m² lies between 3.5 and 4.3 (White 1980), although a few higher values have been reported (Lonsdale and Watkinson 1982). It has also been suggested that the highest values of log K are characteristic of species with erect or needle-leaved foliage and low extinction coefficients (Monsi and Saeki 1953), whereas the lowest values are seen in species with horizontal leaves and high extinction coefficients (Harper 1977; Hutchings and Budd 1981; Westoby 1984). Figure 3 shows the mean ramet weights and densities observed under the three experimental conditions and also indicates, for thinning lines with log K values of 4.5, 4.0 and 3.5, the maximum values of mean ramet weight which can be supported at each of the observed ramet densities. If mean weights exceed these values, ramet thinning would be observed. The results indicate that none of the ramet populations which developed under the experimental treatments should undergo density dependent mortality if the growth of *Glechoma hederacea* is limited by a power equation with a log K value of either 4.5 or 4.0. However, if *G. hederacea* is limited by a thinning line with an intercept of anything lower than 3.75 (which may be the case in view of its horizontal foliage), the ramets in both the combined and severed treatments should display some density dependent mortality. (The data for mean ramet weight and density for the intact clones lies just below the thinning line with log K = 3.5). Although the mean weights of rooted ramets in the treatments involving severing were lower than those in the intact clones, this reduction in mean weight would not have prevented thinning but merely reduced its intensity.

These results indicate the likelihood that ramet thinning would occur in *Glechoma hederacea* under experimental conditions in which integration between ramets is destroyed. *G. hederacea* is a species with a growth form with guerilla characteristics, and it does not always form pure stands of vegetation. If the experiment were repeated on field populations of phalanx clonal species such as *Mercurialis perennis* L. and *Chamerion angustifolium* L. Holub which usually develop dense, monospecific stands, it is probable that density dependent ramet mortality would be more marked. However, the problems of distinguishing mortality caused by physiological shock after severing connections between ramets and mortality caused by high density may be considerable in such experiments.

The size structure and population dynamics of ramet populations in clonal herbs

Whereas competing plants in populations grown from seeds display increasing

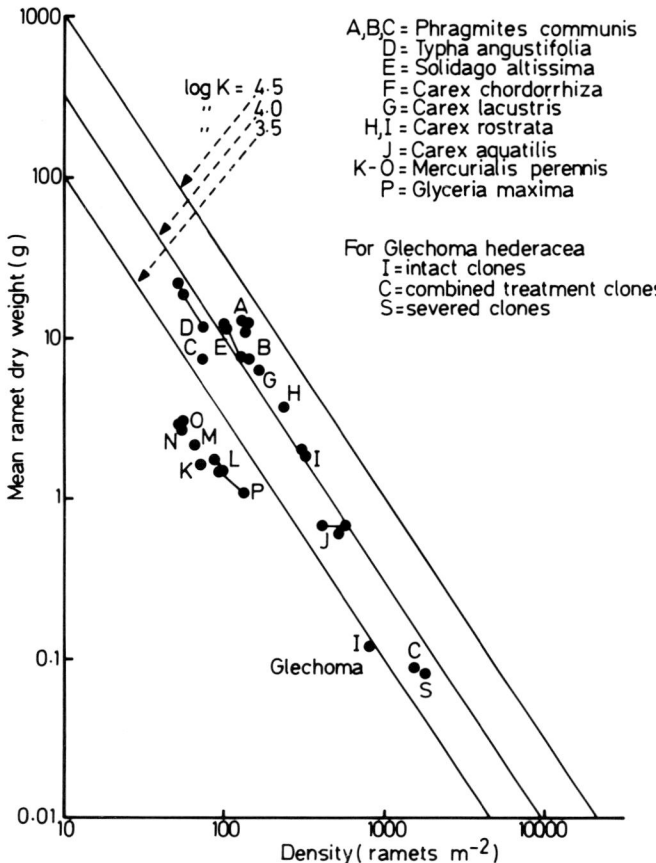

Fig. 3. The relationship between log (mean ramet weight) and log (ramet density) for several natural stands of herbaceous clonal species, compared with −3/2 boundary lines with values of K = 4.5, 4.0 and 3.5 at a density of one ramet per m². Points of closest approach to the boundary lines are plotted for all species, together with experimental data for *Glechoma hederacea* clones growing intact (I), for combined treatment clones (C) and for clones in which all connections between ramets are severed (S). See text for further details.

positive skewness[*] in their size frequency distribution as growth proceeds (Obeid et al. 1986), the same does not appear to be true of the annually-produced populations of ramets which are developed by many clonal herbs. For example, ramet populations of *Mercurialis perennis* develop a limited amount of size (measured as either weight or height) skewness in the first month after shoot emergence, but no further increase in skewness is observed as the shoots continue to develop (Hutchings and Barkham 1976). In *M. perennis* the initial limited development of skewness is more likely to be a result of small differences in the data of commencement of shoot growth rather than of competition between shoots. The development of marked positive skewness in the distribution of genet weights has been

---

[*] The use of skewness as a statistic for measuring the development of size hierarchies in plant populations has been superceded by measurement of size inequality using the Gini coefficient (Weiner and Solbrig 1984). However, use of skewness as a descriptive statistic is retained in this discussion because measurements of the Gini coefficient have not been published for ramet populations.

Table 3. Comparison of mean (± S.D.) flowering characteristics of clones of *Glechoma hederacea* in different sites.

	Grassland		Woodland
	Male sterile clones	Hermaphrodite clones	Hermaphrodite clones
% Flowering			
1984	36 ± 2a	61 ± 10b	70 ± 6b
1985	51 ± 2a	71 ± 19ab	90 ± 1b
% Bearing seeds			
1984	22 ± 1a	42 ± 13b	59 ± 6b
1985	27 ± 0a	59 ± 20ab	77 ± 5b
Flowers per flowering stem			
1984	8.4 ± 0.4a	14.8 ± 2.5b	15.8 ± 2.3b
1985	8.9 ± 1.6a	16.1 ± 4.2b	16.7 ± 2.4b

Significant differences (at least $p < 0.05$) between mean values for the different clones are indicated by different letters.

interpreted as a direct consequence of one-sided competition. Its absence in ramet populations indicates a lack of strong competition between ramets and the limitation of ramet density by correlative inhibition of the growth of many of the ramet initials possessed by the clone. Within ramet populations there is a marked difference between the mean sizes of flowering and sterile shoots, with fertile shoots being taller and heavier in *Solidago canadensis*, *Mercurialis perennis* and *Glechoma hederacea* (Bradbury 1981; Hutchings 1983; Slade 1986).

Recent analyses of ramet performance in *Glechoma hederacea*, which is gynodioecious, have revealed marked differences between hermaphrodite and male sterile clones, and between clones of the same sex in different habitats. Populations of ramets were censused, and aspects of their performance recorded at monthly intervals for a period of two years, in a grassland site in which both hermaphrodite and male sterile clones were found, and in a woodland site containing only hermaphrodite clones. The two sites were 100 m apart. Hermaphrodite clones from both sites produced a higher proportion of flowering ramets in both years of the study (the difference was significant in 1984) than male sterile clones (Table 3), and in both years of the study seeds were borne by a significantly smaller proportion of the ramets from the male sterile clones than by the ramets of either hermaphrodite clone. The woodland hermaphrodite clones consistently produced higher proportions of flowering and seed-bearing ramets than the grassland hermaphrodite clones. In both years the ramets of male sterile clones which flowered produced no more than 60% of the mean number of flowers borne by ramets of hermaphrodite clones in either habitat ($p < 0.05$, Table 3). Hermaphrodite flowers are considerably larger than male sterile flowers, and they may therefore attract a greater number of pollinators. This may explain why a far greater proportion of the flowering ramets of hermaphrodite clones bore seeds compared with the male sterile clones. There was no significant difference in the mean flower production by flowering ramets of hermaphrodite clones from either the grassland or woodland site. If hermaphrodite flowers are self-compatible, maintenance of gynodioecy in a species demands that there should be greater flowering and seed set by male sterile plants or clones (Shore 1978; Webb 1981). If this

Table 4. Mean (± S.E.) half lives (days) of ramets of *Glechoma hederacea*. Half lives for the whole population were calculated using data collected from April 1st to November 1st each year. Half lives of non-flowering, flowering and seed-bearing ramets were calculated using data collected from June 1st to November 1st of each year.

	Grassland site		Woodland site
	Male sterile clones	Hermaphrodite clones	Hermaphrodite clones
Whole ramet population			
1984	215 ± 33	308 ± 40	139 ± 15
1985	212 ± 37	383 ± 34	166 ± 12
Non-flowering ramets			
1984	168 ± 22	177 ± 19	104 ± 9
1985	169 ± 24	256 ± 28	119 ± 10
Flowering ramets			
1984	233 ± 30	463 ± 31	125 ± 8
1985	264 ± 38	433 ± 23	182 ± 11
Seed-bearing ramets			
1984	328 ± 38	416 ± 49	145 ± 16
1985	395 ± 47	395 ± 40	186 ± 14

is not the case, hermaphrodites will eventually come to dominate the population. Gynodioecy can be maintained however, if the hermaphrodite flowers are self-incompatible, and this is the case in *G. hederacea* (Gill 1979).

From initiation of growth until the start of winter frosts, populations of annual ramets appear to exhibit relatively constant and low mortality rates (Hutchings 1975; Slade 1986). Half lives for the whole populations of ramets from the different clones of *G. hederacea*, calculated from data collected between April and November, were markedly different (Table 4). In both years of the study, the ramets of the grassland hermaphrodite clones had the lowest mortality rates (and longest half lives) and the woodland hermaphrodite clones had the highest mortality rates (and shortest half lives). The differences in mortality rates were significant ($p < 0.01$ in 1984 and $p < 0.05$ in 1985). Grassland male sterile clones had intermediate half lives. Different subsets of the ramets produced by given clones also differed in half lives, with flowering and seed-bearing ramets having longer half lives than sterile ramets. In all but one out of six comparisons, mortality rates of sterile ramets from a given clone were significantly greater than mortality rates of flowering ramets from the same clone. Half lives of both flowering and non-flowering ramet subsets from the grassland hermaphrodite clones considerably exceeded those for the corresponding ramet subsets of woodland hermaphrodite and grassland male sterile clones (Table 4). However, although mortality rates of flowering and non-flowering subsets of ramets were significantly different for the hermaphrodite clones from the two sites, the mortality rates of ramet subsets for the grassland hermaphrodite and male sterile clones were not.

Whereas there is virtually no recruitment of new plants from seed in *G. hede-*

racea and many other herbaceous clonal species (Sarukhan and Harper 1973; Lovett Doust 1981b; Hutchings 1975; Thomas and Dale 1975; Angevine 1983; Hartnett and Bazzaz 1985; Pitelka *et al.* 1985), a considerable amount of energy is allocated to sexual reproduction. The results outlined above demonstrate that, for *G. hederacea* at least, ramets engaged in sexual activities have greater survivorship than sterile ramets, increasing the likelihood that seeds will be successfully matured and dispersed. Sexual reproductive effort appears to be greater in hermaphrodite clones; more ramets are sexual and these ramets have longer half lives than the sexual ramets of male sterile clones.

Epilogue

Our knowledge of structure in populations of genets is extensive (Hutchings 1986). The same cannot yet be said of clonal species in which the population consists of genetically uniform ramets. We are conscious that this paper has dealt with data from only a limited range of species and that we have relied extensively on data collected from *Glechoma hederacea* clones. Nevertheless, this paper illustrates what has been achieved in understanding the structure of ramet populations in clonal herbs, and indicates the types of data which are now required from other clonal species.

Acknowledgments

Part of this work was carried out while A.J.S. was in receipt of a Natural Environment Research Council studentship.

References

Alpert, P. and Mooney, H.A. 1986. Resource sharing among ramets in the clonal herb *Fragaria chiloensis*. Oecologia (Berlin) 70: 227–233.
Angevine, M.W. 1983. Variations in the demography of natural populations of the wild strawberries *Fragaria vesca* and *F. virginiana*. J. Ecol. 71: 959–974.
Bradbury, I.K. 1981. Dynamics, structure and performance of shoot populations of the rhizomatous herb *Solidago canadensis* L. in abandoned pastures. Oecologia (Berlin) 48: 271–276.
Clegg, L. 1978. The morphology of clonal growth and its relevance to the population dynamics of perennial plants. Ph.D. Thesis, University of Wales.
Cook, R.E. 1983. Clonal plant populations. Am. Sci. 71: 244–253.
Gill, L.S. 1979. Cyto-taxonomic studies of the tribe Nepeteae (Labiateae) in Canada. Genetica 50: 111–118.
Ginzo, H.D. and Lovell, P.H. 1973a. Aspects of the comparative physiology of *Ranunculus bulbosus* L. and *Ranunculus repens* L. 1. Response to nitrogen. Ann. Bot. 37: 753–764.
Ginzo, H.D. and Lovell, P.H. 1973b. Aspects of the comparative physiology of *Ranunculus bulbosus* L. and *Ranunculus repens* L. 2. Carbon dioxide assimilation and distribution of photosynthates. Ann. Bot. 37: 765–776.
Grindey, E. 1975. Ecological aspects of the runnering habit of plants with special reference to mineral nutrition. Ph.D. Thesis, University of Sheffield.
Guttridge, C.B. 1959. Evidence of a flower inhibitor and growth promoter in the strawberry. Ann. Bot. 23: 351–360.

Harper, J.L. 1977. Population Biology of Plants. Academic Press, London.
Harper, J.L. 1983. A Darwinian plant ecology. In: D.S. Bendall (ed.), Evolution from Molecules to Man, pp. 323–345. Cambridge University Press, Cambridge.
Harper, J.L., Rosen, B.R. and White, J. 1986. The Growth and Form of Modular Organisms. The Royal Society, London.
Hartnett, D.C. and Bazzaz, F.A. 1985. The genet and ramet population dynamics of *Solidago canadensis* in an abandoned field. J. Ecol. 73: 407–414.
Hoshino, M. 1972. Translocation and accumulation of assimilates in forage plants. Jap. Agric. Res. Quart. 6: 165–168.
Hutchings, M.J. 1975. The dynamics of shoot populations of a woodland perennial herb (*Mercurialis perennis* L.) growing in pure stands. Ph.D. Thesis, University of East Anglia.
Hutchings, M.J. 1979. Weight-density relationships in ramet populations of clonal perennial herbs, with special reference to the $-3/2$ power law. J. Ecol. 67: 21–33.
Hutchings, M.J. 1983. Shoot performance and population structure in pure stands of *Mercurialis perennis* L., a clonal perennial herb. Oecologia (Berlin) 58: 260–264.
Hutchings, M.J. 1986. The structure of plant populations. In: M.J. Crawley (ed.), Plant Ecology, pp. 97–136. Blackwell Scientific Publications, Oxford.
Hutchings, M.J. and Barkham, J.P. 1976. An investigation of shoot interactions in *Mercurialis perennis* L., a clonal perennial herb. J. Ecol. 64: 723–743.
Hutchings, M.J. and Bradbury, I.K. 1986. Ecological perspectives on clonal perennial herbs. BioScience 36: 178–182.
Hutchings, M.J. and Budd, C.S.J. 1981. Plant competition and its course through time. BioScience 31: 640–645.
Hutchings, M.J. and Slade, A.J. 1988. Morphological plasticity, foraging and integration in clonal perennial herbs. In: A.J. Davy, M.J. Hutchings and A.R. Watkinson (eds), Plant Population Biology. 28th Symposium of the British Ecological Society. Blackwell Scientific Publications, Oxford.
Jackson, J.B.C., Buss, L.W. and Cook, R.E. 1985. Population Biology and Evolution of Clonal Organisms. Yale University Press, New Haven.
Lonsdale, W.M. and Watkinson, A.R. 1982. Light and self-thinning. New Phytol. 90: 431–445.
Lovett Doust, L. 1981a. Interclonal variation and competition in *Ranunculus repens*. New Phytol. 89: 495–502.
Lovett Doust, L. 1981b. Population dynamics and local specialisation in a clonal perennial (*Ranunculus repens*). 1. The dynamics of ramets in contrasting habitats. J. Ecol. 69: 743–755.
Monsi, M. and Saeki, T. 1953. Über den Lichtfaktor in den Pflanzengesellschaften und seine Bedeutung für die Stoffproduktion. Jap. J. Bot. 14: 22–52.
Newell, S.J. 1982. Translocation of ^{14}C-photoassimilate in two stoloniferous *Viola* species. Bull. Torrey Botanical Club 109: 306–317.
Noble, J.C. and Marshall, C. 1983. The population biology of plants with clonal growth. 2. The nutrient strategy and modular physiology of *Carex arenaria*. J. Ecol. 71: 865–877.
Norton, R.A. and Wittwer, S.W. 1962. Foliar and root absorption and distribution of phosphorus and calcium in the strawberry. Proc. Am. Soc. Hort. Sci. 82: 277–286.
Obeid, M., Machin, D. and Harper, J.L. 1967. Influence of density on plant to plant variation in Fiber Flax, *Linum usitatissimum*. Crop Science 7: 471–473.
Pitelka, L.F. and Ashmun, J.W. 1985. Physiology and integration of ramets in clonal plants. In: J.B.C. Jackson, L.W. Buss and R.E. Cook (eds), Population Biology and Evolution of Clonal Organisms, pp. 399–435. Yale University Press, New Haven.
Pitelka, L.F., Hansen, S.B. and Ashmun, J.W. 1985. Population biology of *Clintonia borealis*. I. Ramet and patch dynamics. J. Ecol. 73: 169–184.
Sarukhan, J. and Harper, J.L. 1973. Studies on plant demography: *Ranunculus repens* L., *R. bulbosus*, L. and *R. acris*. I. Population flux and survivorship. J. Ecol. 61: 675–716.
Shore, B.F. 1978. Breeding systems in *Carpodetus serratus*. New Zealand J. Bot. 16: 179–184.
Silander, J.A. 1985. Microevolution in clonal plants. In: J.B.C. Jackson, L.W. Buss and R.E. Cook (eds), Population Biology and Evolution of Clonal Organisms, pp. 107–152. Yale University Press, New Haven.
Slade, A.J. 1986. The population biology and foraging behaviour of the clonal perennial herb *Glechoma hederacea*. D. Phil. Thesis, University of Sussex.
Slade, A.J. and Hutchings, M.J. 1987. The effects of nutrient availability on foraging in the clonal herb *Glechoma hederacea*. J. Ecol. 75: 95–112.

Slade, A.J. and Hutchings, M.J. 1987. The effects of light intensity on foraging in the clonal herb *Glechoma hederacea*. J. Ecol. 75: 639–650.
Slade, A.J. and Hutchings, M.J. 1987. Clonal integration and plasticity in *Glechoma hederacea*. J. Ecol. 75: 1023–1036.
Slade, A.J. and Hutchings, M.J. 1987. An analysis of the costs and benefits of physiological integration between ramets in the clonal perennial herb *Glechoma hederacea*. Oecologia 73: 425–431.
Slade, A.J. and Hutchings, M.J. 1987. An analysis of the influence of clone size and stolon connections between ramets on the growth of *Glechoma hederacea*. New Phytol. 106: 759–771.
Sutherland, W.J. The foraging behaviour of plants. In: N.C. Stenseth and I.R. Swingland (eds), Living in a Patchy Environment. Oxford University Press, Oxford, in press.
Thomas, A.G. and Dale, H.M. 1975. The role of seed reproduction in the dynamics of established populations of *Hieracium floribundum* and a comparison with that of vegetative reproduction. Can. J. Bot. 53: 3022–3031.
Webb, C.J. 1981. Test of a model predicting equilibrium frequencies of females in populations of gynodioecious angiosperms. Heredity 46: 397–405.
Weiner, J. and Solbrig, O.T. 1984. The meaning and measurement of size hierarchies in plant populations. Oecologia (Berlin) 61: 334–336.
Westoby, M. 1984. The self thinning rule. Advances in Ecological Research 14: 167–225.
White, J. 1980. Demographic factors in populations of plants. In: Demography and Evolution in Plant Populations, O.T. Solbrig (ed.), pp. 21–48. Blackwell Scientific Publications, Oxford.
Yoda, K., Kira, T., Ogawa, H. and Hozumi, K. 1963. Self thinning in overcrowded pure stands under cultivated and natural conditions. J. Biol. Osaka City Univ. 14: 107–129.

NITROGEN AVAILABILITY, OPTIMAL SHOOT/ROOT RATIOS AND PLANT GROWTH

TADAKI HIROSE*
Department of Plant Ecology, University of Utrecht, Lange Nieuwstraat 106, 3512 PN Utrecht, The Netherlands

Abstract

Three simulation models of plant growth were developed from experimentally determined relationships (a) between the nitrogen concentration of the whole plant (PNC) and the partitioning of dry matter and nitrogen between organs, (b) between the specific leaf weight (SLW) and leaf nitrogen concentration (LNC), and (c) between the net assimilation rate (NAR) and LNC on an area basis.
 1. The relative growth rate (RGR) as a function of PNC was derived. A strong dependence of RGR upon PNC resulted from the increase in NAR and LWR (fraction of plant dry weight in the leaves) and the decrease in SLW with increasing PNC.
 2. A model of vegetative growth was developed. The RGR under steady-state exponential growth increased with increasing nitrogen availability although with diminishing returns. Shoot/root ratio increased with increasing nitrogen availability.
 3. Optimal partitioning between shoot and root to give the highest RGR was determined for different nitrogen availabilities. Although actual shoot/root ratios were always higher than the optimal ones, the actual plants attained RGRs which were close to the values of the optimal plants except under the lowest nitrogen availabilities.

Introduction

Nitrogen is one of the major limiting factors for plant growth in many natural and agricultural systems (Bradshaw *et al.* 1964). Nitrogen affects plant growth through its effect on physiological processes such as photosynthesis, respiration and protein synthesis (Thomas *et al.* 1978; Lambers *et al.* 1981; Stulen *et al.* 1981; Evans 1983; Moorby and Besford 1983; Waring *et al.* 1985; Hirose and Werger 1987), and also through its morphogenetic effects such as matter partitioning between organs (Bradshaw *et al.* 1964; Brouwer 1966; Chapin 1980; Novoa and Loomis 1981; McDonald *et al.* 1986).

In a nutrition experiment with *Polygonum cuspidatum* Siebold et Zucc. (Hirose and Kitajima 1986), Hirose (1986) found that the partitioning of dry matter and of nitrogen between organs are highly correlated with the nitrogen concentration of the whole plant (PNC) and that the specific leaf weight (SLW, leaf dry weight per unit leaf area) is negatively correlated with leaf nitrogen concentration (LNC). In another experiment with *P. cuspidatum*, Hirose (1984) showed that the net assimilation rate (NAR, net dry matter productivity per unit leaf area) could be approximated by a rectangular hyperbolic function of LNC on an area basis. From

---

Present address: Biological Institute, Faculty of Science, Tohoku University, Sendai 980, Japan.

these relations Hirose (1987) constructed a simulation model of plant growth, in which the adaptive significance of plasticity in matter partitioning was examined. In the present paper, first, the importance of internal plant nitrogen concentration as a determinant of the relative growth rate (RGR) is shown (Hirose 1988). Second, the plant growth under varying nitrogen availability (Hirose 1987) is presented with concomitant changes in shoot/root ratios. And finally, an optimal partitioning between shoot and roots was determined so as to give the highest growth rate under a given availability of nitrogen and compared with the shoot/root ratio obtained by the model simulation of actual plant growth.

Relationship between RGR and internal nitrogen concentration

To indicate an efficiency of nitrogen use in plant growth, Ingestad (1979) introduced the concept of nitrogen productivity (NP) as

$$NP = \frac{1}{N}\frac{dW}{dt} = \frac{1}{W}\frac{dW}{dt} / \frac{N}{W} \tag{1}$$

where N and W are total plant nitrogen and dry weight, respectively (see Appendix for abbreviations and symbols used in the equations). Ingestad and coworkers showed that if relative growth rates (RGR = 1/W dW/dt) are plotted against total plant nitrogen concentration (PNC = N/W), a linear relationship holds over a wide range of PNC (Ingestad 1979, 1981; Ingestad and Kähr 1985; Ericsson 1981; Ericsson et al. 1982). The y-intercept is usually negative and the NP against PNC gives a rectangular hyperbola (Ågren 1985). However, so far no one seems to have further analyzed the strong correlation between PNC and RGR (Hirose 1988). The RGR is a rather complex concept which involves both functional and structural components of plant growth (Lambers and Dijkstra 1987). The RGR is analyzed as

$$RGR = NAR \times LWR / SLW \tag{2}$$

where NAR represents physiological activity of the plant, while LWR (leaf weight ratio, the fraction of plant dry matter in leaves) and SLW represent structural components of plant growth. From the relationships between nitrogen concentration and the three components of plant growth (NAR, LWR and SLW) presented by Hirose (1984, 1986), the RGR can be formulated as a function of PNC as follows.

The fractions of plant dry weight (f_{LW}) and total nitrogen (f_{LN}) in the leaves are linear functions of plant nitrogen concentration (ν_P) (Hirose 1986):

$$f_{LW} = a_{LW} + b_{LW} \nu_P \tag{3}$$

and

$$f_{LN} = a_{LN} + b_{LN} \nu_P \tag{4}$$

where a and b are constants. Then leaf nitrogen concentration (ν_L) is given by

$$\nu_L = \nu_P f_{LN} / f_{LW}. \tag{5}$$

Specific leaf weight (σ) is a rectangular hyperbolic function of leaf nitrogen concentration (Hirose 1986):

$$\sigma = d / (1 + c \nu_L) \tag{6}$$

where c and d are constants. Net assimilation rate (ρ_W) is a rectangular hyperbolic function of nitrogen concentration on an area basis ($\sigma\nu_L$) (Hirose 1984):

$$\rho_W = \frac{h(\sigma\nu_L - k)}{1 + g(\sigma\nu_L - k)} \tag{7}$$

where g, h, and k are constants.

If PNC is given, the LWR is calculated with equation 3. Equations 3–5 determine the LNC which in turn gives the SLW (equation 6). Once LNC and SLW are determined, the NAR is calculated with equation 7. Thus the three components of RGR, i.e., NAR, LWR and SLW are determined for a given PNC. Equation 2 gives RGR.

Parameter values used in the calculation are presented in Table 1; they were derived from a sand culture experiment with *P. cuspidatum* (Hirose 1984). Figure 1 gives the RGR, NAR, LWR, SLW and NP as functions of PNC. A curvilinear relationship between RGR and PNC was obtained. Steep reduction in NAR at PNC lower than 2% contributed to this curvilinearity. Increases in RGR for the PNC over 2% are ascribed mainly to the increase in LWR and the decrease in SLW with increasing PNCs. Nitrogen productivity (NP), being the ratio of RGR to PNC (equation 1), was observed maximal at 1.5% PNC. The reduction in NP for PNC over 1.5% may reflect some effect of self-shading in the present experiment (Hirose 1988).

Plant growth under varying nitrogen availability

Hirose (1987) developed a model for vegetative growth of the plant which consists of leaves, stem and roots upon the three relationships partly presented above. (a) The partitioning of dry matter and nitrogen between organs is controlled by the nitrogen concentration of the whole plant (PNC). In addition to equations 3 and 4, an equation for the fraction of dry matter in roots should be formulated:

$$f_{RW} = a_{RW} + b_{RW} \nu_P \tag{8}$$

(b) specific leaf weight (SLW) is controlled by the leaf nitrogen concentration (LNC) (equation 6). (c) Net assimilation rate (NAR) is a function of LNC on an area basis (equation 7). The environmental variable considered in the present model is the availability of nitrogen. The effect of nitrogen availability on plant growth is examined through its influence upon the specific absorption rate (SAR, net nitrogen uptake rate per unit root dry weight).

Now suppose that the amount of dry matter and total nitrogen in the plant at time t are W and N, respectively. Then,

Table 1. Parameter values used in the model simulations. Derived from a sand-culture experiment with *Polygonum cuspidatum*.

Parameter	Unit	Value	Equations
a_{LW}	–	0.352	3, 16
a_{LN}	–	0.533	4, 18
a_{RW}	–	0.589	8, 17
b_{LW}	–	8.46	3, 16
b_{LN}	–	6.17	4, 18
b_{RW}	–	−9.63	8, 17
c	–	18.2	6
d	g DW m^{-2}	56.6	6
g	m^2 (g N)$^{-1}$	16.5	7
h	g DW (g N)$^{-1}$ d^{-1}	101	7
k	g N m^{-2}	0.60	7

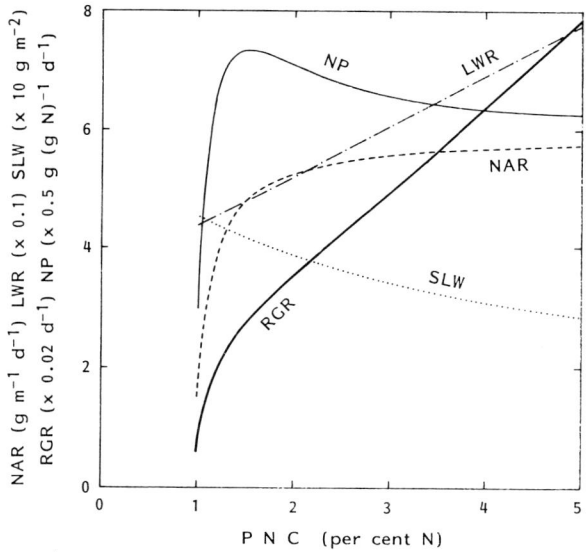

Fig. 1. The dependence on plant nitrogen concentration (PNC) of relative growth rate (RGR), net assimilation rate (NAR), leaf weight ratio (LWR), specific leaf weight (SLW), and nitrogen productivity (NP), where RGR = NAR × LWR / SLW and NP = RGR / PNC.

$$\nu_P = N / W. \tag{9}$$

The dry matter and nitrogen partitioning in the leaves are calculated by substituting equation 9 into equations 3 and 4. The amount of dry matter (W_L) and nitrogen (N_L) in the leaves are

$$W_L = f_{LW}W \text{ and } N_L = f_{LN}N \tag{10}$$

respectively. Leaf nitrogen concentration (ν_L) is given by

$$\nu_L = N_L / W_L. \tag{11}$$

Nitrogen availability, optimal shoot/root ratios and plant growth

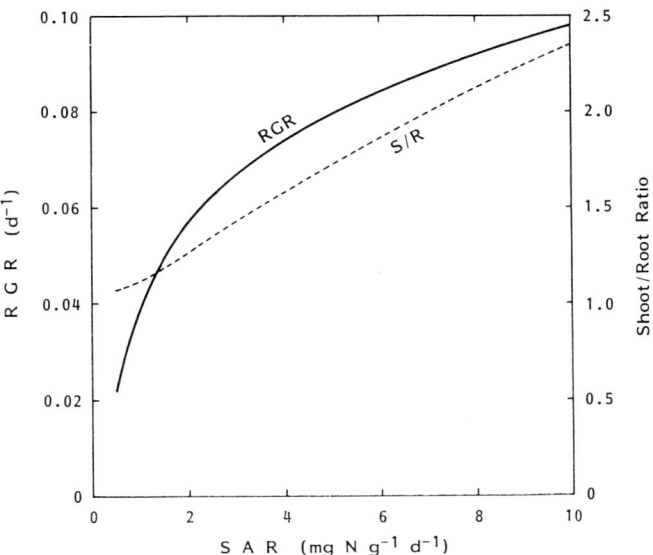

Fig. 2. Changes in relative growth rate (RGR) and shoot/root ratio with varying specific absorption rates (SAR) under steady-state exponential growth.

Specific leaf weight (σ) is determined by equations 6 and 11, and leaf area (A_L) is

$$A_L = W_L / \sigma. \tag{12}$$

The absolute growth rate of the plant is the product of net assimilation rate (ρ_W, given in equation 7) and leaf area (equation 12):

$$dW/dt = \rho_W A_L. \tag{13}$$

Equations 8 and 9 give the root dry weight at time t as

$$W_R = f_{RW} W. \tag{14}$$

The SAR is given as an independent variable (ρ_N), and the rate of nitrogen uptake is calculated as the product of SAR and root dry weight:

$$dN/dt = \rho_N W_R. \tag{15}$$

Thus if the amounts of dry matter and total nitrogen in the plant at time t are given, those at t + dt are determined using equations 13 and 15. If these calculations are repeated under a given SAR, the whole process approaches a steady-state exponential growth (Thornley 1976), which defines a steady-state value of relative growth rate.

The RGRs under steady-state exponential growth are plotted against varying SARs in Fig. 2. Increase in the availability of nitrogen, which was supposed to increase the SAR, increased the RGR, although with diminishing returns. Hirose (1988) showed that this pattern of increase in RGR was caused by the increase in NAR and LWR and the decrease in SLW with increasing SAR. In Fig. 2, changes in shoot/root ratio are also shown. The shoot/root ratio increased more than twofold with the increase in SAR from 0.05 to 10 mg N g^{-1} d^{-1}. Many authors observed an increase in shoot/root ratio with increasing availability of nitrogen (Bradshaw et al. 1964; Brouwer 1966; Chapin 1980; Lambers et al. 1981).

Changes in shoot/root ratio with nitrogen availability have been considered adaptive because they compensate for potentially growth limiting conditions (Robinson 1986). Under nutrient-poor conditions, a relatively large mass of roots compensates for the low nutrient uptake rate per unit root weight; and under nutrient-rich conditions, greater allocation of biomass to the shoot ensures the plant to obtain a large amount of photosynthates for further growth.

Hirose (1987) demonstrated the adaptive significance of phenotypic plasticity in matter partitioning, comparing the growth of actual plants which modify their partitioning according to the availability of nitrogen with those less plastic in this respect. He showed that the plant with plasticity attained a higher growth rate under varying availabilities of nitrogen. But it is still an open question to what extent this plastic change in matter partitioning of the actual plant is adaptive. Or, more specifically: Is there any optimal partitioning of biomass between shoot and root for a particular availability of nitrogen in the environment? And if there is, does the partitioning observed in actual plants approach the optimal one? And finally, what is the effect of suboptimal partitioning of biomass on plant growth? These questions will be addressed in the following section.

Optimal partitioning between shoot and root

Suppose a plant which can change its dry matter and nitrogen allocation freely within the following constraints:

$$f_{LW} = a_{LW} + b_{LW}\delta \qquad (16)$$

$$f_{RW} = a_{RW} + b_{RW}\delta \qquad (17)$$

and

$$f_{LN} = a_{LN} + b_{LN}\delta. \qquad (18)$$

These equations are the same as equations 3, 8 and 4, respectively, except for δ in these equations instead of ν_p. A 'dummy' variable, δ, generates a wide variation in the partitioning of dry matter and nitrogen between organs, independent of the nitrogen concentration of the whole plant. The same assumptions as before were made on SLW and NAR; i.e., they are functions of leaf nitrogen concentration per unit weight and per unit area, respectively (equations 6, 7). Because the NAR contains not only leaf photosynthesis, but also respiratory loss by non-photosynthetic organs such as roots, the assumption of NAR as a function only of LNC may give some bias in the results. But at present, no data is available on the regulation of respiration for P. cuspidatum. Surveying root respiration of several species, however, Lambers (1979) showed a tendency that root growth respiration decreases with the decrease in the shoot/root ratio. If this is the case, some compensation is expected between the root mass and the root respiration.

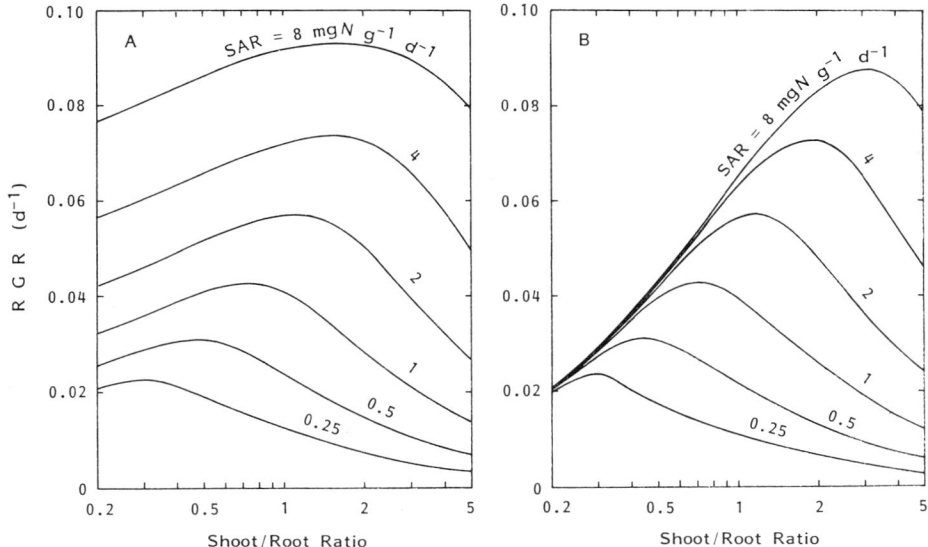

Fig. 3. (A) Changes in relative growth rate (RGR) with varying shoot/root ratios for different specific absorption rates (SAR). (B) The same as (A) except that SLW was fixed at 40 g m^{-2}.

Upon the above assumptions, changes in RGR of plants with varying shoot/root ratios were examined for different values of SAR (Fig. 3). The shoot/root ratio was changed by controlling δ in equations 16 and 17. This δ simultaneously determines the partitioning of nitrogen to the leaves according to equation 18. The RGR under the steady-state exponential growth was calculated following the same algorithm as described before.

Figure 3A gives RGRs as a function of shoot/root ratio for different nitrogen availabilities which were supposed to change the SAR of the plant. The shoot/root ratio is plotted on a logarithmic scale. Optimum curves were observed. They were skewed to lower shoot/root ratios, *i.e.*, reduction of RGR was smaller at lower shoot/root ratios than at higher ratios. The optimal shoot/root ratio that gave the maximum RGR changed with different SARs: it shifted to higher values with increasing nitrogen availabilities. In Fig. 3B, another simulation is shown in which the SLW was fixed at 40 g m^{-2}, irrespective of leaf nitrogen concentration. A stronger reduction in RGR for the lower shoot/root ratios was observed. This reduction in RGR was so strong that virtually no difference in RGR was observed between different SARs at a shoot/root ratio of e.g., 0.2. If the SLW was changed according to the leaf nitrogen concentration, the plant with low shoot/root ratios could also respond positively to higher nitrogen availabilities. This is because the lower allocation of biomass to leaves could be compensated for by making the leaf area per unit dry weight larger. The plants could have a relatively high LAR (= LWR /SLW) even though the LWR was low (Hirose 1987). Thus, the plastic change in SLW favors the plant's ability to maintain high growth rates under varying availabilities of nitrogen in the environment.

Growth response at varying nitrogen availabilities (SAR) is shown in Fig. 4A for

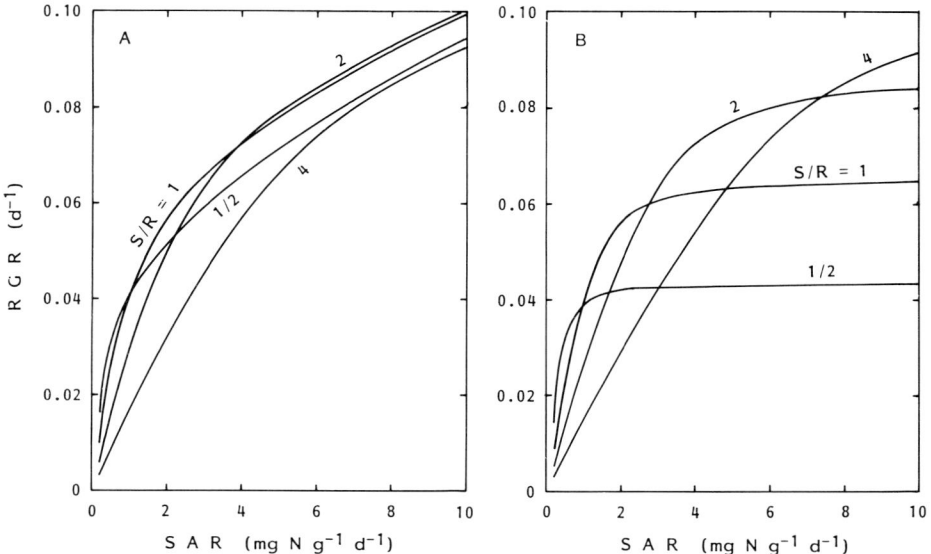

Fig. 4. (A) Changes in relative growth rate (RGR) with specific absorption rates (SAR) in plants with different shoot/root ratios. (B) The same as (A) except that SLW was fixed at 40 g m^{-2}.

plants with different shoot/root ratios. Figure 4B shows the responses when the SLW was fixed at 40 g m^{-2}. Quite different patterns resulted from the two simulations, particularly for low shoot/root ratios. When the SLW was fixed, larger differences between plants with different shoot/root ratios were observed. Both simulations, however, showed that under high nitrogen availabilities, a plant with a shoot/root ratio of 2 grew better than a plant with a low shoot/root ratio (1/2). The reverse was true under low nitrogen availabilities. An extremely high shoot/root ratio (4) gave lower RGRs even under high SARs in the simulation of Fig. 4A.

In Table 2, optimal shoot/root ratios and RGRs are summarized against different SARs, together with 'actual' shoot/root ratios and RGRs. The 'actual' plant was supposed to change the partitioning of dry matter and nitrogen between shoot and root according to its internal nitrogen concentration following equations 3, 4 and 8 (see Fig. 2). The optimal plant is a hypothetical one in which the partitioning between shoot and root was assumed to change so as to maintain the highest RGR throughout varying availabilities of nitrogen (see Fig. 3A). Optimal shoot/root ratios were always lower than the 'actual' ones, particularly under low nitrogen availabilities (Table 2). However, even though shoot/root ratios differed, the 'actual' plants attained RGRs which were close to the values of the optimal plants, except under low nitrogen availabilities. Under the lowest availability of nitrogen in the simulation experiment, the 'actual' plant attained an RGR less than half of that of the optimal plant.

P. cuspidatum has a wide distribution in Japan. It grows well in nutrient-rich environments such as river banks, roadsides, etc., and also in nutrient-poor habitats such as volcanic deserts (Hirose and Tateno 1984). The results of the present simulation, however, suggest that this species is more adapted to the

Table 2. Optimal and 'actual' shoot/root ratios for different SARs and the RGRs of optimal and 'actual' plants attained under steady-state exponential growth for respective SARs. Optimal plants change the partitioning between shoot and root to maintain the highest RGR for respective SARs, while 'actual' plants change the partitioning according to their internal nitrogen concentration (see text for further explanation).

SAR	Shoot/root ratio		RGR		
	Optimal	'Actual'	Optimal	'Actual'	'Actual'/Optimal
mg N g^{-1}d^{-1}			—— d^{-1} ——		%
8	1.55	2.12	0.0932	0.0921	98.8
4	1.50	1.57	0.0739	0.0739	99.9
2	1.07	1.27	0.0573	0.0566	98.8
1	0.74	1.11	0.0426	0.0390	91.5
0.5	0.49	1.07	0.0309	0.0220	71.2
0.25	0.30	1.04	0.0226	0.0100	44.2

environment of high nitrogen availability because it attained near-optimal growth rates only under high nitrogen availabilities (Table 2).

Acknowledgments

I thank Hans Lambers and Marinus J.A. Werger for comments. The Netherlands Organization for the Advancement of Pure Research (Z.W.O.) allowed me to continue this study during my stay at the Department of Plant Ecology, University of Utrecht. This study was also financed in part by the Grant-in-Aid from the Ministry of Education, Science and Culture, Japan.

Appendix

Abbreviations

LNC	Leaf nitrogen concentration
LWR	Leaf weight ratio
NAR	Net assimilation rate
NP	Nitrogen productivity
PNC	Plant nitrogen concentration
RGR	Relative growth rate
SAR	Specific absorption rate
SLW	Specific leaf weight

Symbols used in the equations

A_L	Leaf area (equation 12), m^2
N	Plant nitrogen (equations 1, 9), g N
N_L	Nitrogen in leaves (equation 10), g N
W	Plant dry weight (equations 1, 9), g DW
W_L, W_R	Dry weight of leaves and roots (equations 10, 14), g DW
f_{LN}	Fraction of plant nitrogen in leaves (equation 4), dimensionless
f_{LW}, f_{RW}	Fraction of plant dry weight in leaves (LWR) and roots (equations 3, 8), dimensionless

t	Time (equations 1, 13, 15), d
δ	A 'dummy' variable (equations 16–18), dimensionless
ν_P	Nitrogen concentration of the whole plant (PNC) (equations 3, 4, 8, 9), g N (g DW)$^{-1}$
ν_L	Nitrogen concentration of leaves (LNC) (equations 5, 11), g N (g DW)$^{-1}$
ρ_N	Specific absorption rate (SAR) (equation 15), g N (g DW)$^{-1}$ d^{-1}
ρ_W	Net assimilation rate (NAR) (equations 7, 13), g DW m^{-2} d^{-1}
σ	Specific leaf weight (SLW) (equations 6, 12), g DW m^{-2}

References

Ågren, G.I. 1985. Theory for growth of plants derived from the nitrogen productivity concept. Physiol. Plant. 64: 17–28.

Bradshaw, A.D., Chadwick, M.J., Jowett, D. and Snaydon, R.W. 1964. Experimental investigations into the mineral nutrition of several grass species. IV. Nitrogen level. J. Ecol. 52: 665–677.

Brouwer, R. 1966. Root growth of grasses and cereals. In: F.L. Milthorpe and D.J. Irvins (eds), The Growth of Cereals and Grasses, pp. 153–166. Butterworth, London.

Chapin, F.S. III. 1980. The mineral nutrition of wild plants. Ann. Rev. Ecol. Syst. 11: 233–260.

Ericsson, T. 1981. Effects of varied nitrogen stress on growth and nutrition in three *Salix* clones. Physiol. Plant. 51: 423–429.

Ericsson, T., Larsson, C.-M. and Tillberg, E. 1982. Growth responses of *Lemna* to different levels of nitrogen limitation. Z. Pflanzenphysiol. 105: 331–340.

Evans, J.R. 1983. Nitrogen and photosynthesis in the flag leaf of wheat (*Triticum aestivum* L.). Plant Physiol. 72: 297–302.

Field, C. and Mooney, H.A. 1986. The photosynthesis-nitrogen relationship in wild plants. In: T.J. Givnish, (ed.), On the Economy of Plant Form and Function, pp. 25–55. Cambridge University Press, London.

Hirose, T. 1984. Nitrogen use efficiency in growth of *Polygonum cuspidatum* Sieb. et Zucc. Ann. Bot. 54: 695–704.

Hirose, T. 1986. Nitrogen uptake and plant growth. II. An empirical model of vegetative growth and partitioning. Ann. Bot. 58: 487–496.

Hirose, T. 1987. A vegetative plant growth model: adaptive significance of phenotypic plasticity in matter partitioning. Functional Ecol. 1: 195–202.

Hirose, T. 1988. Modelling the relative growth rate as a function of plant nitrogen concentration. Physiol. Plant. 72: 185–189.

Hirose, T. and Kitajima, K. 1986. Nitrogen uptake and plant growth I. Effect of nitrogen removal on growth of *Polygonum cuspidatum*. Ann. Bot. 58: 479–486.

Hirose, T. and Tateno, M. 1984. Soil nitrogen patterns induced by colonization of *Polygonum cuspidatum* on Mt. Fuji. Oecologia (Berlin) 61: 218–223.

Hirose, T. and Werger, M.J.A. 1987. Nitrogen use efficiency in instantaneous and daily photosynthesis of leaves in the canopy of a *Solidago altissima* stand. Physiol. Plant. 70: 215–222.

Ingestad, T. 1979. Nitrogen stress in birch seedlings. II. N, K, P, Ca, and Mg nutrition. Physiol. Plant. 45: 149–157.

Ingestad, T. 1981. Nutrition and growth of birch and grey alder seedlings in low conductivity solutions and at varied relative rates of nutrient additions. Physiol. Plant. 52: 454–466.

Ingestad, T. and Kähr, M. 1985. Nutrition and growth of coniferous seedlings at varied relative nitrogen addition rate. Physiol. Plant. 65: 109–116.

Lambers, H. 1979. Efficiency of root respiration in relation to growth rate, morphology and soil composition. Physiol. Plant. 46: 194–202.

Lambers, H. and Dijkstra, P. 1987. A physiological analysis of genotypic variation in relative growth rate: Can growth rate confer ecological advantage? In: J. van Andel, J.P. Bakker and R.W. Snaydon (eds), Disturbance in Grasslands. Causes, Effects and Processes, pp. 239–253. Dr Junk, Dordrecht.

Lambers, H., Posthumus, F., Stulen, I., van de Dijk, S.J. and Hofstra, R. 1981. Energy metabolism of *Plantago lanceolata* as dependent on the supply of mineral nutrients. Physiol. Plant. 51: 85–92.

McDonald, A.J.S., Lohammar, T. and Ericsson, A. 1986. Growth response to step-decrease in nutrient availability in small birch (*Betula pendula* Roth). Plant Cell Environ. 9: 427–432.

Moorby, J. and Besford, R.T. 1983. Mineral nutrition and growth. In: A. Lauchli and R.L. Bieleski (eds), Encyclopedia of Plant Physiology, New Series, Vol. 15B, pp. 481–527. Springer Verlag, Berlin.

Natr, L. 1975. Influence of mineral nutrition on photosynthesis and the use of assimilates. In: J.P. Cooper (ed.), Photosynthesis and Productivity in Different Environments, pp. 537–555. Cambridge University Press, London.

Novoa, R. and Loomis, R.S. 1981. Nitrogen and plant productivity. Plant Soil 58: 177–204.

Robinson, D. 1986. Compensatory changes in the partitioning of dry matter in relation to nitrogen uptake and optimal variations in growth. Ann. Bot. 58: 841–848.

Stulen, I., Lanting, L., Lambers, H., Posthumus, F., van de Dijk, S.J. and Hofstra, R. 1981. Nitrogen metabolism of *Plantago lanceolata* as dependent on the supply of mineral nutrients. Physiol. Plant. 51: 93–98.

Thomas, S.M., Thorne, G.N. and Pearman, I. 1978. Effect of nitrogen on growth, yield and photorespiratory activity in spring wheat. Ann. Bot. 42: 827–837.

Thornley, J.H.M. 1976. Mathematical Models in Plant Physiology. Academic Press, London.

Waring, R.H., McDonald, A.J.S., Larsson, S., Ericsson, T., Wiren, A., Arwidsson, E., Ericsson, A. and Lohammar, T. 1985. Differences in chemical composition of plants grown at constant relative growth rates with stable mineral nutrition. Oecologia (Berlin) 66: 157–160.

PHENOTYPIC RESPONSES OF PLANTS TO ENVIRONMENTAL CONDITIONS

TOSHIHIKO HARA and AKIRA HARAGUCHI
Laboratory for Plant Ecological Studies, Faculty of Science, Kyoto University, Kyoto 606, Japan

Abstract

Experiments with *Impatiens balsamina* L. were carried out to investigate phenotypic responses of plants to density and nutrient levels. Changes in specific leaf area and leaf:stem:root ratio in response to environmental conditions are analysed by a mathematical model.

Introduction

The plant changes its phenotype in response to environmental conditions. Many workers have focused on root:shoot ratio and found that it increases under low-nutrient or high-light levels (Davidson 1969; Hunt and Burnett 1973; Hunt 1975; Richards *et al.* 1979; Chapin 1980; Morris and Myerscough 1985; Hunt and Nicholls 1986). This phenomenon has been accounted for by mechanistic simulation models (Thornley 1972a, b; Reynolds and Thornley 1982; Johnson 1985), empirical models (Hirose 1984; Hirose 1986; Hunt and Nicholls 1986) or models considering adaptive strategies (Orians and Solbrig 1977; Mooney and Gulman 1979; Iwasa and Roughgarden 1984; Robinson 1986). These models take the leaf and stem as one functional system, *i.e.*, shoot. However, the leaf as a photosynthesizing system and the stem as a supporting system have different functions and should not be taken as one system except for some species, *e.g.*, graminoids. In this paper we deal with leaf, stem and root as three independent functional systems for understanding the plants' phenotypic responses to environments. We introduce three parameters in our model: specific leaf area (SLA), ratio of stem weight to whole plant weight (k_1), and ratio of leaf weight to weight of the productive system (leaf and root) (k_2).

Materials and methods

Seeds of *Impatiens balsamina* L., an annual herb, were sown in plastic flowerpots (59 cm long, 33 cm wide, 22 cm deep) filled with 30 l of coarse vermiculite on 22 April 1986 and harvests were made four times on 3 June (H1), 17 June (H2), 1 July (H3) and 15 July (H4). Pots of same treatment were set in a greenhouse, packed closely to avoid edge effect.

Five levels of density and four levels of nutrients were set, *i.e.*, 5 × 4 treatments. Density levels were as follows: D1, 20 per m²; D2, 46 per m²; D3, 140 per m²; D4, 416 per m²; D5, 1250 per m². Nutrient level were controlled by different con-

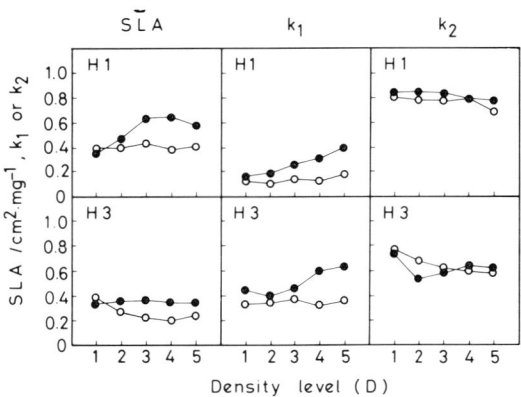

Fig. 1. Averaged values of SLA, k_1 and k_2 of *Impatiens balsamina* populations against density level for nutrient levels N1 (open circle) and N4 (closed circle) at H1 (above) and H3 (below).

centrations of nutrient solution 'HYPONEX 5-10-5' (5% total N(NH$_4$-N, 2%; NO$_3$-N, 1%; ureal N, 2%), 10% P$_2$O$_5$ and 5% K$_2$O) as follows: N1, × 1/20000; N2, × 1/5000; NS, × 1/1250; N4, × 1/313.

All harvested plants for the D_1, D_2 and D_3 treatments and 40–50 plants selected randomly from harvested ones for the D_4 and D_5 treatments were used for measurements. Shoot height, stem diameter at ground level, leaf area and dry weights of stems, leaves and roots of each individual plant were measured for each treatment. Photosynthetic rates of leaves cut from a plant of average size were measured by the infrared gas analyser (Hitachi-Horiba ASSA-2) for treatment combinations of all nutrient levels and density levels D1, D3 and D5. Nitrogen and carbon contents were determined for well-ground samples of leaves using the N-C analyzer (Sumigraph NC-80).

Results

We introduce three parameters which describe phenotypic properties of a plant; k_1 (= stem weight / whole plant (stem + leaf + root) weight, k_2 (= leaf weight/leaf + root weight) and SLA (specific leaf area). Averaged k_1, k_2 and SLA of populations N1 and N4 at H1 and H3 are given against the density level in Fig. 1. Values of SLA and k_1 generally increased with density at high-nutrient levels, while those values were almost constant at low-nutrient levels. Values of k_2 decreased a little with density at both nutrient levels. The values of SLA, k_1 and k_2 of high-nutrient populations were generally higher than those of low-nutrient populations at each density. k_1, k_2 and SLA of each plant in populations N4D1 and N4D4 at H1 are given against the plant size (whole plant weight) in Fig. 2. Values of SLA and k_1 decreased with plant size in the high-density population, while these values decreased a little or were almost constant in the low-density population. Values of k_2 were almost constant irrespective of plant size in both populations. Fig. 3 shows light-photosynthetic rate curves for different nutrient treatments of population D1 at H2. Nitrogen content per unit leaf area was positively correlated with the maximal photosynthetic rate per unit leaf area but was not

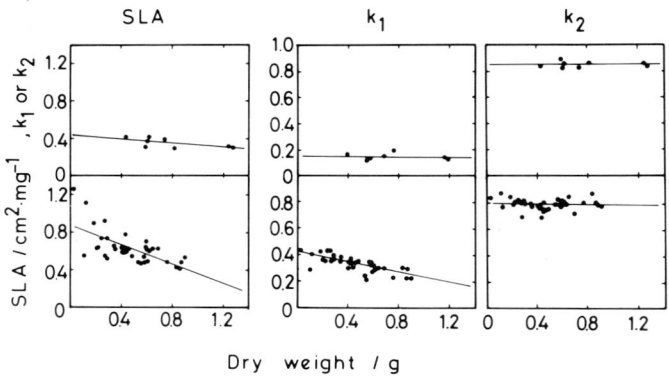

Fig. 2. SLA, k_1 and k_2 of each individual plant in *Impatiens balsamina* populations N4D1 (above) and N4D4 (below) against plant size (whole plant weight) at H1. Regression lines are also given.

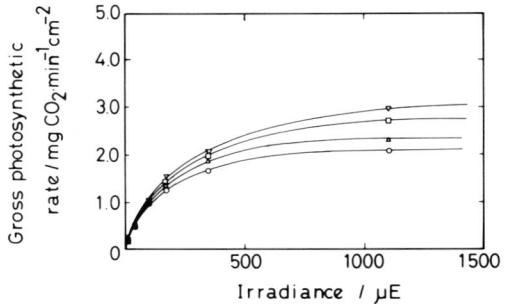

Fig. 3. Light-photosynthetic rate curves of unit leaf area of *Impatiens balsamina* for different nutrient treatments of D1 populations at H2: N1, circle; N2, triangle; N3, square; N4, inverse triangle. Nitrogen contents per unit leaf area are: N1, 0.093; N2, 0.087; N3, 0.108; N4, 0.100 mg.cm^{-2}.

significantly correlated with the photosynthetic rate at low light intensity (Fig. 3). The former relationship was given by a rectangular hyperbola curve. Nitrogen content per unit leaf area was affected only a little by density and nutrient levels, while nitrogen content per unit leaf dry weight was affected markedly by these.

Discussion

Many studies have focused on the root:shoot ratio which gives the ratio of belowground system to aboveground system. Thus it ignores the difference in function of stem and leaf and is not adequate to describe the plants' phenotypic responses to environmental conditions. In this paper we divide a plant into three independent functional systems, root, stem and leaf. Each of these three systems have significance of their own in dry matter production by photosynthesis: the root absorbs nutrients which are necessary for photosynthesis; the leaf is an organ where photosynthesis takes place; the stem supports the leaves so as to gain favorable light conditions in a crowded stand. The leaf and root are productive systems and the stem is a supporting system. Therefore, we presented a model introducing two ratios, k_1 and k_2, and SLA to describe the plants' phenotypic responses in more detail (Hara and Haraguchi unpublished). Our model is based on the photosynthetic activity of a leaf which depends on light intensity and leaf nitrogen content. $(1-k_1) \times (1-k_2)$ determines the proportion of root, thus the amount of absorbed nutrients (in terms of nitrogen in this study). The amount of absorbed nitrogen, k_2 and SLA determine nitrogen content per unit leaf area which determines the photosynthetic ability of a leaf. k_1 gives the stature of a plant in a stand, thus determines the light conditions the plant enjoys. Light intensity and nitrogen content per unit leaf area determine the photosynthetic rate per unit leaf area which together with $(1-k_1) \times k_2 \times$ SLA give the photosynthetic rate of a plant. Thus the photosynthetic rate of a plant is a function of k_1, k_2 and SLA. Further, we proposed the hypothesis that a plant keeps these three variables at the optimal values that produce the maximal photosynthetic rate of the plant under given environmental conditions. A plant changes its phenotype by adjusting these three variables to the optimal values when environmental conditions are changed.

Our model together with the hypothesis predicts that the following features are all advantageous in dry matter production: small values of k_2 and SLA under low-nutrient conditions; a small value of k_2 and a large value of SLA under low-light conditions; a large value of k_1 under crowded conditions; and a constant nitrogen content per unit leaf area irrespective of the nutrient level under given light conditions.

Our experimental results can be explained well by these theoretical predictions. The averaged SLA of population N4 at H1 increased as the population became more crowded, *i.e.*, as leaf area index (LAI) increased from 0.41 to 6.48 with density level (Fig. 1). That is because small suppressed plants in a crowded population are subject to low-light conditions and tend to have higher values of SLA as compared with large dominant plants (Fig. 2). N1 populations at H1 were fairly sparse (LAI = 0.75 even at D5), and N1 and N4 populations at H3 had almost constant LAIs irrespective of the density level (LAI of N1 populations was about 0.95 (0.78 – 1.34), and LAI of N4 populations about 5.21 (4.09 – 6.31)). This is why averaged SLAs of N1 populations at H1 and N1 and N4 populations at H3 were almost constant irrespective of the density level. SLA was higher when nutrient level was higher at each density level (Fig. 1). These trends of SLA have been reported by many authors (*e.g.*, Gulmon and Chu 1981; Hirose 1986), although some exceptions have been reported concerning the nutrient effects on SLA (Hirose 1984; Andrews *et al.* 1985).

Averaged k_1 of a population increased with density level under the high-nutrient level (Fig. 1). This is due to large values of k_1 of small suppressed plants in crowded populations under high-nutrient level N4 (H1, LAI = 0.41 – 6.48; H3, LAI = 4.09 – 6.31) (Fig. 2). Populations with small LAIs under low-nutrient level N1 (H1, LAI = 0.09 – 0.75; H3, LAI = 0.78 – 1.34) showed little effect of density on k_1.

Averaged k_2 of a population decreased as density level increased at H1 and H3 or as nutrient level decreased at H1. The nutrient effect on k_2 was not clear at H3. Effects of density and nutrient levels on k_2 were not so marked as compared with these effects on SLA and k_1.

Many models so far proposed do not describe changes in the partitioning ratio (*e.g.*, root:shoot) and SLA simultaneously, but our model can do. The degree of changes in k_1, k_2 and SLA will depend on species characteristics (*e.g.*, photosynthetic rate curve per unit leaf area as affected by leaf nitrogen content and light intensity, nutrients uptake rate per unit root weight, etc.) and strength of environmental stress (see also Hunt and Nicholls 1986). The adaptive strategies and life histories of plants under different environmental conditions should be investigated in more detail in this context.

References

Andrews, M., MacFarlane, J.J. and Sprent, J.I. 1985. Carbon and nitrogen assimilation by *Vicia faba* L. at low temperature: the importance of concentration and form of applied-N. Ann. Bot. 56: 651–658.

Chapin, F.S. 1980. The mineral nutrition of wild plants. Ann. Rev. Ecol. Syst. 11: 233–260.

Davidson, R.L. 1969. Effects of root/leaf temperature differentials on root/shoot ratios in some pasture grasses and clover. Ann. Bot. 33: 561–569.

Gulmon, S.L. and Chu, C.C. 1981. The effects of light and nitrogen on photosynthesis, leaf characteristics, and dry matter allocation in the chaparral shrub, *Diplacus aurantiacus*. Oecologia (Berl.) 49: 207–212.

Hirose, T. 1984. Nitrogen use efficiency in growth of *Polygonum cuspidatum* Sieb. et Zucc. Ann. Bot. 54: 695–704.

Hirose, T. 1986. Nitrogen uptake and plant growth II. An empirical model of vegetative growth and partitioning. Ann. Bot. 58: 487–496.

Hunt, R. 1975. Further observations on root-shoot equilibria in perennial ryegrass (*Lolium perenne* L.). Ann. Bot. 39: 745–755.

Hunt, R. and Burnett, J.A. 1973. The effects of light intensity and external potassium level on root/shoot ratio and rates of potassium uptake in perennial ryegrass (*Lolium perenne* L.). Ann. Bot. 37: 519–537.

Hunt, R. and Nicholls, A.O. 1986. Stress and the coarse control of growth and root-shoot partitioning in herbaceous plants. Oikos 47: 149–158.

Iwasa, Y. and Roughgarden, J. 1984. Shoot/root balance of plants: optimal growth of a system with many vegetative organs. Theor. Pop. Biol. 25: 78–105.

Johnson, I.R. 1985. A model of the partitioning of growth between the shoots and roots of vegetative plants. Ann. Bot. 55: 421–431.

Mooney, H.A. and Gulmon, S.L. 1979. Environmental and evolutionary constraints on the photosynthetic characteristics of higher plants. In: O.T. Solbrig, S. Jain, G.B. Johnson and P.H. Raven (eds), Topics in Plant Population Biology, pp. 316–337. Columbia University Press, New York.

Morris, E.C. and Myerscough, P.J. 1985. Effects of nutrient level on thinning and non-thinning crowding in even-aged populations of subterranean clover. Aust. J. Ecol. 10: 469–479.

Orians, G.H. and Solbrig, O.T. 1977. A cost-income model of leaves and roots with special reference to arid and semiarid areas. Am. Nat. 111: 677–690.

Reynolds, J.F. and Thornley, J.H.M. 1982. A shoot:root partitioning model. Ann. Bot. 49: 585–597.

Richards, D., Goubran, F.H. and Collins, K.E. 1979. Root-shoot equilibria in fruiting tomato plants. Ann. Bot. 43: 401–404.

Robinson, D. 1986. Compensatory changes in the partitioning of dry matter in relation to nitrogen uptake and optimal variations in growth. Ann. Bot. 58: 841–848.

Thornley, J.H.M. 1972a. A model to describe the partitioning of photosynthate during vegetative plant growth. Ann. Bot. 36: 419–430.

Thornley, J.H.M. 1972b. A balanced quantitative model for root:shoot ratios in vegetative plants. Ann. Bot. 36: 431–441.

EFFECT OF FOLIAGE DISTRIBUTION WITHIN TREE CROWNS ON INTERCEPTED RADIANT ENERGY AND PHOTOSYNTHESIS

J.C. GRACE
Forest Research Institute, Ministry of Forestry, Rotorua, New Zealand

Abstract

A climate-driven model which simulates the interception of solar radiant energy and canopy photosynthesis for stands of *Pinus radiata* growing on fertile sites is presented. The crown of each tree is represented by an ellipsoid, which may be truncated at the base. Within the crown the foliage is distributed between four shells allowing for non-random foliage distributions.

Defoliating all trees by 48%, leaving an outer shell containing the youngest age-class of foliage, reduced annual intercepted photosynthetically active radiant energy (PAR) by 8% in a stand with 1841 stems ha^{-1} and by 18% in a stand with 500 stems ha^{-1}. Canopy photosynthesis was reduced by 16% and 28% respectively.

At 500 stems ha^{-1}, removing 48% of the foliage from alternate trees, increased tree photosynthesis of unaffected trees by 6%, and decreased tree photosynthesis of affected trees by 30% compared with a tree in an unaffected stand.

With green-crown pruning, the percentage reduction in intercepted PAR and canopy photosynthesis increased exponentially with percentage foliage removed. At 500 stems ha^{-1}, when 48% of the foliage was removed by pruning, intercepted PAR and canopy photosynthesis were reduced by 29% and 35% respectively.

These results and other studies suggest that parameters describing crown shape should be measured when developing models of tree growth for forest managers.

Introduction

Traditionally, models of forest growth have been derived from large mensurational data sets covering a range of management options and site conditions. Such models can be assumed to give reliable predictions of growth for stands growing on similar sites and subjected to similar management conditions, provided that the growing conditions were well represented in the data set. Outside these ranges the model may not be reliable. For example, Manley (1986) showed that the basal area increment of heavily thinned stands of *Pinus radiata* D. Don with less than 225 stems ha^{-1} was overestimated using a model which had been derived from moderately thinned stands.

An alternative approach which will allow the consequences of different management options as well as defoliation due to disease and insect attack to be explored without recourse to large databases is to develop models which simulate the biological processes controlling tree growth. Such a model is being developed to simulate the growth of *P. radiata* on sites where water and nutrients are non-limiting. This model should be applicable to the extensive commercial plantings of *P. radiata* in the central North Island of New Zealand where there are no significant water and nutrient limitations (Beets and Brownlie 1987). Currently the model simulates interception of solar radiant energy by a stand of trees, and canopy photosynthesis. Sub-models simulating respiration and allocation of carbon to different parts of the tree are being developed.

This paper reports two simulation studies in which the model was used to investigate the effect of defoliation, from the center of the crown outwards, and the effect of green crown pruning from the base upwards on yearly intercepted PAR and canopy photosynthesis.

Methods

1. Simulation model

As trees are often widely spaced in New Zealand (*e.g.*, in newly-planted stands; heavily thinned stands; and in agroforestry systems), the sub-model simulating the interception of solar radiant energy has been developed from the model of Norman and Welles (1983) which assumes that the crown of each tree can be represented by an ellipsoid. Other features of the current model (Grace *et al.* 1987a) are that the position of each tree is specified, and that the crown can be divided into four shells by specifying three smaller ellipsoids within the ellipsoid representing the crown shape. Within each shell the foliage is assumed to be randomly distributed, however the foliage area density (foliage surface area per unit volume) can vary between shells allowing for non-random distribution of foliage within the crown. A specified length of the crown, from the base upwards, may be removed allowing green-crown pruning regimes to be simulated.

For each hour of daylight, the model calculates the hourly average intercepted solar radiant energy per m^2 in photosynthetically active (PAR, 400–700 nm) and near infra-red (NIR, 700–3000 nm) wavebands taking into account the position of the sun, proportions of diffuse and direct solar radiant energy and scattering. To estimate yearly intercepted PAR, the year is split into 3 periods, namely: (1) January, February, November, December; (2) March, April, September, October; (3) May, June, July, August.

For each period the model is run for selected days covering the range of measured weather conditions and an equation predicting daily intercepted PAR from daily incoming PAR is derived. These equations are used to predict daily intercepted PAR for each day of the year. Over a six month period (January–June) the percentage difference between using this method rather than running the model for each day was less than 12%, and on a monthly basis the percentage difference was less than 5% (Grace *et al.* 1987a).

Using this model Grace *et al.* (1987a) showed that for stands of *P. radiata* growing on a fertile site, above ground dry matter production was linearly related to simulated yearly intercepted PAR.

Net photosynthesis for a tree, on an hourly basis, is simulated by splitting the tree crown into a maximum of 52 segments and estimating the rate of net photosynthesis at a fixed point within each segment. This rate is assumed to apply to all foliage within that segment. Canopy photosynthesis is obtained by summing over all segments of all trees within the stand. Yearly photosynthesis was estimated using the same mathematical procedure as that used to obtained yearly intercepted PAR.

Table 1. Details of stands for which the effects of defoliation have been simulated.

Stand	Stand Structure
1.	1841 SPH no trees affected by *Cyclaneusma*
2.	1841 SPH foliage on each tree reduced by 48%, remaining foliage randomly distributed throughout the tree crown
3.	1841 SPH foliage on each tree reduced by 48%, remaining foliage randomly distributed in the outer shell of crown
4.	500 SPH no trees affected by *Cyclaneusma*
5.	500 SPH foliage on each tree reduced by 48%, remaining foliage randomly distributed throughout the crown
6.	500 SPH foliage on each tree reduced by 48%, remaining foliage distributed in the outer shell of crown

The rate of net photosynthesis is assumed to increase asymptotically with increasing incident PAR, and decrease with increasing water vapor saturation deficit (VPD) and specific leaf area (SLA) (Grace et al. 1987b).

2. Defoliation

In stands of *P. radiata* the pathogen, Cyclaneusma (*Cyclaneusma minus* (Butin) Di Cosmo *et al.*), causes premature needle-cast of 1-year and older needles during spring while the current years needles are resistant. In the field infection does not usually show up until trees are six years old and the symptoms are most apparent in stands aged 11–20 years (Gadgil 1985).

The effect of removing older foliage from the center of the crown leaving just the youngest age-class in the outer shell was simulated by estimating yearly intercepted PAR and canopy photosynthesis for stands with open and closed canopies (500 and 1841 stems ha^{-1} (SPH) respectively) (Table 1). Each tree was assumed to have the crown shape given in Table 2. This is the average crown shape of a six-year old *P. radiata* tree growing in a stand with 1841 SPH at Puruki (38° 30' S, 176° 15' E), an experimental forest growing on a fertile site about 40 km southwest of Rotorua. To investigate the importance of the shape of the outer shell, yearly intercepted PAR and canopy photosynthesis were also calculated assuming that the remaining foliage was evenly distributed throughout the crown.

Tree photosynthesis was also calculated for unaffected and partially defoliated trees in a stand with 500 SPH where 48% of the foliage on a surface area basis had been removed from the center of the crown of alternate trees. The crown shape given in Table 2 was used in this simulation.

3. Pruning

Green-crown pruning is carried out in stands of *P. radiata* in New Zealand with the objective of producing clearwood. It is carried out in conjunction with thinning, and up to 70% of the green crown length may be removed.

The effect of green-crown pruning was simulated by calculating yearly intercepted radiant energy and canopy photosynthesis for Stand 4 (Table 1) with 9%, 29%, 48% and 70% of the foliage, on a surface areas basis, removed due to green crown pruning.

Table 2. Average crown shape of a six-year old *Pinus radiata* tree growing on a fertile site in stand with 1841 stems ha^{-1}.

Tree Height (m)	7.4
Crown Length (m)	6.7
Crown Width (m)	2.5
Surface area of foliage (m^2)	45.4
Surface area of 1-year-old foliage (m^2)	23.4
Foliage area density (surface area per unit volume, m^{-1})	2.1

Note: foliage surface areas are on a one-sided basis.

Table 3. Estimated yearly intercepted PAR and canopy photosynthesis for stands given in Table 1.

Stand	Yearly intercepted PAR (GJ m^{-2} y^{-1})	Canopy Photosynthesis (t C ha^{-1} y^{-1})
1	2.27	24.3
2	2.14	21.2
3	2.09	20.5
4	1.25	11.6
5	1.04	8.5
6	1.03	8.3

Stands assumed to be growing at 38° 30' S, 176° 15' E.

Table 4. Estimated yearly photosynthesis for trees growing in stands with 500 stems ha^{-1}. A. Unaffected stand. B. Stand where alternate trees have had the inner 48% of their foliage removed. Stands assumed to be growing at 38° 30' S, 176° 15' E. Crown shape given by Table 2.

			Tree Photosynthesis (kg C tree^{-1} y^{-1})
A.	Unaffected Stand		23.3
B.	Affected Stand	– Unaffected tree	24.7
		– Affected tree	16.3

Results

Defoliation

At 1841 SPH, removing 48% of the foliage leaving just the youngest age-class reduces simulated yearly intercepted PAR by 6% if the remaining foliage is evenly distributed throughout the crown and 8% if the remaining foliage is distributed in the outer shell of the crown (Table 3). Canopy photosynthesis is reduced by 13% if the remaining foliage is evenly distributed throughout the crown and 16% if the remaining foliage is distributed in the outer shell (Table 3).

At 500 SPH, defoliation has a greater effect. Simulated yearly intercepted PAR and canopy photosynthesis are reduced by 18% and 28% respectively if the remaining foliage is distributed in the outer shell; and by 17% and 27% respectively if the remaining foliage is distributed throughout the crown (Table 3).

Table 5. Estimated yearly intercepted PAR and canopy photosynthesis for stands where a given percentage of the foliage has been removed through green crown pruning. (Unpruned stand corresponds to Stand 4, Table 1). Stands assumed to be growing at 38° 30′ S, 176° 15′ E.

% foliage removed	Leaf area index	Yearly intercepted PAR (GJ m^{-2} y^{-1})	Canopy photosynthesis (t C ha^{-1} y^{-1})
0	2.3	1.25	11.6
9	2.1	1.20	11.0
29	1.6	1.07	9.4
48	1.2	0.89	7.5
70	0.7	0.64	4.9

Removing 48% of the foliage, from the center of the crown outwards, from alternate trees increased yearly tree photosynthesis of unaffected trees by 6%, and decreased yearly tree photosynthesis by 30% compared with trees in an unaffected stand. The difference in tree photosynthesis between unaffected and partially defoliated trees was 34% (Table 4).

Pruning

The percentage reduction in both yearly intercepted PAR and canopy photosynthesis increases exponentially with increasing percentage of foliage removed by pruning (Table 5 and Fig. 1). The ratio of canopy photosynthesis to intercepted PAR also changes with the amount of foliage removed (Fig. 1).

Discussion

The results (Table 3) indicate the maximum effect that defoliation due to *Cyclaneusma* would have on stands as the same amount of foliage was removed from each tree. In practice rarely more than 50% of trees are affected (Gadgil 1985). The results (Table 4) indicate that the growth of unaffected trees in a stand affected by *Cyclaneusma* is likely to be greater than in a stand where no trees are affected. The results also indicate that defoliation will have a greater impact on tree growth in stands where the canopy is open compared to stands where the canopy is closed. As susceptibility to *Cyclaneusma* is heritable, these results suggest that growth losses could be minimised by leaving stands unthinned till there are signs of *Cyclaneusma*, and then removing susceptible trees as suggested by van der Pas *et al.* (1984) and Gadgil (1985).

The results (Table 3) indicate that defoliation causes a greater reduction in canopy photosynthesis than in intercepted PAR. Assuming that above-ground dry matter production is linearly related to intercepted PAR (*e.g.*, Grace *et al.* 1987), this suggests that defoliation should affect allocation patterns and respiration.

There have been no studies on the effect of defoliation on biomass allocation in *P. radiata*, however other studies indicate that defoliation causes a greater reduction in basal area growth than in height growth (*e.g.*, Rook and Whyte 1976; van der Pas *et al.* 1984). For example Rook and Whyte (1976) showed that re-

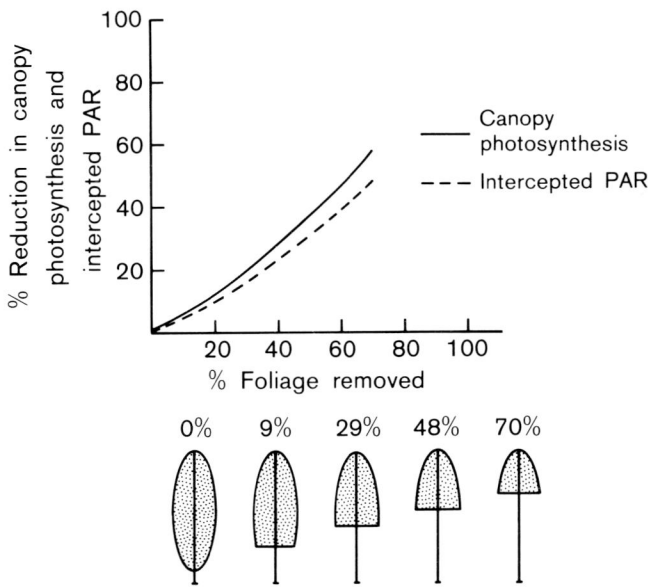

Fig. 1. Estimated percentage reduction in yearly intercepted PAR and canopy photosynthesis due to removing a given percentage of foliage by green-crown pruning.

moving the oldest 25% of foliage in a 5-year old stand, with 2500 SPH, reduced height growth by 5% and basal area growth by 16% in the first year after defoliation.

While yearly intercepted PAR and canopy photosynthesis were slightly higher when the foliage was distributed throughout the crown compared to foliage distributed in the outer shell, the differences are likely to be unimportant and the results suggest that it is the shape of the crown which is important.

Figure 1 indicates light pruning is likely to have little effect on tree growth, but that as the percentage of foliage removed increases the reduction in intercepted PAR and canopy photosynthesis increases exponentially.

Removing 48% of the foliage by pruning leaves the same amount of foliage within the crown as in stand 6 (Table 1). The density of the foliage within the part of the crowns occupied by foliage is also the same; however, the values of yearly intercepted PAR and canopy photosynthesis are very different, being reduced by 18% and 28% respectively in the case of defoliation and by 29% and 35% in the case of pruning. This indicates that the shape of the crown is important as has been shown in other studies. For example Jahnke and Lawrence (1965) indicate that taller crowns are more efficient at intercepting light. It is not the volume that is important: Kellomäki *et al.* (1986) show that for a given crown volume the radiant energy absorbed by a cone is dependent on the ratio of height to base radius. Good correlations have been obtained between growth and crown surface area (*e.g.*, Kramer 1986). Also functions of crown surface area have been used with reasonable success in a number of individual tree growth models (*e.g.*, Hatch *et al.* 1975; Grace 1980). These results suggest that the effect of crown surface area on tree growth should be investigated further.

This study indicates approximate reductions in photosynthesis due to defoliation and green-crown pruning. However, without a reliable theory for allocating carbon to different parts of the tree the long-term consequences of pruning and defoliation on tree growth cannot be realistically simulated. It does, however, indicate important interactions between stocking, leaf area and crown shape. It would be worthwhile for forest mensurationists to measure leaf area and crown shape as manipulation of these variables may improve traditional mensurational models.

References

Beets, P.N. and Brownlie, R.K. 1988. Puruki experimental catchment: site, climate, forest management and research. New Zealand J. For. Scie. 17 (in press).

Gadgil, P.D. 1985. Cyclaneusma needle-cast. Forest pathology in New Zealand No 11. Forest Research Institute, Rotorua. 4pp.

Grace, J.C. 1980. Computer modelling of individual tree growth. D. Phil. thesis, Oxford University. 183 pp. (unpublished).

Grace, J.C., Jarvis, P.G. and Norman, J.M. 1987a. Modelling the interception of solar radiant energy in intensively managed forests. New Zealand J. For. Scie. 17 (in press).

Grace, J.C., Rook, D.A. and Lane, P.M. 1987b. Modelling canopy photosynthesis in *Pinus radiata* stands. New Zealand J. For. Scie. 17 (in press).

Hatch, C.R., Gerrard, D.J. and Tappeiner II, J.C. 1975. Exposed crown surface area: a mathematical index of individual tree growth potential. Can. J. For. Res. 5: 224–228.

Jahnke, L.S. and Lawrence, D.B. 1965. Influence of photosynthetic crown structure on potential productivity of vegetation, based primarily on mathematical models. Ecology 46: 319–326.

Kellomäki, S., Kuuluvainen, T. and Kurttio, O. 1986. Effect of crown shape, crown structure, and stand density on the light absorption in a tree stand. In: T. Fujimori and D. Whitehead, (eds), Crown and Canopy Structure in Relation to Productivity. Forestry and Forest Products Research Institute, Ibaraki, Japan.

Kramer, H. 1986. Relation between crown parameters and volume increment of *Picea abies* stands damaged by environmental pollution. Scand. J. For. Res. 1: 251–263.

Manley, B. 1986. Performance of the Kaingaroa growth model for low-stocking regimes. New Zealand Forest Service, FRI Bulletin No. 113, 21 pp.

Norman, J.M. and Welles, J.M. 1983. Radiative transfer in an array of canopies. Agron. J. 75: 481–488.

Rook, D.A. and Whyte, A.G. 1976. Partial defoliation and growth of 5-year-old Radiata pine. New Zealand J. For. Scie. 6: 40–56.

van der Pas, J.B., Slater-Hayes, J.D., Gadgil, P.D. and Bulman, L. 1984. Cyclaneusma (Naemacyclus) needle-cast of *Pinus radiata* in New Zealand. 2: Reduction in growth of the host, and its economic implication. New Zealand J. For. Scie. 14: 197–209.

DYNAMICS OF THE BRANCH POPULATION IN THE CANOPY OF YOUNG SCOTS PINE STANDS BASED ON MODULAR GROWTH

SEPPO KELLOMÄKI
University of Joensuu, Faculty of Forestry, P.O. Box 111, SF-80101 Joensuu, Finland

Abstract

The paper describes the dynamics of the branch number and biomass in crowns of young Scots pines (*Pinus sylvestris*) assuming that the crown system could be treated as a population of branches occupying the crown volume or the crown projection. The processes underlying the life tables developed for Scots pine branches were analyzed on bases of the birth and death of branches using a population model. The total number and biomass in the crown system was incorporated using the $-3/2$ power model where the branch number per crown projection area (branch density) formed the feedback system controlling the size of the branch population.

Introduction

A tree is traditionally divided into crown, stem and root system. In this context crown structure and crown processes are of primary importance for the productivity of trees. For example, crown shape is claimed to be responsible for differences in the productivity between forest trees (Jahnke and Lawrence 1965; Oker-Blom and Kellomäki 1982). Obviously, crown structure could regulate within-crown competition between modules of the crown structure and, consequently, production of stem wood.

The crown systems of most trees are composed of branches, branchlets and twigs. This is the result of modular growth which is characterized as the growth of a genetic individual by repeated iteration of multicellular parts (Harper *et al.* 1986). In conifers in particular this kind of structure is pronounced due to their monopodial growth pattern. The tree's structure in terms of modular growth is a consequence of the growth with the modules remaining attached to each other. The whole tree as an organism is, thus, an accumulation of dead modules supporting living modules necessary for obtaining resources through branches and roots connected by the vascular system (Shinozaki *et al.* 1964a, b; Harper *et al.* 1986). This growth pattern seems to allow one to treat the tree as a system characterized by the modular structure of roots, stem, branches and shoots or leaves (needles) (Harper *et al.* 1986).

A tree as a whole and its modules are fixed in a certain position. This makes them susceptible to depletion of local resources. Light absorption, especially in the upper crown parts, limits the growth of branches with the consequent death of branches in the lower crown parts due to shading by other branches. However, living branches occur deep into the crown if the light conditions are sufficient to maintain the photosynthetic functions. These effects of light conditions on branch growth show that the depletion of local resources has a major effect on the survival of any individual branch and that the growth of a branch is only loosely dependent on the growth of other branches. This allows one to assume that the tree crown can be treated as a population of branches with mutual competition for available resources when they occur close enough to each other to be able to exploit each other's resources.

A biological population is traditionally defined as a group of individual organisms of the same species within a particular area or community. The branches occupying the crown of a particular tree fall outside this definition, since branches are normally not regarded as individuals but as organs of an individual. This conceptual difficulty can be avoided if one constructs a tree from proper structural modules, *i.e.*, a branch or a shoot. In this context the selected module is an organism which forms a community called a tree. Consequently, a tree crown is assumed to be a population of branches occupying a land area equal to the crown projection area of the tree; its branches with their necessary connections to the soil are seen as a basic module of the tree's structure. The stem of a tree is, thus, a bunch of pipes which connect the branches to the soil. The rationale of this approach lies in the pipe model theory of plant form expressed by Shinozaki and his associates (1964a, b).

The size of any population is determined by birth and death rates and the time over which these rates are integrated. The birth rate of the branch population is determined by the number of branches formed annually around the stem apex. Similarly, the number of branches dying annually determines the death rate of the branch population. Obviously, the death of branches in a tree crown is closely related to the total biomass of all branches, since the accumulation of branches cannot exceed the capacity of the crown space available for a tree.

This paper applies the concepts of population ecology for a study of the dynamics of branch number and biomass in the crown system of young Scots pines (*Pinus sylvestris* L.) assuming that the crown structure of Scots pine could be analyzed on the basis of modular growth. Approaches based on life tables for branches, birth and death of branches and the $-3/2$ power model for self-thinning are applied.

Life tables for the branch population in Scots pine

The dynamics of the crown system of Scots pine can be characterized in terms of the birth and death rates of branches due to the monopodial growth pattern of this tree species. Each year one whorl (branch cohort) with a particular number of branches will be born at the stem's apex. As this is repeated annually the preceeding cohort will shift deeper into the canopy where it is exposed to increased shading by the surrounding trees and the crown itself. Consequently, the life expectancy of the branches and whole cohort reduces with increasing mortality of the branches deeper in the crown (Table 1).

In a young Scots pine crown (age 15–25 years) the average number of branches in the uppermost whorl (the birth rate of the branches) is four to six with substantial variation between trees and years (Table 1) (Flower-Ellis *et al.* 1976; Kellomäki and Väisänen 1988). Branches already begin to die in the second year after their

Table 1. Distribution of surviving (s) and dead (d) branches per whorls in the total material (Kellomäki and Väisänen 1988).

Whorl	Branches per whorl, number of cases																				Total	
	1		2		3		4		5		6		7		8		9		10			
	s	d	s	d	s	d	s	d	s	d	s	d	s	d	s	d	s	d	s	d	s	d
1	2		5		19		46		77		39		6		4						198	
2	2	1	3		13		44		81		40		8		4		2				198	1
3		6	3	4	14	1	27	1	65		49		28		9		2		1		198	11
4	2	2	3		23		47		65		36		11		8		1				196	3
5	1	9	5	3	16		50	1	78		24		20		2						198	12
6	3	12	1	5	13	1	46	3	81	3	34	1	15		3		1	1			197	20
7		27	10	11	20	7	48	9	64	12	28		18		3						192	51
8	17	44	11	23	15	6	27	17	51	25	34	9	8	6	7	2					170	111
9	21	38	17	24	26	21	36	18	35	41	9	9	4	2		2		3			148	138
10	16	26	17	19	11	9	21	32	29	46	9	25	7	15		7		4		1	110	165
11	10	11	9	14	11	15	13	36	12	53	7	26	1	20		13		4		2	63	184
12	12	9	8	16	4	14	3	33	3	65		31		20		11		4		2	30	196
13	8	5	5	7		23		32		54	1	29		12		17		1		1	14	193
14	1	11	1	19		13		33		47		39		14		9					2	191
15		14		17		26		28		26		23		5		4				2		173
16		18		28		21		21		18		17		7								145
17		23		21		19		8		12		12		3				1				118
18		17		17		16		7		4		1				1						72
19		9		13		9		4		2		1										43
20		6		8		3		4		2												23
21		4		3				1		1												9
22		1																				1
Total	96	293	98	252	185	204	408	284	641	409	310	223	126	104	40	66	6	13	2	8	1914	1860

Table 2. Survivorship and mortality in branch populations in the canopies of young *Pinus sylvestris* stands in the total material (Kellomäki and Väisänen 1988).

Age interval (yr)	Survivorship o/oo	Average mortality o/oo
0-1	1000	0
1-2	1000	0
2-3	1000	0
3-4	1000	0
4-5	1000	0
5-6	992	8
6-7	853	147
7-8	714	286
8-9	575	425
9-10	437	563
10-11	298	702
11-12	159	841
12-13	21	979
13-14	0	1000

birth. In the upper crown (above the zone where branches overlap) the probability of death is small compared with that in the middle and lower zones of the canopy. Branch death in Scots pine crowns before canopy closure, *i.e.*, in whorls younger than five to six years, is very improbable (Flower-Ellis *et al.* 1976).

Kellomäki and Väisänen (1988) found that the regression between the size of a branch cohort and the age of the cohort (t) was nearly linear when the age exceeded six years, but was less than 14 years ($6 < t < 14$). This indicates a negligible death in the six uppermost whorls and exceptional survival of branches in whorls older than 11-13 years. In this process the branch cohort per tree is reduced at an average rate of 0.7 ± 0.1 branches per year, the initial size of the branch cohort being reduced by half in four to five years (*cf.* also Flower-Ellis *et al.* 1976).

The life span of branches of young Scots pines can be described using life tables which give the rate of survival and mortality of branches representing different age classes (Table 2) (Flower-Ellis *et al.* 1976; Kellomäki and Väisänen 1988). The rate of survival gives the life expectancy of a branch at a particular age just as the mortality indicates the risk for the same branch to die. As discussed above the life expectancy of branches younger than five to six years experience no risk of death in healthy trees. Thereafter the mortality of the branches increases so that the age of branches of young Scots pines only exceptionally exceeds 14 years.

Effects of birth and death on the total number of branches

A model for branch number

The processes underlying the life tables could be analyzed with the help of population models incorporating the birth and death rates of the branches as suggested by Kellomäki and Väisänen (1988). Let $N(t)$ denote the total number of branches in the crown at moment t, $N(t-1)$ the respective number at the preceeding moment and ΔN the change in branch number at the time interval $(t-1, t)$. The change in branch number is

$$N(t) = N(t-1) + (B-D), \tag{1}$$

where B is the birth and D the death rate of branches (per crown.yr^{-1}). The per capita rate of increase of branches during the period $(t-1, t)$ will be (Berryman 1985; pp. 36–52).

$$R = \frac{\Delta N}{N(t-1)} = \frac{B-D}{N(t-1)} = \frac{N(t)-N(t-1)}{N(t-1)}, \tag{2}$$

where R is the per capita increase rate (dimensionless). The size of the branch population can be written as follows

$$N(t) = N(t-1) + RN(t-1). \tag{3}$$

Let R_m denote the maximum per capita increase in branch number, *i.e.*, the surrounding trees have no effect on the birth and death of branches. Then

$$R = R_m - sN(t), \tag{4}$$

where s indicates the strength of the interaction between trees in a stand (Berryman 1985). Consequently,

$$N(t) = N(t-1) + (R_m - sN(t-1))N(t-1). \tag{5}$$

It is reasonable to assume that in a closed stand the birth and death of branches occur with equal frequency, *i.e.*, the crown space is fully utilized with no more space for additional growth of the branch population. In other words

$$R_m = sK, \tag{6}$$

where K indicates the maximum number of branches able to occupy the crown space (carrying capacity) (per crown). Consequently,

$$N(t) = N(t-1) + R_m(1-N(t-1)/K)N(t-1). \tag{7}$$

The parameters R_m and K achieve values in the range of R_m = 0.47–0.65 and K = 36–52 for young Scots pine plantations on sites with *Myrtillus* and *Vaccinium*. Values of both parameters were on the average greater on sites of the *Myrtillus* type as compared to the *Vaccinium* type. The estimates of branch number in individual tree crowns given by Equation (7) are close to the realized ones, but Equation (7) underestimates the number of young branches and overestimates the number of old ones (Fig. 1).

Rate of increase and carrying capacity of the branch population

The realized rate of increase (R) can be written as a function of parameter K using Equations (4) and (5) as follows

$$R(t) = s(K-N(t)). \tag{8}$$

The following relations exist between these quantities

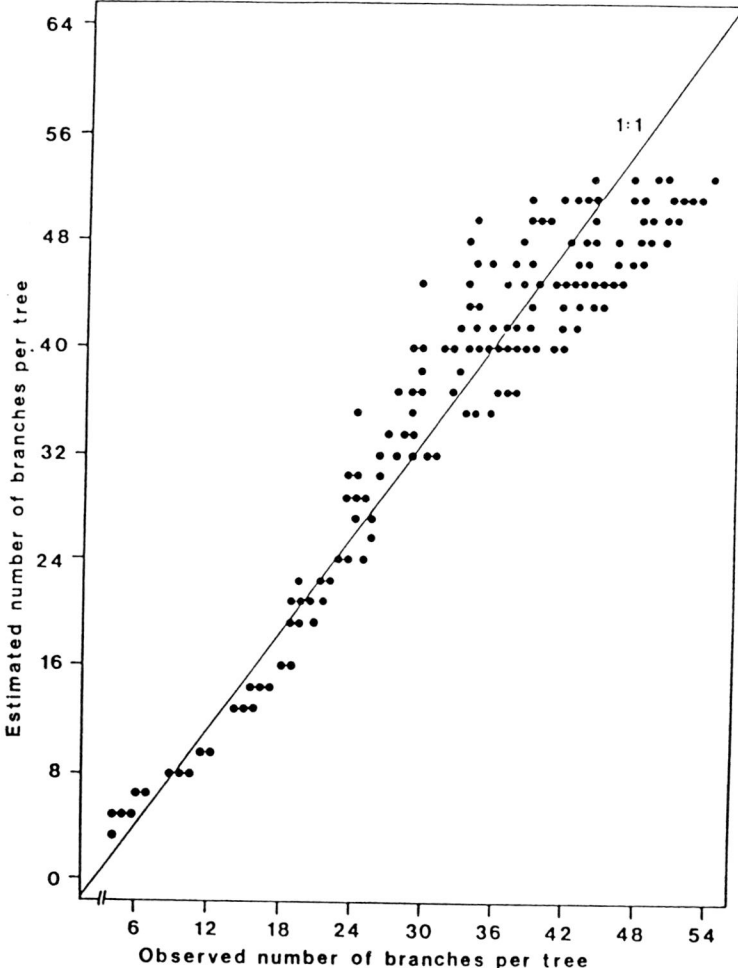

Fig. 1. Example of the relationship between observed and predicted values of the number of surviving branches on sites of the *Vaccinium* type as calculated with Equation (7) (Kellomäki and Väisänen 1988).

i. When N(t) → 45.4 of the *Vaccinium* site and 51.9 at the *Myrtillus* site, then R → 0, *i.e.*, the above values of N(t) indicate the carrying capacity (K) of the branch population in the present material (Fig. 2).
ii. When N(t) → 0, then R → R_m, *i.e.*, the realized and potential per capita rates of increase are nearly equal when the total number of branches is low (Fig. 2).

Based on the above consideration, one could expect that the competition stress between branches on sites of the *Vaccinium* type is greater than on the *Myrtillus* type. Consequently, there are 6–7 branches more on young Scots pines on sites of the *Myrtillus* type than on the *Vaccinium* type. This results on average in one to two more whorls on sites of the *Myrtillus* type as compared to the *Vaccinium* type.

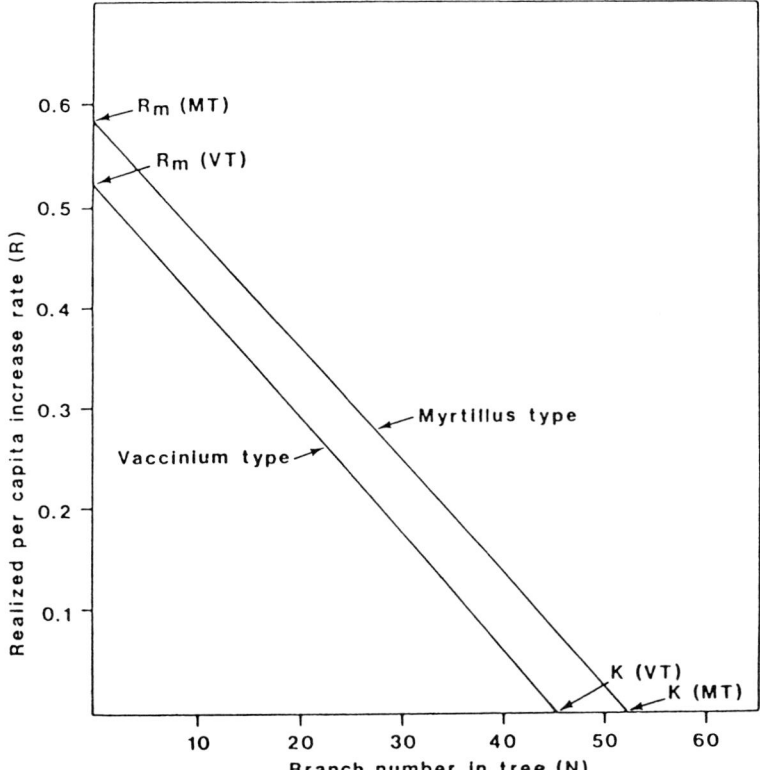

Fig. 2. Regression between the actual values of the per capita growth rate (R) and the number of branches (N) per tree. K indicates the carrying capacity of branches per mean tree (Kellomäki and Väisänen 1988).

Number of dead branches

The change in population size $\Delta N = B - D$ incorporates the death rate (D) which can be written as a function of the birth rate, per capita rate of increase and the population size as follows

$$D(t) = B(t) - RN(t-1) = B(t) - (R_m - sN(t))N(t-1). \tag{9}$$

The estimated values for dead branches using Equation (9) remained lower than the observed ones (Fig. 3) and the death of the branches was initiated later than expected. It was, however, evident that

i. The increasing value of B enhances the death of branches, *i.e.*, the increasing birth rate of branches accelerates the accumulation of branches in the crown, causing the increasing death of branches due to increasing competitive stress (s).
ii. The death of branches is approximately relative to the second power of the actual number of branches, which also emphasizes the role of birth (B) and competitive stress (s) as factors controlling the death of branches.

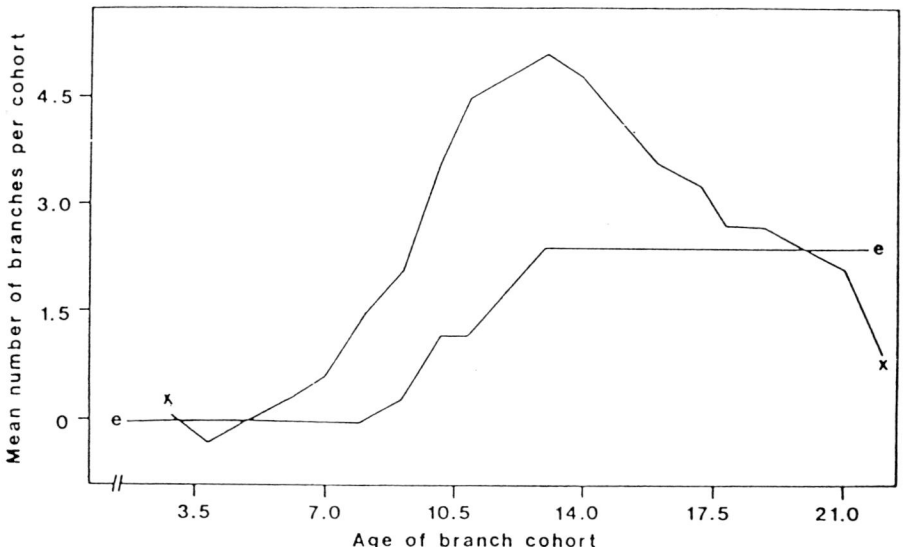

Fig. 3. Observed (x) and predicted (e) number of dead branches per mean tree as a function of the age of the branch cohort in years, calculated with Equation (9). (Kellomäki and Väisänen 1988).

iii. The increasing R_m decreases the death of branches, *i.e.*, potentially branchy trees are actually branchy due to a decreased death rate.

Obviously, most dead branches in absolute and relative terms occur on trees which represent branchy genotypes or experience high interaction stress owing to competition. For example, narrow spacing or a poor site, independently of genotype, realize these premises.

Branch number and the biomass of the crown

Obviously, the death of branches is related to the accumulation of branch biomass as indicated by the relationship between the death of branches and the total number of branches. Based on this fact Kellomäki (1986) applied the $-3/2$ power model by Yoda *et al.* (1963) when he described the relations between branch number and biomass in crowns of Scots pines as follows

$$w = C\rho_B^k \tag{10}$$

$$Y = \rho w = C\rho_B^{k+1} \tag{11}$$

where w is the mean branch biomass, Y the total branch biomass, ρ_B the branch density (number of branches per crown projection area) and C and k parameters.

The $-3/2$ power model was originally developed for whole stands at the stage of self-thinning. In the case of a single tree there is only one particular tree per crown area, *i.e.*, the stand density (ρ) in the projected crown area (A_c) of a partic-

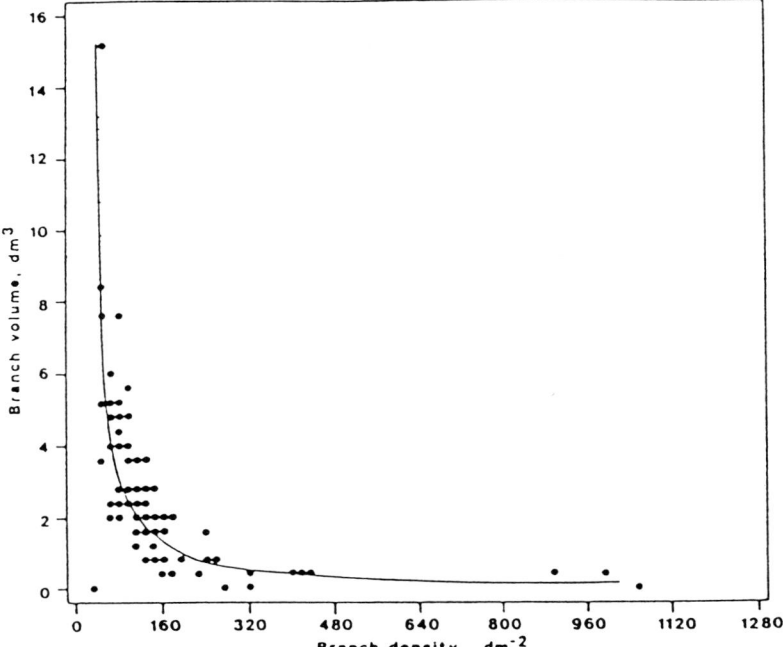

Fig. 4. Relationship between branch density and branch biomass (Kellomäki 1986).

ular tree is one tree per unit land area ($\rho = 1A_c^{-1}$). Consequently, the mean biomass (w) of the tree inside the crown projection area equals the total biomass (Y) of this particular tree (w = Y).

Assuming that any tree is composed of a number of modules indicated by the number of branches on a tree (N), Kellomäki (1986) claimed *a priori* that the stand density in the crown projection area is approximately equal to the branch density (ρ_B), *i.e.*, the number of branches per crown projection area ($\rho_B = N \cdot A_c^{-1}$). Thus, it was assumed that the total branch biomass in a tree crown is proportional to the k:th power of the branch number per crown projection area of the tree, *i.e.*,

$$Y = C\rho_B^k \qquad (12)$$

The −3/2 power model gave satisfactory results in the material representing young Scots pine stands (age 15–20 years) of varying stand density (stand density 1300–4300 stem.ha^{-1}) (Fig. 4). A lower branch density yielded a greater branch volume than a higher branch density, while stand levels with narrow spacing resulted in a greater total stem or branch volume than wide spacing (Yoda *et al.* 1963).

Stand level and tree level seemed to be comparable, if in the former the mean tree biomass is related to the stand density according to the −3/2 power model. Therefore it seems possible that the −3/2 power model indicates at the individual level the basic relations of plant geometry associated with the modular structure of Scots pine (Kellomäki 1986). In this context branch density represents the feedback mechanism which controls the size of the branch population. This control mechanism allows one to integrate the above population model and the $-3/2$ power model, since it is reasonable to assume that the parameter K is equal to the maximal number of branches that can survive in the self-pruning crown. The justification of this assumption is still under study.

Conclusions

The paper describes the dynamics of the crown structure of young Scots pines (*Pinus sylvestris*) in population biological terms where the birth and death of branches are the key concepts determining the actual size of the branch population occupying the crown and canopy space. This treatment emphasizes the branch as the basic unit of the canopy. The canopy is seen as consisting of individual branches interacting with each other. It is assumed that the tree structure can be derived on basis of the branches and their connections (stem) to the soil as is done in the pipe model theory of tree structure (Shinozaki et al. 1964a, b).

In this approach branches are organisms which form a community called a tree. Evidence to justify this kind of re-evaluation of the tree concept is still negligible but it stimulates studies that try to integrate structural and functional properties of trees. In particular, the derivation of a tree's structure on basis of the simple principles needed for modular growth could provide a useful framework for modelling the physiological processes in tree crowns. The population approach in itself might be able to tackle, for example, the vague interaction between trees which are difficult to describe in an ecophysiological approach.

References

Berryman, A.A. 1985. Population systems. A general introduction. Plenum Press. New York. 222 pp.
Flower-Ellis, J., Albrektsson, A. and Olsson, L. 1976. Structure and growth of some young Scots pine stands: (1) dimensional and numerical relationship. Swedish Coniferous Project. Tech. Rep. 3/1976:1–98.
Harper, J.L., Rosen, B.R. and White, J. 1986. The growth and form of modular organism. Preface. Phil. Trans. R. Soc. Lond. B 313: 3–5.
Jahnke, L.S. and Lawrence, D.B. 1965. Influence of photosynthetic crown structure on potential productivity of vegetation, based primarily on models. Ecology 46: 291–311.
Kellomäki, S. 1986. A model for the relationship between branch number and biomass in *Pinus sylvestris* crowns and the effect of crowns shape and stand density on branch and stem biomass. Scand. J. For. Res. 1: 455–472.
Kellomäki, S. and Väisänen, H. 1988. Dynamics of branch population in the canopy of young Scots pine stands. Forest Ecology and Management. (In press).
Oker-Blom, P. and Kellomäki, S. 1982. Theoretical computations on the role of crown shape in absorption of light by forest trees. Mathem. BioScience 59: 291–311.
Shinozaki, K., Yoda, K., Hozumi, K. and Kira, T. 1964a. A quantitative analysis of plant form – the pipe model theory. I. Basic analysis. Jap. J. Ecol. 14: 97–105.
Shinozaki, K., Yoda, K., Hozumi, K. and Kira, T. 1964b. A quantitative analysis of plant form – the pipe model theory. II. The evidence of the theory and its application in forest ecology. Jap. J. Ecol. 14: 133–139.
Yoda, K., Kira, T., Ogawa, H. and Hozumi, K. 1963. Self-thinning in overcrowded pure stand under cultivated and natural conditions. J. Biol. Osaka City Univ. 14: 107–129.

EFFECTS OF LIGHT CLIMATE AND NITROGEN PARTITIONING ON THE CANOPY STRUCTURE OF STANDS OF A DICOTYLEDONOUS, HERBACEOUS VEGETATION

M.J.A. WERGER[1] and T. HIROSE[2]*

[1]Department of Plant Ecology, University of Utrecht, Lange Nieuwstraat 106, 3512 PN Utrecht, The Netherlands; [2]Department of Botany, Faculty of Science, University of Tokyo, Hongo, Tokyo 113, Japan

Abstract

Light climate, canopy structure and nitrogen allocation were studied in dense and open stands of the tall, herbaceous *Solidago altissima*, during the late growing and the flowering stages. In the late growing stage relative investment of dry matter and nitrogen in leaves is independent of stand density, but light climate and leaf arrangement strongly differ with stand density. In the flowering stage relative investment of dry matter and nitrogen in leaves has diminished and leaf distribution is more attenuated higher in the canopies.

We conclude that the pattern of dry matter investment in leaves and stems is determined by the amount of nitrogen available to the plant and that leaf distribution in the canopy is mainly affected by the light climate in the canopy. The reduced availability of nitrogen to the leaves in the flowering stage results in a correspondingly reduced leaf area index, so that the nitrogen concentration in the leaves remains the same. The pattern of decreasing nitrogen allocation to leaves from the top to the bottom of the canopy is more narrowly defined in dense stands than in open stands.

Introduction

Most terrestrial herbaceous plants show clear seasonal growth rhythms: in spring they emerge from a root mat or seed bank; they subsequently grow up, often followed by flowering and seed ripening, and in summer or autumn they die back above-ground or completely. This growth rhythm brings about strong seasonal changes in the above-ground stand structure of herbaceous vegetation. The stand's canopy structure is determined by the spatial distribution of stems, leaves and, less importantly, flowers and fruits. Since this spatial distribution pattern, particularly that of the leaves, determines the light penetration into the canopy, it affects the photosynthetic activity of the leaves and as a consequence the photosynthetic performance of the whole canopy (Monsi and Saeki 1953; Saeki 1960; Björkman 1981; Johnson and Thornley 1984). Regulation of the canopy structure in stands of herbaceous vegetation might therefore be adaptive: whole canopy photosynthetic returns depend on canopy structure.

However, not only light availability determines photosynthetic performance; so does nitrogen concentration in the leaves (Natr 1975; Field 1983; DeJong and Doyle 1985; Field and Mooney 1986; Hirose and Werger 1987a, b). The nitrogen concentration in the leaves depends on the availability of nitrogen to the plant and on the pattern of allocation of nitrogen to the various plant parts. Furthermore, particularly when nitrogen availability is limited the allocation of nitrogen to the

---

* *Present address*: Biological Institute, Faculty of Science, Tohoku University, Sendai 980, Japan.

leaves at various positions in the canopy can be expected to follow a pattern that is related to the pattern of light distribution in the canopy, as this ensures near maximum photosynthetic performance (Hirose and Werger 1987b). This pattern should be more narrowly defined in dense stands, where there is a stronger gradient of diminishing light intensity inside the canopy, than in open stands.

Such strong, parallel patterns in allocation of nitrogen to the leaves and in light availability should be expected to go together with a pattern of decreasing values of specific leaf weights (SLW) from top to bottom in a dense canopy. Shading often leads to reduced SLW values (Nobel et al. 1975; Chabot et al. 1979; Jurik et al. 1979; Jurik 1986; Jurik and Chabot 1986) and under such light conditions SLW was shown to be correlated positively with leaf nitrogen concentration on a dry weight basis (Field 1983; Seemann et al. 1987; Hirose et al. 1988). In highly illuminated leaves, however, Hirose et al. (1988) showed a negative correlation between SLW and leaf nitrogen concentration.

It has been shown before that total plant nitrogen correlates with the matter partitioning between stems and leaves. The more nitrogen is available, the higher the proportion of matter invested in leaves (Chapin 1980; Hirose 1986, this volume). It is thus to be expected that in a growing vegetation the matter partitioning between stems and leaves is adjusted when less nitrogen becomes available for leaves because of the development of a nitrogen sink, e.g., as a result of flowering. Such an adjustment might also lead to a change in the spatial distribution of leaves in order to maintain optimum light utilization in the canopy.

We determined the canopy structures, light climates and nitrogen allocation patterns in open and dense stands of the tall dicotyledonous herb *Solidago altissima* during the late growing stage and the flowering stage. By comparing the results for the two phenological stages we studied the effects of the development of a strong nitrogen sink in the plants. Because of earlier findings discussed above, we expected that

1. between the open and dense stands the light climates in the canopies would be different, and this would go together with different spatial patterns of leaf distribution in the canopies, the leaves being spread over a shorter stretch of the stems in the dense stands than in the open stands;
2. between the open and dense stands the matter partitioning between leaves and stems would be similar, since the total plant nitrogen can be expected to be similar in our stands;
3. between the phenological stages the matter partitioning between leaves and stems would show great differences since flowering implies the development of a strong nitrogen sink;
4. because of this change in matter partitioning there would be differences in the light climates in the canopies of stands of the same density but of different phenological stages;
5. the vertical distribution pattern of nitrogen allocation to the leaves would be more narrowly defined in the dense stands than in the open stands;
6. the dense stands, having a stronger gradient in their light climates, would also show a strong, decreasing vertical gradient of SLW values from top to bottom in their canopies. The open stands, allowing for a favorable light climate throughout their canopies, would not show such a decreasing vertical gradient

in SLW values and, in correspondence with their pattern in leaf nitrogen concentration, might even show a reverse trend.

Material and methods

We sampled stands of *Solidago altissima* L. (Compositae) in the flood plain of the Arakawa River, Urawa, Japan (35°50'N, 139°37'E). *Solidago altissima* has been introduced from North America. It forms dense, pure stands on the fertile soils of the flood plain; on disturbed sites stands are open and mixed with much shorter grasses and herbs. *Solidago altissima* is a perennial herb that develops an extensive rhizome system. Plants normally form wintering rosettes which bolt in early April. Stems grow rapidly and reach up to 3 m high in September when inflorescences are initiated. They flower in October–November. Plants usually stay unbranched, except for their inflorescences. They have their narrow, elliptic leaves arranged along a wide section of their stems and can build fairly simply structured canopies. Longevity of individual leaves is usually 1–3 months (Hirose 1971). While flowering the plants start to loose all their green leaves.

We measured stand structure and light climate in open and dense stands around the middle of September 1985, just before the plants initiated inflorescences and apparently had reached their maximum standing crop. We repeated such measurements on 18 October 1985 when plants were in full bloom.

In a dense stand a 2 × 1 m quadrate was established. The distribution of photon flux density (PFD, 400–700 nm) was measured with a photon flux meter (LI 185B, LiCor, USA) every 20 cm from ground level to the top of the canopy under diffuse light conditions. PFD was read 50 times for every horizontal level and the mean was calculated. PFD above the leaf canopy was monitored throughout the measurement to obtain the distribution of relative PFD within the canopy.

After light climate measurements in September the quadrate in the dense stand was divided into five 2 × 0.2 m subquadrates. In each of these subquadrates, and in four open stands, all plants were cut at ground level and, sealed in plastic bags, were brought to the laboratory. Here plants were clipped every 20 cm from the bottom and separated in leaves and stems. In October the same stratified clipping method was followed, but the total quadrate measured 0.5 × 0.5 m. At that sampling also flowering parts were distinguished. Leaf area of all samples was measured with a leaf area meter (AAM-7, Hayashi Denko, Japan). Dry weights of all collected plant parts were determined after oven drying at 80°C for at least three days. Subsamples of the dried material were ground in a Wiley mill and their nitrogen content was determined with an NC-analyser (NC-80, Shimadzu, Japan).

Results

The light climate in the stands is very different. During the late growing stage in the dense stands light is strongly intercepted high in the canopies and the lowest leaves receive only 7–10% of the PFD above the canopies. Much more light penetrates deep into the canopies of the open stands where the lowest leaves on average receive still 53% of full PFD.

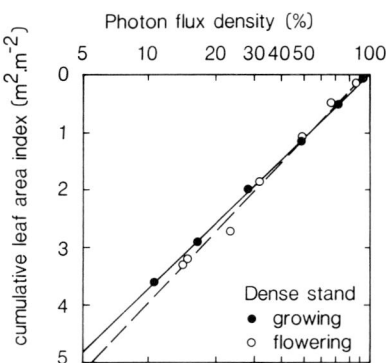

Fig. 1. Relative photon flux density as a function of leaf area index cumulated from the top of the canopy in dense stands of *Solidago altissima* in the late growing and the flowering stages.

About 14% of full PFD reaches ground surface in the dense stands during the flowering stage, so that light conditions for the lowest leaves are better than in the late growing stage. In the late growing and flowering stages the regression lines for proportional PFD received at various height levels with accumulating LAI from the top of the canopy do not differ significantly (Fig. 1).

During the late growing stage the above-ground biomass in the dense stands is 1074 g.m^{-2} and the LAI is 4.24 m^2.m^{-2}. That in the open stand cannot be expressed on a ground area basis because of the large spatial heterogeneity in the disturbed sites: stems emerge in open population patches of greatly different size and at strongly varying interdistances. The relative investment of dry matter in leaves and stems in the dense and open stands is very similar: leaves comprise 23 and 22% of above-ground dry matter respectively, and stems 77 and 78% (Fig. 2; Table 1). Leaf area ratio (LAR, leaf area per unit of above-ground plant dry weight) is 3.9 and 3.2 m^2.kg^{-1} respectively. However, leaf distribution along the stems greatly differs: in the dense stands leaves are spread between 0.6 and 2.4 m above ground surface, with about 20% of total leaf area amassed near median height (at about 1.5 m above ground surface), while in the open stands leaves are spread between 0.4 and 2.4 m above ground surface with only about 13% of total leaf area near median height. Thus, leaves are more evenly spread along the length of the stems in the open stands. In the dense stands SLW values vary from 67 g.m^{-2} (on a dry weight basis) in the top to 50 g.m^{-2} deep in the canopies, while in the open stands they vary from 64 in the top to 75 g.m^{-2} at the bottom of the canopies.

Nitrogen partitioning between leaves and stems is rather similar in both types of stands during the late growing stage. In the dense stands leaves contain about 68% of total above-ground plant nitrogen, and in the open stands this amounts to 62% (Fig. 2B, D). Total nitrogen content in the above-ground plant parts of the dense stands is 8.2 g N.m^{-2} ground area. Nitrogen concentrations in the total above-ground plant parts are very similar in plants of the dense stands (0.78%). In plants of the open stands they vary between 0.60 and 0.95% (see Fig. 3). Of course, the vertical distribution of leaf nitrogen is strongly, though not solely, determined by the vertical distribution pattern of the leaves. The younger tissues

Table 1. Comparison of stand structures and light conditions for lowest leaves. For the open stands quantities cannot be expressed on a ground area basis (see text).

Stage	Late growing		Flowering	
	Open	Dense	Open	Dense
% of dry matter invested in				
leaf	22	23	22	14
stem	78	77	68	79
flower	0	0	10	7
100% = (g.m^{-2})	–	1074	–	1497
% of nitrogen invested in				
leaf	62	68	52	45
stem	38	32	23	27
flower	0	0	25	28
100% = (g.m^{-2})	–	8.24	–	10.27
LAR (m^2.g^{-1}) (× 10^{-4})	32	39	28	22
SLW (g.m^{-2}) top of canopy	64	67	83	69
bottom of canopy	75	50	83	50
% of PFD at lowest leaves	53	7–10	–	14

near the top of the canopies show higher and the older tissues deep inside the canopies lower nitrogen concentrations (Fig. 2A–D).

In the flowering stage dry matter partitioning between plant organs has changed in the dense stands: about 7% of above-ground biomass is invested in flowers, only 14% in leaves and 79% in stems (Fig. 2E; Table 1). Total above-ground biomass is 1497 g.m^{-2}, LAI is 3.31 m^2.m^{-2} and LAR is 2.2 m^2.kg^{-1}. In the open stand 10% of the above-ground biomass is invested in flowers, 22% in leaves and 68% in stems (Fig. 2G, Table 1). LAR is 2.8 m^2.kg^{-1}. Leaf distribution in the dense stands has become more attenuated and is concentrated in the top layers of the canopy but the leaf distribution profile still resembles that in the dense stands of the late growing stage. Around median height (about 1.9 m above ground surface) 24% of total leaf area is amassed. SLW varies from 69 to 50 g.m^{-2} from the top to the bottom of the canopy. In the open stand leaves are again widespread along the length of the stems, though there is some concentration around the middle. Perhaps this pattern is partly influenced by the circumstance that this open flowering stand was not as open as the open stands sampled in the late growing stage. SLW values are about 83 g.m^{-2} throughout the canopy.

Flowering means a strong nitrogen sink to the plants: 25–28% of total aboveground plant nitrogen is contained in the flowering heads, as against 45–52% in the leaves and 23–27% in the stems (Table 1). Total nitrogen in the above-ground plant parts amounts to 10.3 g.m^{-2} in the dense stands. Nitrogen concentrations for the total above-ground plant parts vary over the same narrow range as in the late growing stage (Fig. 3). Again the younger leaves contain proportionally more nitrogen than the older ones (Fig. 2F, H).

In the late growing stage, taking plants of the dense and open stands together, there is a significant positive correlation between the nitrogen concentration of the whole above-ground plant and the fraction of dry matter invested in leaves

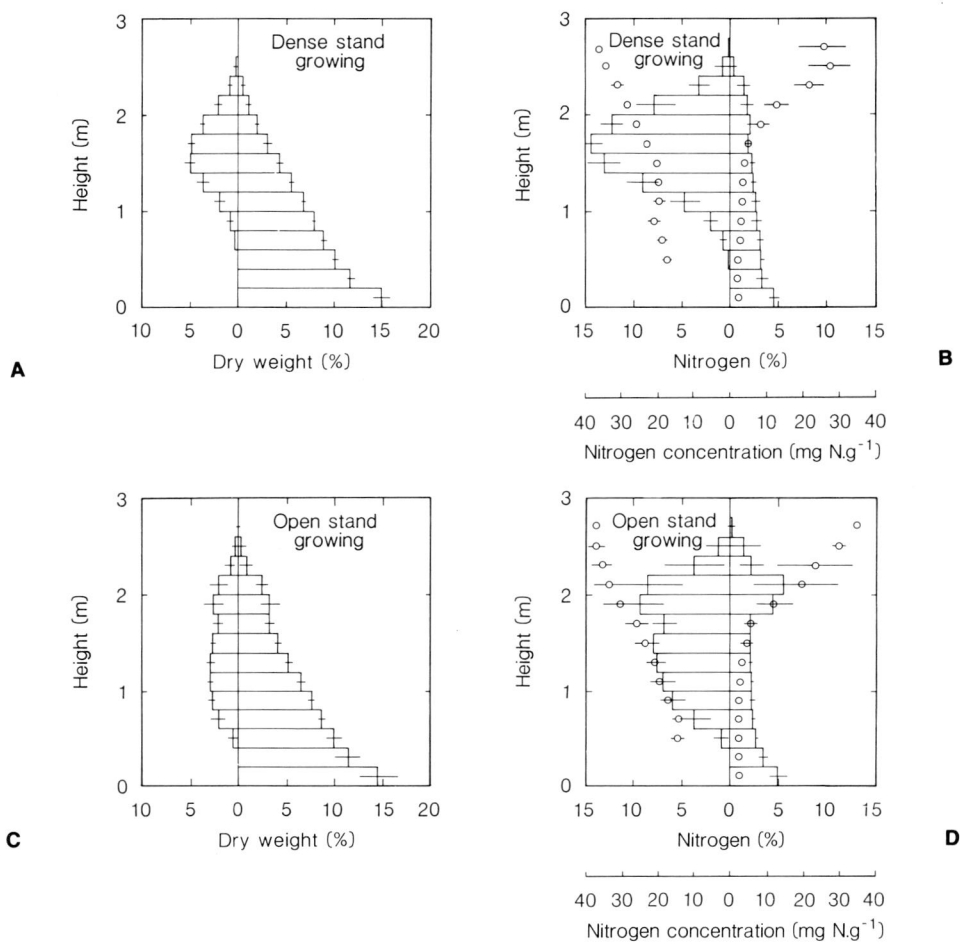

Fig. 2. Vertical distribution of dry matter (A, C, E, G) and nitrogen (B, D, F, H) in dense (A, B, E, F) and open (C, D, G, H) stands of *Solidago altissima* in the late growing (A, B, C, D) and flowering (E, F, G, H) stages. Bars indicate mean percentages (with s.d., A, B, C, D) of total above-ground plant dry weights or total above-ground plant nitrogen in layers of 20 cm high. Leaves are shown on the left side, stems (blank bars) and flowering parts (stippled bars) on the right in each diagram. Circles (B, D, F, H) indicate mean nitrogen concentrations per plant part per layer with the values for flowering parts furthest right (F and H). Where in B and D no standard deviations for nitrogen concentrations are shown they are smaller than the diameter size of the circles.

(Fig. 3). In the flowering stage a similar trend is shown.

The vertical distribution of leaf nitrogen concentration values per unit leaf area shows a narrowly defined range in both the late growing and the flowering stages of dense stands with highest concentrations in the uppermost leaves and gradually diminishing values lower down in the canopies. In the open stands the pattern of vertical leaf nitrogen distribution varies over a broad range, though values are invariably lower deeper in the canopies (Fig. 4).

Effects of light climate and nitrogen partitioning

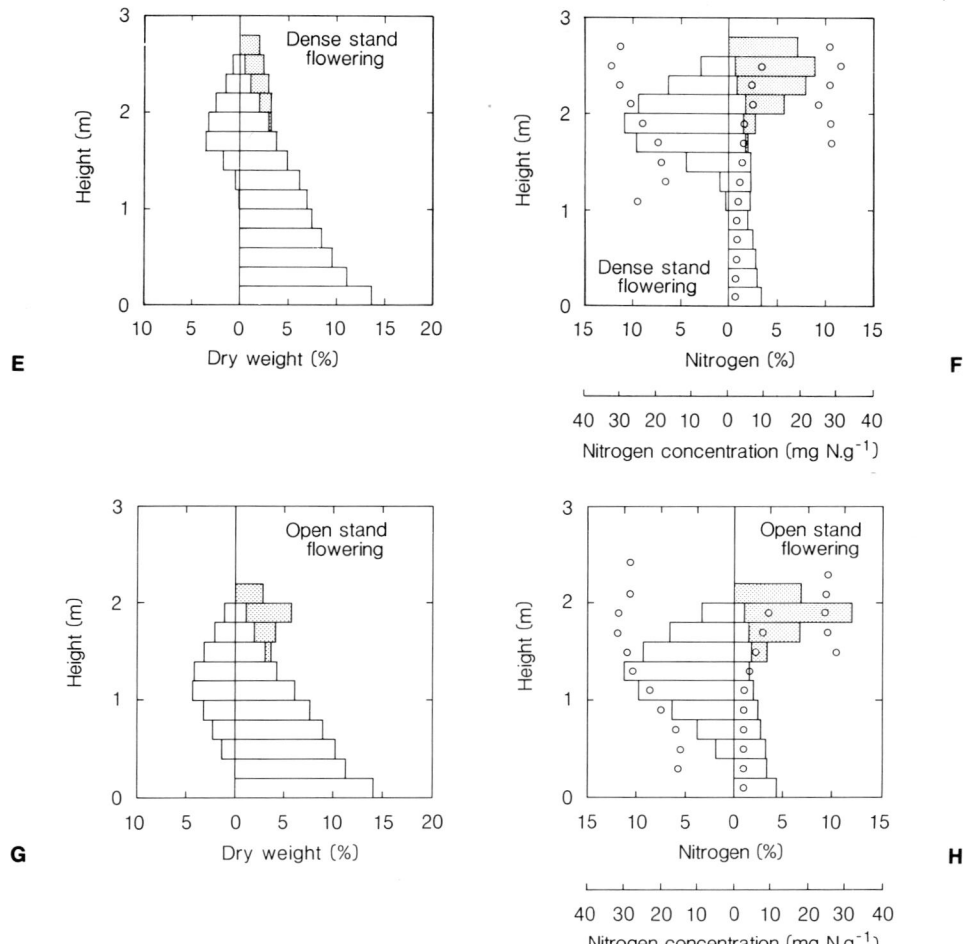

Fig. 2.

Discussion

The light climates in the canopies of the dense and open stands differed strongly from one another. This obviously was to be expected. The spatial distribution of the leaves is strongly determined by the light climate within the stand. In dense stands leaves are concentrated higher in the canopy, while in open stands they are evenly spread along most of the length of the stems.

As expected, the light climates also differed in the canopies of the dense stands in their different phenological stages with somewhat more light penetrating deep into the canopies of the flowering stands. Leaves were even more narrowly concentrated in the top parts of the canopies of the dense stands during flowering. However, the relationship between the proportion of total PFD available at any point inside the canopies and the cumulative LAI above that point was the same at the two phenological stages (Fig. 1).

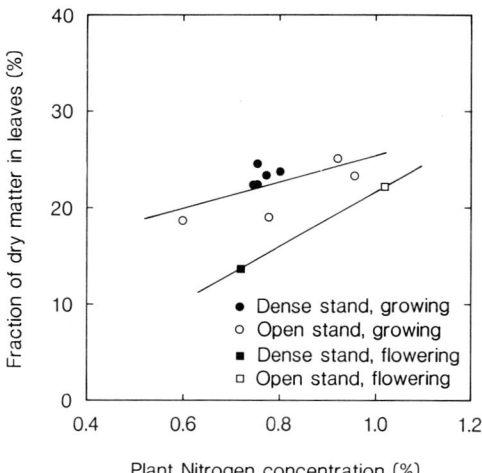

Fig. 3. Fraction of total above-ground dry matter invested in leaves as a function of the nitrogen concentration in the total above-ground plant in stands of *Solidago altissima*.

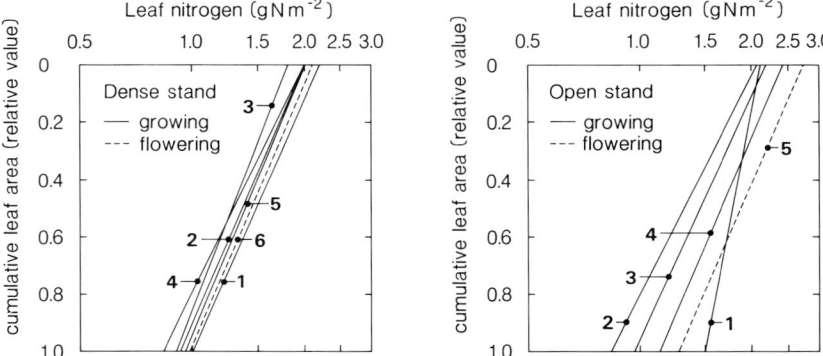

Fig. 4. Leaf nitrogen per unit leaf area as a function of the relative value of the leaf area index cumulated from the top of the canopy in dense and open stands of *Solidago altissima*. Figures identify lines for different plots. Each line is based on 8 to 12 observation points depending on the lengths of the leaf-bearing stretches of the stems.

Dense stands: line 1 r^2 = 0.937, line 2 r^2 = 0.961, line 3 r^2 = 0.939, line 4 r^2 = 0.949, line 5 r^2 = 0.943, line 6 r^2 = 0.927.

Open stands: line 1 r^2 = 0.824, line 2 r^2 = 0.966, line 3 r^2 = 0.989, line 4 r^2 = 0.961, line 5 r^2 = 0.988.

The effect of nitrogen availability to the plant on the partitioning of matter between roots and shoots is well known (Brouwer 1962, 1983; Schulze 1982; Werger 1983; Hirose this volume). Our data show that in the late growing stage on average dry matter partitioning between leaves and stems is equal in the dense and open stands, and the relative investment of plant nitrogen in these plant fractions is also comparable. However, once a substantial nitrogen sink has developed in the form of flowers and initial fruits, the dry matter partitioning between leaves and stems dramatically changes in the dense stands (Fig. 2, Table 1). The plants decrease their investment of dry matter in leaves by dropping them, diminishing total leaf weight and total leaf area, but they keep the nitrogen concentrations in the remaining leaves at approximately the same level (Fig. 2). This suggests that the plants keep the nitrogen concentrations of their leaves above a certain minimum level. The proportional investment of dry matter in leaves is controlled by the nitrogen concentration of the total above-ground plant (Fig. 3). In the flowering stage in the dense stands this ratio lies below that in the late growing stage. However, in flowering plants with a rather high concentration of nitrogen, like in our open stand, the fraction of dry matter invested in leaves may be as high as in plants of the late growing stage (Fig. 3, Table 1).

Hirose and Werger (1987a) showed for *Solidago altissima* that with increasing PFD the optimum values of nitrogen content per unit leaf area with respect to photosynthetic returns per unit leaf nitrogen also increased. This implies that it is beneficial for a plant to have a non-uniform distribution of nitrogen in the canopy, particularly in a dense stand, and to allocate more nitrogen per unit leaf area to leaves higher in the canopy. Such a pattern in vertical leaf nitrogen distribution as a result of redistribution of nitrogen by the plant is indeed found (Figs 2 and 4). This pattern is much more narrowly defined in the dense stands than in the open stands, both in the late growing and flowering stages, and inside the canopies of the open stands leaf nitrogen concentration may vary considerably. Based on a model study Hirose and Werger (1987b) predicted a larger relative advantage in terms of photosynthetic returns of the whole canopy with redistribution of nitrogen in canopies of dense stands than in those of open stands. Thus a more narrowly defined distribution pattern in dense canopies was expected. The lack of a difference between the distribution patterns in the late growing stage and the flowering stage of the dense stands suggests that the leaf nitrogen concentrations at various levels in the profile are kept near optimum levels in terms of nitrogen use efficiency (Hirose and Werger 1987a).

LAR is defined as LWR/SLW, where LWR is the leaf weight ratio, the fraction of plant dry matter invested in leaves (here with reference to both LAR and LWR plant dry weights refer to the above-ground plant parts only). With similar LWR values for both phenological stages of the open stands, the lower LAR value for the flowering stage must be reflected in higher SLW values there (Table 1). Similarly, lower LWR values for the flowering stage as compared to the late growing stage of the dense stands with similar SLW values must result in a lower LAR value. The pattern of diminishing SLW values from top to bottom of the canopies is strong in dense stands, as was expected, and parallels the pattern of leaf nitrogen concentration. In the open stands such a pattern is absent and during the late growing stage SLW values are even higher low in the canopy than at the top (Table 1). As expected for these well-illuminated leaves (Hirose *et al.* 1988), their nitrogen concentrations are lower than those in the top of the canopy (Fig. 2). However, there is no clear difference in leaf nitrogen concentrations between the late growing and flowering stages of the open stands that would explain the higher SLW values at the flowering stage. This suggests that apart from light climate and nitrogen availability to each leaf there might be another factor involved in the control of SLW within the canopy.

Conclusions

1. The light climates inside the canopies of open and dense stands of the tall herb *Solidago altissima* during the late growing stage differed strongly. In the open stands 53% of total PFD reached the lowest leaves; in the dense stands this was 7–10% of total PFD.
2. Light climate regulates the spatial distribution of the leaves in the canopy. In the open stands leaves are spread over a longer stretch of the stems than in the dense stands, where they are concentrated towards the top of the canopies.
3. Light climates inside dense canopies differed between late growing and flowering stages with more light penetrating inside the canopy to the lowest leaves during the flowering stage. However, the relation between the proportion of total PFD available at any point inside the canopy and the accumulative LAI above that point was equal in both phenological stages.
4. The fraction of above-ground plant matter invested in leaves is determined by the nitrogen concentration of the total above-ground plant. This fraction is often smaller during the flowering stage than during the late growing stage, because flowering represents an important nitrogen sink for the plant.
5. The vertical pattern of nitrogen allocation to the leaves is more narrowly defined in dense stands than in open stands.
6. The vertical pattern of SLW values is strong in dense stands and parallels the pattern of leaf nitrogen concentration. In open stands no such pattern of SLW values exists or it is even reversed. The SLW values in open stands do not entirely correlate, as was expected, with leaf nitrogen concentrations.

Acknowledgment

MJAW gratefully acknowledges the support of the Japanese Society for the Promotion of Science and the hospitality and facilities provided by T. Saeki and his staff during the field work. TH acknowledges partly support by Grant-in-Aid No. 60480004 from the Ministry of Education, Science and Culture, Japan, and Grant No. B 84-270 of the Netherlands Organization for Scientific Research (NWO). J.T. Lambers and H.J. During made useful comments on the manuscript.

References

Björkman, O. 1981. Responses to different quantum flux densities. In: O.L. Lange et al. (eds), Encyclopaedia of plant physiology, N.S., Vol. 12 A: 57–107. Springer, Berlin.
Brouwer, R. 1962. Nutritive influences on the distribution of dry matter in the plant. Neth. J. Agric. Sci. 10: 399–408.
Brouwer, R. 1983. Functional equilibrium: sense or nonsense? Neth. J. Agric. Sci. 31: 335–348.
Chabot, B.F., Jurik, T.W. and Chabot, J.F. 1979. Influence of instantaneous and integrated light flux density on leaf anatomy and photosynthesis. Amer. J. Bot.66: 940–945.
Chapin, F.S. 1980. The mineral nutrition of wild plants. An. Rev. Ecol. Syst. 11: 233–260.
DeJong, T.M. and Doyle, J.F. 1985. Seasonal relationships between leaf nitrogen content (photosynthetic capacity) and leaf canopy light exposure in peach (*Prunus persica*). Plant Cell Environ. 8: 701–706.

Field, C. 1983. Allocating leaf nitrogen for the maximization of carbon gain: leaf age as a control on the allocation program. Oecologia (Berl.) 56: 341–347.

Field, C. and Mooney, H.A. 1986. The photosynthesis-nitrogen relationship in wild plants. In: T.J. Givnish (ed.), On the economy of plant form and function, pp. 25–55. Cambridge U.P., London.

Hirose, T. 1971. Nitrogen turnover and dry-matter production of a *Solidago altissima* population. Jap. J. Ecol. 21: 18–32.

Hirose, T. 1986. Nitrogen uptake and plant growth. II. An empirical model of vegetative growth and partitioning. Ann. Bot. 58: 487–496.

Hirose, T. 1988. Nitrogen availability, optimal shoot/root ratios and plant growth. pp. 135 ff. this volume.

Hirose, T. and Werger, M.J.A. 1987a. Nitrogen use efficiency in instantaneous and daily photosynthesis of leaves in the canopy of a *Solidago altissima* stand. Physiol. Plantarum 70: 215–222.

Hirose, T. and Werger, M.J.A. 1987b. Maximizing daily canopy photosynthesis with respect to the leaf nitrogen allocation pattern in the canopy. Oecologia (Berl.) 72: 520–526.

Hirose, T., Werger, M.J.A., Pons, T.L. and Van Rheenen, J.W.A. 1988. Canopy structure and leaf nitrogen distribution in a stand of *Lysimachia vulgaris* L. as influenced by stand density. Oecologia (Berl.) 77: (in press).

Johnson, I.R. and Thornley, J.H.M. 1984. A model of instantaneous and daily canopy photosynthesis. J. Theor. Biol. 107: 531–545.

Jurik, T.W. 1986. Temporal and spatial patterns of specific leaf weight in successional northern hardwood tree species. Amer. J. Bot. 73: 1083–1092.

Jurik, T.W. and Chabot, B.F. 1986. Leaf dynamics and profitability in wild strawberries. Oecologia (Berl.) 69: 296–304.

Jurik, T.W., Chabot, J.F. and Chabot, B.F. 1979. Ontogeny of photosynthetic performance in *Fragaria virginiana* under changing light regimes. Plant Physiol. 63: 542–547.

Monsi, M. and Saeki, T. 1953. Über den Lichtfaktor in den Pflanzengesellschaften und seine Bedeutung für die Stoffproduktion. Jap. J. Bot. 14: 22–52.

Natr, L. 1975. Influence of mineral nutrition on photosynthesis and the use of assimilates. In: J.P. Cooper (ed.), Photosynthesis and productivity in different environments, pp. 537–555. Cambridge U.P., London.

Nobel, P.S., Zaragoza, L.J. and Smith, W.K. 1975. Relation between mesophyll surface area, photosynthetic rate and illumination level during development for leaves of *Plectranthus parviflorus* Henckel. Plant Physiol. 55: 1067–1070.

Saeki, T. 1960. Interrelationships between leaf amount, light distribution and total photosynthesis in a plant community. Bot. Mag. Tokyo 73: 55–63.

Schulze, E.D. 1982. Plant life forms and their carbon, water and nutrient relations. In: O.L. Lange et al. (eds), Encyclopaedia of plant physiology, N.S., Vol. 12 B: 615–676. Springer, Berlin.

Seemann, J.R., Sharkey, T.D., Wang, J. and Osmond, C.B. 1987. Environmental effects on photosynthesis, nitrogen-use efficiency, and metabolic pools in leaves of sun and shade plants. Plant Physiol. 84: 796–802.

Werger, M.J.A. 1983. Functional equilibrium between shoots and roots. Introduction. Neth. J. Agric. Sci. 31: 287–289.

VERTICAL DISTRIBUTION OF PHOTOSYNTHETIC AND NON-PHOTOSYNTHETIC PHYTOMASS IN *ULEX EUROPAEUS*

I. OJEA, J. PEREIRAS and M. BASANTA
Dpto. Ecología, Fac. Biología, Univ. Santiago de Compostela, Spain

Abstract

A detailed study of vertical structure at the species level is necessary prior to any attempt to evaluate existing relationships between structure and environmental factors. In this work we have described the vertical distribution of the phytomass of different structures of *Ulex europaeus* L., the dominant species in many of the shrubland communities of Galicia (NW Spain).

Ten a priori height classes were established and 54 individuals were collected and classified according to these classes. Phytomass data were obtained by 6 cm strata for photosynthetic, woody and dead structures.

Analysis of the results permitted a reclassification into three types of vertical distribution: (1) low plants (0–30 cm in height), with green structures (66 gr/m² on average) concentrated in the lowest strata; (2) medium plants (31–90 cm in height) with green structures (293 gr/m² on average) concentrated in the middle strata and woody structures (301 gr/m² on average) concentrated in the lowest strata; and (3) tall plants (> 90 cm in height) with woody structures (1769 gr/m² on average) concentrated in the lowest strata.

The ratio P_p/P_T (photosynthetic phytomass/total phytomass) decreases as plants grow, due to an increase in woody and dead phytomass and a slight decrease in green phytomass in the tallest individuals. The relationship between green, woody or dead phytomass and total phytomass was analyzed using allometric equations. A good fit was obtained for green and woody structures, indicating their potential usefulness as predictive equations in these cases.

Introduction

Vegetation architecture and environmental factors are closely related, and the detailed study of vertical structure at the species level is one way to detect such relationships. In this paper we present some of the first results obtained for the characterization of the vertical structure of *Ulex europaeus* L. These results are part of a wider study that includes other species of shrubland communities in Galicia (NW Spain). We have selected this species because it is the most representative one of these communities, which are composed mainly of Papilionaceae and Ericaceae and grow in acid, low-nutrient soils. These features coincide with those of similar shrub communities found in other European Atlantic regions.

The aims of the present article are the study of: the vertical distribution of phytomass as regards living and dead tissue, the change in the ratio between photosynthetically active and inactive portions in relation to height of the plants, and the development of allometric equations for living and dead tissue.

Materials and methods

The study was carried out on a site whose physical characteristics are typical of

Table 1. Total phytomass and phytomass of green, woody and dead structures (mean and standard deviation, in g.m^{-2}), for each height class.

Height class	Green		Woody		Dead		Total	
	X	S.D.	X	S.D.	X	S.D.	X	S.D.
I (0–15 cm)	48.4	21.0	9.7	10.9	22.9	23.9	81.0	48.0
II (16–30 cm)	84.1	27.6	31.3	19.4	33.9	18.5	149.4	52.0
III (31–45 cm)	224.6	71.8	105.3	28.2	69.9	54.6	399.8	59.2
IV (46–60 cm)	220.3	61.6	185.7	89.5	83.1	82.7	489.1	188.4
V (61–75 cm)	369.0	144.6	366.9	182.3	186.9	181.0	922.7	311.5
VI (76–90 cm)	359.0	153.8	548.1	198.6	311.9	220.2	1218.9	299.6
VII (91–105 cm)	409.6	98.3	1149.1	409.4	761.5	332.7	2320.1	667.3
VIII (106–120 cm)	254.3	44.3	1318.4	201.6	977.1	344.6	2549.8	512.0
IX (121–135 cm)	473.2	234.0	2328.7	631.1	1156.3	336.3	3958.2	991.3
X (136–150 cm)	364.7	120.5	2693.7	1014.0	1059.1	516.5	4104.2	1624.6

Table 2. Total phytomass and phytomass of green, woody and dead structures (mean and standard deviation, in g.m^{-2}), in plants of 0–30 (Group 1), 31–90 (Group 2) and 91–150 (Group 3) cm in height.

	Green		Woody		Dead		Total	
	$\bar{X}$	S.D.	$\bar{X}$	S.D.	$\bar{X}$	S.D.	$\bar{X}$	S.D.
Group 1	66.3	30.1	20.5	18.9	28.4	21.5	115.2	60.0
Group 2	293.2	128.7	301.5	219.0	162.9	170.6	757.6	406.4
Group 3	380.3	141.3	1769.1	861.1	956.1	370.3	3105.5	1200.1

the areas preferred by *U. europaeus*: a hill-side with a 10° slope and an acid, sandy soil on granite bedrock.

Fifty-four individuals were selected and classified by height in ten classes defined a priori by 15 cm intervals between 0 and 150 cm. Phytomass data were obtained by cutting each individual in 6 cm strata, separating green (spines and photosynthetic stems), woody (non-photosynthetic stems) and dead (unshed necromass) structures, obtaining dry weight after desiccation in a forced-air oven at 120°C until no further reduction in weight was observed.

Results and discussion

Vertical distribution of phytomass

Mean values obtained for green, woody and dead phytomass (g.m^{-2}) in the ten height classes are presented in Table 1.

The evaluation of values of phytomass (total and fractions) and their distribution over the strata permitted a reclassification of individuals into three groups (Table 2), as follows:

1. Low plants (0–30 cm in height), with green structures (66 g.m^{-2} on average)

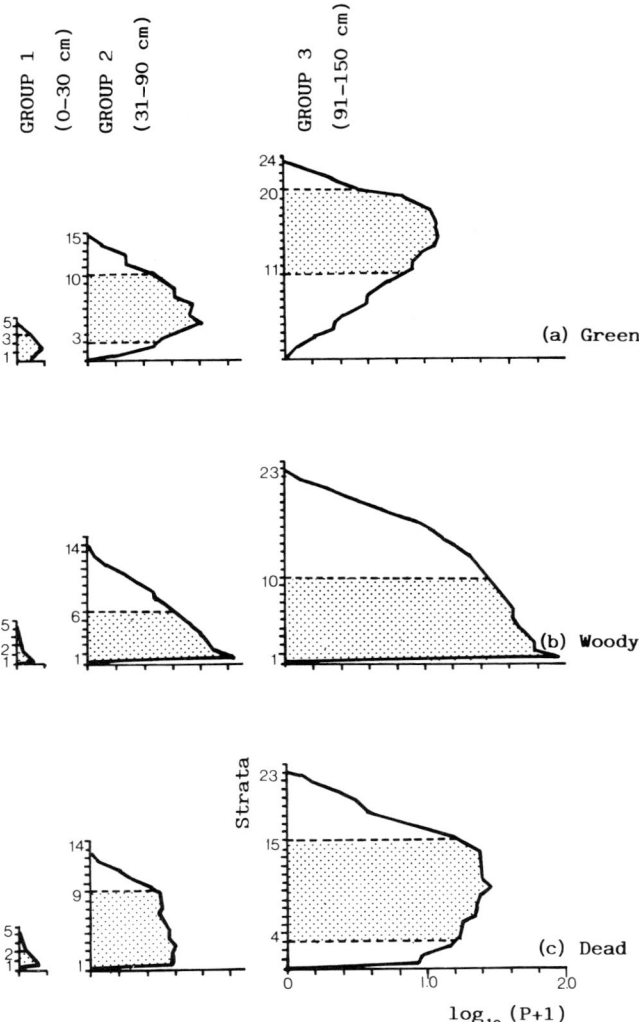

Fig. 1. Vertical distribution of phytomass of (a) green, (b) woody and (c) dead structures in plants of 0–30, 31–90 and 91–150 cm in height.

concentrated in the lowest strata (83%); they account for 58% of the total phytomass.

2. Medium plants (31–90 cm in height), with green structures (293 g.m^{-2} on average) concentrated in the middle strata (82%) and woody structures (302 g.m^{-2} on average) concentrated in the lowest strata (80%); photosynthetic phytomass accounts for 39% and woody phytomass accounts for 40% of the total phytomass.

3. Tall plants (91–150 cm in height) with woody structures (1769 g.m^{-2} on average) concentrated in the lowest strata (79%); they account for 57% of the total phytomass.

Figure 1 shows that woody phytomass always presents a peak near ground level, independent of the size group, whereas the peak of green phytomass shifts to higher strata as the plant grows. Dead phytomass, which is initially concentrated in the lower strata, becomes evenly distributed, accounting for 22% to 31% of total weight; it never becomes the dominant fraction.

The above results coincide with those of Debussche (1978), Debussche et al. (1980), Gray (1982) and Diaz Barradas and Garcia Novo (1987), who also report that the green tissue peak rises as the plant grows, whereas woody phytomass always tends to accumulate in the lowest strata. Similarly, Giliberto et al. (1977), Debussche (1978) and Gray and Schlesinger (1981) point out that support organs displace photosynthetic organs in older or higher plants of various shrub species. Finally, the tendency for the green phytomass to accumulate more slowly in tall plants, observed in our data, is also reported by Schlesinger and Gill (1978, 1980) and Gray (1982) for other shrub species.

Ratios P_p/P_t and P_p/P_l

The ratios between photosynthetic phytomass and total phytomass (P_p/P_t) and between photosynthetic phytomass and live phytomass (P_p/P_l) were calculated for each individual and the data fitted to linear regression equations (Fig. 2). The F values are high, indicating a good fit in both cases.

Several authors (Debussche 1978; Merino and Martin 1981; Basanta Alves 1982; Diaz Barradas and Garcia Novo 1987) have employed these or similar ratios to evaluate how the plant allocates photosynthates in each stage of development: to increase production or to create and maintain support structures. In general, the phytomass corresponding to non-photosynthetic structures (wood, roots) increases as the plant ages or grows (Giliberto et al. 1977; Debussche 1978; Gray and Schlesinger 1981), and the same trend can be observed in our data.

The ratio P_p/P_t decreases sharply from the lowest to the tallest individuals, in accordance with an increase in woody and dead phytomass and a slight decrease in green phytomass in the tallest individuals. The ratio P_p/P_l also decreases, but dispersion values are smaller than in the first case due to the variability of the data for dead structures. The amount of dead phytomass is strongly influenced by microclimatic conditions. Also, P_p/P_t is smaller than P_p/P_l for any individual, although the difference between both values decreases as height increases, because dead phytomass is not as important as woody phytomass in the tallest individuals.

Thus, the morphology of the tallest individuals is characterized by a small percentage of peripherally located photosynthetic material, a large percentage of wood and a variable amount of dead material attached to the living stems in the central part of the shrub. This distribution model is the same as reported by other authors (Debussche et al. 1980; Houssard et al. 1980; Gray 1982; Diaz Barradas and Garcia Novo 1987) for various shrub species.

Allometric equations

The relationship between the different phytomass fractions and total phytomass was analyzed by means of nonlinear regression methods. Linear and nonlinear

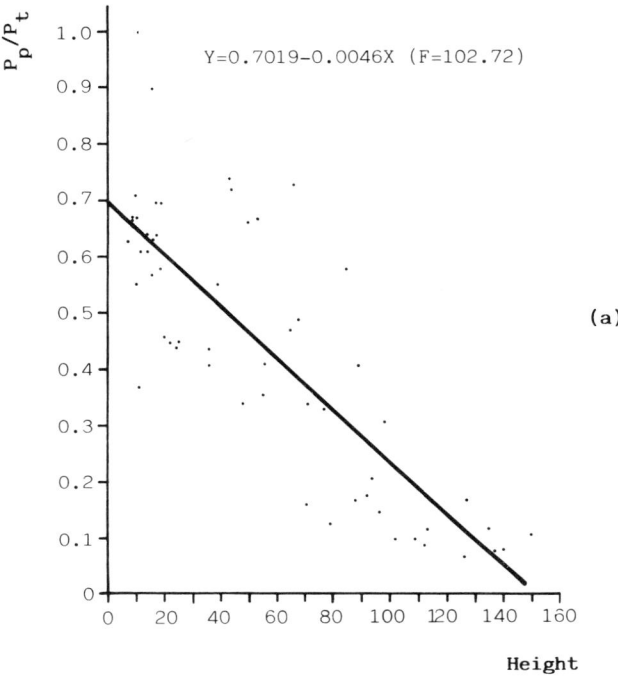

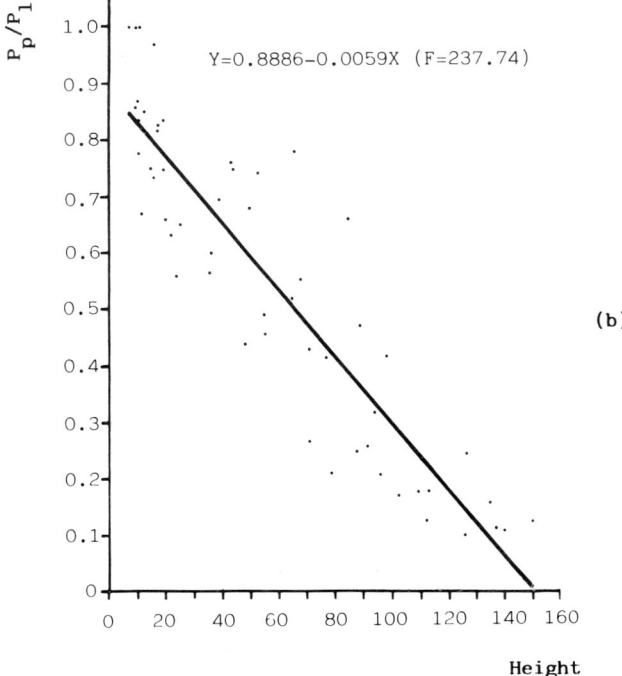

Fig. 2. Regression equations for (a) P_p/P_t values and (b) P_p/P_l values with height (cm).

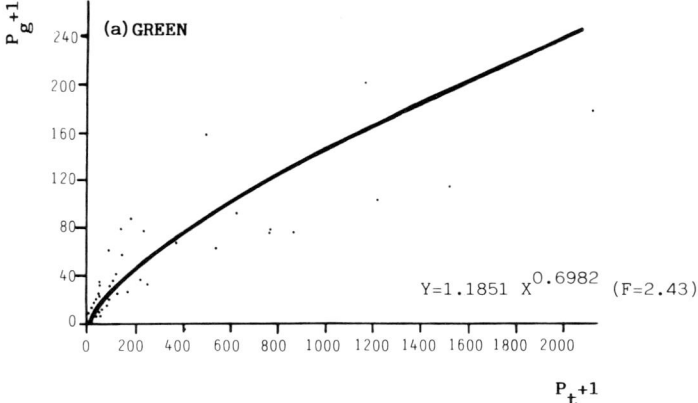

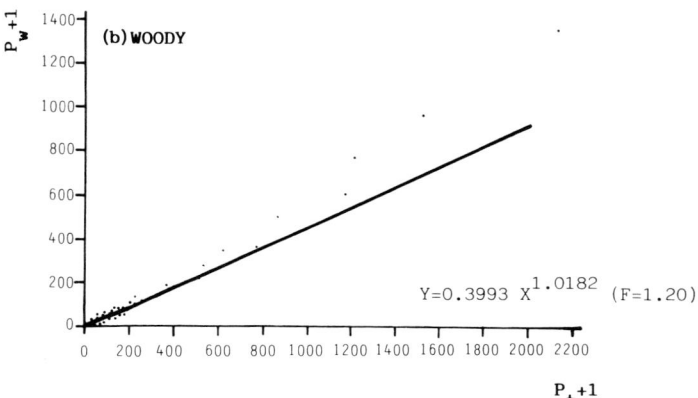

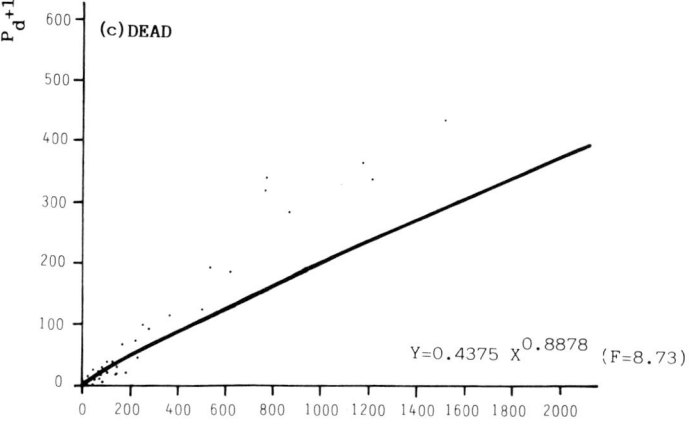

Fig. 3. Allometric relationships between total phytomass and (a) green phytomass, (b) woody phytomass and (c) dead phytomass.

regression techniques have been widely used to predict phytomass values from different kinds of data. Gray and Schlesinger (1981), Gray (1982), Brand and Smith (1985) and Campbell et al. (1985) relate phytomass to diameter for trees or shrubs; Schlesinger and Gill (1980) and Strauss and Ledig (1985) relate total phytomass to phytomass fractions, and Pyke and Zamora (1982) relate phytomass to cover.

We have used the allometric equation $Y = aX^b$, where a and b are estimated parameters, X is total phytomass and Y is green, woody or dead phytomass. The resulting equations were:

	Equation	r^2	F
Green	$Y = 1.1851\ X^{0.6982}$	0.95	2.43
Woody	$Y = 0.3993\ X^{1.0182}$	0.99	1.20
Dead	$Y = 0.4375\ X^{0.8878}$	0.92	8.73*

(* $p = 0.05$)

The curves and equations are presented in Fig. 3. The relationships between total phytomass and green or woody fractions were highly significant ($r^2 = 0.95$ and 0.99 respectively). The coefficient of determination (r^2) is also high for the dead fraction, but the F-test demonstrates that a good fit is achieved only in the first two cases. This can be interpreted as a reflection of a greater dependence of live structure on internal regulation processes.

The good fit obtained for green and woody structures indicates the potential usefulness of these equations to predict green or woody phytomass from total phytomass values. However, more extensive data, especially those corresponding to high total phytomass values, would permit the determination of new equations and possibly a better fit, as many of the points used for the present calculations come from plants with low total phytomass values.

Acknowledgments

We are indebted to J. Fariña for his valuable help in the statistical treatment of the data.

References

Basanta Alves, A. 1982. Vegetación seral en Sierra Morena. Estudio ecológico de las respuestas del matorral a distintas intervenciones humanas en el Coto Nacional 'La Pata del Caballo' (Huelva). Ph.D. thesis, University of Sevilla.
Brand, G.J. and Smith, W.B. 1985. Evaluating allometric shrub biomass equations fit to generated data. Can. J. Bot. 63: 64–67.
Campbell, J.S., Lieffers, V.J. and Pielou, E.C. 1985. Regression equations for estimating single tree biomass of trembling aspen: assessing their applicability to more than one population. For. Ecol. Man. 11: 283–295.
Debussche, M. 1978. Etude de la dynamique de la végétation sur le versant nord-ouest du Mont Aigoual. Ph.D. thesis, University of Languedoc.
Debussche, M., Escarré, J. and Lepart, J. 1980. Changes in mediterranean shrub communities with *Cytisus purgans* and *Genista scorpius*. Vegetatio 43: 73–82.

Díaz Barradas, M.C. and García Novo, F. 1987. Vertical structure of *Halimium halimifolium* shrubs in Doñana National Park (SW Spain). In: J.D. Tenhunen, F.M. Catarino, O.L. Lange and W. Oechel (eds), Plant Response to Stress. Functional Analysis in Mediterranean Ecosystems, pp. 531–545. Springer-Verlag, Berlin.

Giliberto, J., Mooney, H.A. and Kummerow, J. 1977. Shrub structure analysis. In: N.J.W. Thrower and D.E. Bradbury (eds), Chile-California Mediterranean Scrub Atlas, pp. 144–147. Dowden, Hutchinson & Ross, Stroudsburg.

Gray, J.T. 1982. Community structure and productivity in *Ceanothus* chaparral and coastal sage scrub of southern California. Ecol. Monogr. 52: 415–435.

Gray, J.T. and Schlesinger, W.H. 1981. Biomass, production, and litterfall in the coastal sage scrub of southern California. Amer. J. Bot. 68: 24–33.

Houssard, C., Escarré, J. and Romane, F. 1980. Development of species diversity in some mediterranean plant communities. Vegetatio 43: 59–72.

Merino, J. and Martín, A. 1981. Biomass, productivity and succession in the scrub of the Doñana Biological Reserve in southwest Spain. In: N.S. Margaris and H.A. Mooney (eds), Components of Productivity of Mediterranean-climate Regions. Basic and Applied Aspects, pp. 197–203. Junk, The Hague.

Pyke, D.A. and Zamora, B.A. 1982. Relationship between overstory structure and understory production on the Grand Fir/Myrtle Boxwood habitat type of Northcentral Idaho. J.R. Manag. 35: 769–773.

Schlesinger, W.H. and Gill, D.S. 1978. Demographic studies of the chaparral shrub, *Ceanothus megacarpus*, in the Santa Ynez Mountains, California. Ecology 59: 1256–1263.

Schlesinger, W.H. and Gill, D.S. 1980. Biomass, production and changes in the availability of light, water and nutrients during the development of pure stands of the chaparral shrub, *Ceanothus megacarpus*, after fire. Ecology 61: 781–789.

Strauss, S.H. and Ledig, F.T. 1985. Seedling architecture and life history evolution on pines. Am. Nat. 125: 702–715.

IS EXPOSURE-RELATED VARIATION IN LEAF CHARACTERISTICS OF TROPICAL RAIN FOREST SPECIES ADAPTIVE?

F. BONGERS and J. POPMA
Department of Plant Ecology, University of Utrecht, Lange Nieuwstraat 106, 3512 PN Utrecht, The Netherlands; Laboratorio de Ecología, Facultad de Ciencias, Universidad Nacional Antónoma de México, 04510 México D.F., México.

Abstract

Tropical rain forest presents contrasting light environments. A considerable variation in leaf characteristics with respect to this environmental variation is often found. In order to identify leaf traits which probably have adaptive value in shaded or in exposed habitats we investigate a series of different leaf traits. Twenty leaf traits (regarding morphology, nutrient levels and anatomy) of sun-grown and shade-grown leaves of 61 woody tropical rain forest species were studied at Los Tuxtlas, Mexico. We determined (1) within-species shade value/sun value ratios, and (2) differences in among-species variance between sun values and shade values. Within species, sun-grown leaves have a smaller area, a higher specific leaf weight, a higher specific leaf water content, higher nutrient concentrations (N, P, K) on an area basis but lower ones on a weight basis, and thicker laminae (palisade and spongy mesophyll), than shade-grown leaves. Between species, sun-grown leaves had a lower variance in leaf area, in [K]/weight and in the ratio spongy/palisade mesophyll thickness, but a higher variance in SLW, [N]/area and stomatal density, than shade-grown leaves. The leaf traits which most probably show adaptive responses (leaf area, specific leaf weight, [K]/weight and [N]/area) were checked for correlations, in order to evaluate whether these traits vary independently or not. Probable adaptations to low or to high light availability are discussed in terms of the leaf's functioning.

Introduction

Leaves are the energy capturing organs of a plant. Simultaneously they are an energy sink, because carbohydrates are needed for their construction, maintenance and functioning. Leaf characteristics may be phylogenetically determined. However, many different leaf traits may influence the energetic cost/benefit balance of a given leaf in a given environment, and thus may be subject to selective pressures for optimization of the leaf's cost/benefit balance in relation to both availability of resources and environmental constraints (Givnish 1987). Consequently selective pressures can result in an optimization of certain traits which may at the same time put constraints on the optimization of other traits. As a result characters that represent a 'trade-off' may be selected for. Presumably, each leaf type is a resultant of all traits under selective pressures and represents a functional solution of the cost/benefit balance.

A given leaf may be optimized for a specific set of environmental conditions. The environment itself, however, is in general highly variable, in space as well as in time. This may explain why large differences in some traits are found even between leaves from different parts of the crown of a single individual, and partly

Correspondence address: F. Bongers, Department of Plant Ecology, University of Utrecht, Lange Nieuwstraat 106, 3512 PN Utrecht, The Netherlands.

Ashton 1978; Fetcher *et al.* 1983; Roth 1984; Fisher 1986; Oberbauer and Strain 1986; Bongers *et al.* 1988b; Popma and Bongers 1988).

In tropical rain forest important differences on a local scale are those found between leaves from shaded and those from exposed habitats. Based on their appearance the leaves are called hygromorphic, respectively scleromorphic. This distinction is not clear-cut however, as both scleromorphism and hygromorphism each comprise a whole complex of leaf characters and are generally poorly defined (*cf.* Roth 1984). It seems worthwhile therefore to investigate more specifically the differences in a number of leaf traits in shaded and in exposed habitats, to estimate the probability that the state of, and the phenotypic plasticity in a given leaf trait provides an adaptation to the environmental conditions of shaded or exposed habitats. Also relationships between leaf traits and possible trade-offs can be studied. Only relative probabilities of adaptive values can be estimated, however, since phylogenetic constraints and selective pressures cannot be distinguished easily (*cf.* Givnish 1987). The tropical rain forest provides an excellent opportunity to investigate such probabilities by means of convergence arguments. It is very rich in species with highly different phylogenetic origins, but with similar growth forms (*i.e.*, trees), and ecologies. Furthermore, it provides strong differences in micro-environments related to differences in light exposure (Chazdon and Fetcher 1984; Chiariello 1984).

Several arguments, focusing on the probability of adaptation of a specific leaf trait to shady or sunny environments, may be used in an additive way:

1. If in one site the within-species difference in a leaf trait between sun-grown and shade-grown leaves is significant over a wide taxonomical range, this suggests that the variation in that trait is the result of evolutionary convergence. This in turn suggests that the character states of that trait reflect adaptations to variation in the environment.
2. Traits which show large within-species variation related to the differences between the shady and the sunny environments may be assumed to be more critical with respect to the functioning of a leaf in those environments than traits which show no, or only small variation.
3. If in one site the among-species variability in a given leaf trait is smaller in the shady (sunny) than in the sunny (shady) environment, it may be assumed that the state of this trait is more critical for the functioning of a leaf in the shady (sunny) environment. A larger variability in a specific environment suggests that the state of the trait is less subject to selective pressures.

Application of these three arguments could identify leaf traits which show relatively high probabilities of being adaptative to shady or sunny rain forest environments. A fourth argument may be used to identify possible trade-offs:

4. Certain leaf traits may be associated within a range of species, considering only sun leaves or only shade leaves. If such a trait changes in relation to a shift from one environment to the other, it may be expected that the associated traits will change also. In that case it is probable that this trait is of adaptive value in conjunction with its associated traits, and there is a fair probability that it changes as

a trade-off with other traits. The alternative is that a trait changes between environments while its associated traits do not.

In the present study we apply these arguments to data concerning the morphology, anatomy, and foliar mineral levels of 61 woody rain forest species, in order to identify traits which are probable adaptations to shady or sunny environments.

Methods

Data were collected in the tropical rain forest of Los Tuxtlas, Mexico (95°06′W, 18°35′N). Annual precipitation is 4639 mm, and mean annual temperature is 24.6°C (climate data of the Coyame station). Soils are of volcanic origin, rich in nutrients (N, P, K) and contain high amounts of organic matter. A description of the floristic composition, structure and dynamics of this forest is given by Bongers *et al.* 1988a, and by Popma *et al.* (1988).

The species sampled in this study may be considered representative for the Los Tuxtlas rain forest (Bongers and Popma 1988). Sun-leaves and shade-leaves of 61 woody species were sampled from a well-exposed and from a shaded position in the crown (height mostly between 5–10 m) of mature individuals. Every sample contained 21 mature leaves of one individual. Leaves with epiphylls or herbivore damage were avoided. One leaf of each sample was conserved in FAA (5% Formaldehyde, 48% Ethanol, 4% Acetic acid) for anatomical studies, 20 leaves were used for morphological measurements and subsequent nutrient analyses.

Leaf area was measured using a leaf area meter (LiCor, Lincoln, Nebraska) or by drawing the leaf circumference on paper and tracing it afterwards on a computer digitizer which calculated the area. Both methods did not differ significantly in their results. Length and width of the leaf blade were measured. After measurement of the leaf area the leaves were left to saturate with water overnight (24 hours between wet paper at 18°C), then dried superficially and weighed. Afterwards, leaves were oven-dried (24 hours, 105°C) and weighed again.

From these data the following parameters were derived: leaf shape index (leaf length/leaf width), specific leaf weight (SLW, leaf dry weight/leaf area, $g.m^{-2}$), and specific leaf water content (SLWC, (fully saturated leaf weight – leaf dry weight)/leaf area, $g.m^{-2}$).

All 20 leaves of a sample were grouped and ground for nutrient analyses. Total nitrogen, phosporus and potassium were determined using a modified Kjeldahl method. Na_2SO_4 was used to raise the boiling temperature of the digestion. [N] and [P] were determined colorimetrically in an auto-analyser (Skalar). [K] was determined by flame spectrophotometry. Every sample was measured twice and mean values were used for calculations, as replicates generally did not deviate more than 5% from the mean.

Stomatal prints (Bayer Xantopren Plus) were made of the lower epidermis of one leaf. Actual counts of stomata (five microscope fields of 0.0967 mm²) and measurements of stomatal length (five measurements per field) were made from positive nail-varnish impressions of the Xantopren negatives. A stomatal area index was calculated as the product of stomatal length and stomatal density. In four species no stomatal prints could be made due to pubescence or to the fact that stomata were located in crypts.

Of the leaves conserved in FAA, a part of the leaf between midrib and leaf margin was embedded in paraffin. Transversal leaf cuts were made with a microtome, and stained with safranine and fast green following standard techniques. Thicknesses of the lamina, the upper epidermis, the palisade mesophyll, the spongy mesophyll, and the lower epidermis were measured. 11 species were not included in the analyses of anatomy, because either sun- or shade-leaf sections were not available.

The mean shade-value/sun-value ratio for each leaf trait was tested for significant deviation from unity. This ratio ranges from 0 to 1 when the sun value is the larger one, and from 1 to infinity when the shade value is the larger one. This nonlinear response range was linearized using an arctangens transformation of the ratio before testing. The null hypothesis

$$H_o: SHSU(x) = 0$$

where $SHSU(x) = ARCTAN(shade(x)/sun(x)) - ARCTAN (1)$

was tested for all leaf characters (x) using a two-tailed T-test. The absolute deviation of $SHSU(x)$ from 0 is the same, whether the sun-value is a times as large as the shade-value or whether the shade-value is a times as large as the sun-value. It is important to note that the shade/sun ratio as used here is not the same as the quotient of two population means, but rather the mean of the shade/sun ratios of all species which could be sampled in both conditions.

Differences in variance in a trait between sun and shade leaves were tested using a two-tailed F-test (Sokal and Rohlf 1981).

Results

The values for several characteristics of sun and shade leaves are summarized in Table 1. Shade leaves are larger and have higher [N], [P], and [K] on a weight basis than sun leaves. Shade leaves have lower SLW, SLWC, [N] and [P] on an area basis, and are thinner (less palisade as well as less spongy mesophyll) than sun leaves. Leaf shape, [K] on an area basis basis, [N]/[P] ratio, stomatal density, stomatal size, stomatal area index, thickness of upper and lower epidermis, and ratio of spongy/palisade mesophyll does not vary with the different environments.

The leaf traits which showed a shade-sun ratio significantly deviating from unity were ordered in terms of this deviation. Those that increased from sun to shade were [K]/weight (with a deviation of 54%), leaf area (50%), [P]/weight (15%), [N]/weight (10%). The traits decreasing from sun to shade were SLW (-30%), [N]/area (-24%), thickness of palisade mesophyll (-22%), [P]/area (-20%), thickness of lamina (-18%), thickness of spongy mesophyll (-14%), and SLWC (-14%).

Only a few leaf traits show significant differences in variance between sun-grown and shade-grown leaves (Table 1). The variance in leaf area, [K]/weight, and the ratio spongy/palisade mesophyll is higher within shade leaves, while the vari-

Table 1. Characteristics of sun leaves and shade leaves, and differences between sun and shade leaves of trees and shrubs from the lowland tropical rain forest of Los Tuxtlas, Mexico.

		Sun-leaves		Shade-leaves		Shade/sun ratio			Variance test
	N	Mean	SE	Mean	SE	Mean	SE	p[1]	[2]
Leaf area (10^{-1} m^2)	61	.60 ±	.08	.81 ±	.11	1.50 ±	.082	*	sun < shade
Shape index ($m.m^{-1}$)	61	3.18 ±	.42	3.00 ±	.33	0.98 ±	.022	.	.
SLW ($g.m^{-2}$)	61	81.41 ±	3.51	55.34 ±	2.55	0.70 ±	.025	*	sun > shade
SLWC ($g.m^{-2}$)	61	171.24 ±	6.41	146.81 ±	5.69	0.86 ±	.014	*	.
N/weight ($mg.g^{-1}$)	61	16.63 ±	.58	17.88 ±	.61	1.10 ±	.035	*	.
P/weight ($mg.g^{-1}$)	61	1.28 ±	.06	1.43 ±	.08	1.15 ±	.044	*	.
K/weight ($mg.g^{-1}$)	61	10.38 ±	.53	15.10 ±	.75	1.54 ±	.078	*	sun < shade
N/area ($g.m^{-2}$)	61	1.30 ±	.06	.95 ±	.04	0.76 ±	.029	*	sun > shade
P/area ($g.m^{-2}$)	61	.100 ±	.005	.075 ±	.0004	0.80 ±	.035	*	.
K/area ($g.m^{-2}$)	61	.79 ±	.04	.80 ±	.05	1.03 ±	.049	.	.
N/P ($g.g^{-1}$)	61	14.06 ±	.57	13.67 ±	.50	1.03 ±	.051	.	.
Stomatal density (mm^{-2})	57	373 ±	58	318 ±	39	1.12 ±	.117	.	sun > shade
Stomatal length (μm)	57	21.52 ±	1.09	21.02 ±	.90	1.08 ±	.066	.	.
Stomatal index (mm^{-1})	57	6.13 ±	.47	5.87 ±	.54	1.24 ±	.180	.	.
Thickness upp. epid. (μm)	50	31.16 ±	2.15	26.80 ±	1.76	0.94 ±	.049	.	.
Thickness palisade mes. (μm)	50	71.21 ±	4.01	52.20 ±	3.33	0.78 ±	.040	*	.
Thickness spongy mes. (μm)	50	97.00 ±	6.00	78.00 ±	5.65	0.86 ±	.057	*	.
Thickness low. epid. (μm)	50	14.44 ±	.75	14.76 ±	.64	1.09 ±	.056	.	.
Leaf thickness (μm)	50	213.8 ±	9.35	171.76 ±	8.40	0.82 ±	.027	*	.
Ratio spongy/palisade ($m.m^{-1}$)	50	1.48 ±	.09	1.65 ±	.13	1.28 ±	.128	.	sun < shade

[1] Test of mean within-species shade value/sun value ratios for deviation from unity. T-test for Ho: SHSU(x) = 0, * = $p < 0.0001$, further explanation see text.
[2] Test for differences in among-species variance between sun values and shade values. Two-tailed F-test, $p < 0.05$; . = not significant; sun < shade = sun-variance smaller than shade-variance, sun > shade = sun-variance larger than shade-variance.

ance in SLW, [N]/area, and stomatal density is larger in sun leaves.

Discussion

Differences between sun and shade leaves

Differences between sun and shade leaves are often mentioned. In general, sun leaves have a smaller surface area (cf. Napp-Zinn 1984; Roth 1984; Chiariello 1984; Fisher 1986; Popma and Bongers 1988), a higher specific leaf weight (Blackman and Wilson 1951; Evans 1972; Björkman 1981; Fetcher et al. 1983; Fisher 1986; Oberbauer and Strain 1986; Popma and Bongers 1988), higher nutrient concentrations on an area basis (Björkman 1981; Field and Mooney 1986) and more and smaller stomata per unit area (cf. Boardman 1977; Fetcher et al. 1983; Napp-Zinn 1984; Oberbauer and Strain 1986; Bongers et al. 1988b). Lamina thickness is larger in sun leaves (Wylie 1951; Jackson 1967; McClendon and McMillen 1982; Napp-Zinn 1984; Roth 1984; Oberbauer and Strain 1986; Bongers et al. 1988b), usually through an increase in mesophyll thickness, particularly palisade mesophyll (Wylie 1951; Jackson 1967, Napp-Zinn 1984; Fisher 1986). With respect to morphology, nutrient concentrations and anatomy, the differences between sun and shade leaves of the Los Tuxtlas rain forest species (cf. Table 1) conform well to this general pattern.

Identification of probable adaptive traits

Additive application of the first three arguments mentioned in the introduction should lead to the identification of probable adaptive traits. However, if a trait does not vary with differences in environment, no conclusions about its adaptive value can be made. The absence of a significant response over a wide range of species only indicates that there is no common solution to a given problem. The specific responses shown in individual species may well be adaptive.

Responses to differences in light exposure may be defined both in terms of the mean within-species variability in the state of a trait (qualitatively as well as quantitatively) and in terms of among-species variability in that trait. Seven traits did not respond significantly in any way in relation to differences in light exposure. Thirteen traits showed a significant response, either in mean within-species shade/sun ratio, or in among-species variance. Variation in these traits thus may be of adaptive significance. Four traits (leaf area, SLW, [K]/w and [N]/a) responded significantly in both ways. These four showed also the largest mean deviation comparing shade to sun leaves ($+54\%$, -30%, $+50\%$, -27% respectively). It is concluded that the variation in these four traits in relation to differences in light exposure is significant and pronounced. Therefore the probability that this variation is of adaptive significance is relatively high. We further assumed that the state of a trait is more crucial for leaf functioning in the condition where it shows the lesser variance. This assumption leads to the conclusion that a reduction in leaf area and in [K]/w are probably adaptations to a sunny environment, whereas a reduction in SLW and in [N]/a are probably adaptations to a shady environment.

Table 2. Correlations between leaf traits in sun and in shade leaves (A), and between the transformed shade/sun ratios (B).

A.

		sun		
	Leaf area	SLW	[K]/w	[N]/a
shade				
Leaf area	–	–.04 n.s.	+.11 n.s.	+.09 n.s.
SLW	–.03 n.s.	–	–.52 ****	+.64 ****
[K]/w	+.07 n.s.	–.29 *	–	–.34 **
[N]/a	+.21 n.s.	+.64 ****	+.03 n.s.	–

B.

	SHSU (leaf area)	SHSU (SLW)	SHSU ([K]/w)	SHSU ([N]/a)
SHSU (leaf area)	–	–.25 n.s.	+.22 n.s.	–.14 n.s.
SHSU (SLW)		–	–.49 ****	+.70 ****
SHSU ([K]/w)			–	–.20 n.s.
SHSU ([N]/a)				–

Pearson correlation, * = $p < .05$; ** = $p < .01$; *** = $p < .001$; **** = $p < .0001$; n.s. = not significant.

Our fourth argument concerned the relations between leaf traits and the identification of possible trade-offs. The relationships between the four leaf traits with the highest probabilities of having adaptive significance are shown in Table 2. Leaf area is not correlated to SLW, [K]/w and [N]/a, neither in sun nor in shade leaves. Also the change from shade to sun (the shade/sun ratios) in leaf area is independent from the other three parameters. Both [K]/w and [N]/a are correlated to SLW in shade and in sun leaves, and changes in both parameters are correlated to changes in SLW as well. [N]/a and [K]/w are only correlated in sun leaves, and changes in both parameters are not correlated. Accordingly, this indicates that leaf area is probably independently adaptive, whereas [K]/w and [N]/a are probably trade-offs related to SLW and changes therein. SLW itself however, most likely forms part of a larger complex of factors (Bongers and Popma 1988).

Functional aspects

Small leaves have a lower heat load and therefore a lower transpirational water loss than large leaves, due to a lower boundary layer conductance for heat transfer (Vogel 1968; Parkhurst and Loucks 1972; Taylor 1975; Givnish and Vermey 1976; Chiariello 1984). The adaptive significance of a reduction in size of fully exposed leaves may therefore lie in the optimization of the heat and water balances of the leaf.

SLW is correlated to various morphological, anatomical and physiological leaf traits (Napp-Zinn 1984; Field and Mooney 1986; Bongers and Popma 1988). Changes in SLW thus may be the consequence of changes in other traits. In sun and in shade leaves SLW is positively correlated with the thicknesses of the lamina and of both mesophyll layers (Bongers and Popma 1988). The decrease in shade-sun ratio in SLW is larger than that of the leaf tissue layers however, indicating that the variation in SLW is not completely explained by the variation in the thickness of the leaf (tissues). SLW is mainly used as a measure of the investment in dry matter per unit photosynthesizing area. SLW and rate of photosynthetically fixed carbon per unit dry weight are negatively correlated (Small 1972; Mooney et al. 1978; Medina 1984; Field and Mooney 1986). Under shady conditions it thus is favorable to produce low SLW leaves as these leaves have a lower cost/benefit ratio due to a higher productivity per unit of dry weight. Our results support this: shade leaves have a lower SLW than sun leaves, and the smaller variance in the shade leaves suggests an adjustment to shade.

Nitrogen content of a leaf is linearly related to photosynthetic capacity, on an area basis as well as on a weight basis (Mooney et al. 1978; Field and Mooney 1986; Hirose and Werger 1987). This relationship might explain, at least partly, the relationship between SLW and photosynthetic capacity, since SLW is positively related to [N]/a. The shade leaves have a lower [N]/a and a smaller variance in that trait than the sun leaves. In a shaded environment [N]/a is probably low because potential maximum photosynthetic rates per unit of leaf area are limited by low light availability.

[K]/w was found to be a probable adaptation to shade. However, the lack of ecophysiological studies on the importance and functioning of K makes an ecological interpretation of the results impossible. The large differences between shade and sun leaves indicate that it might be rewarding to conduct more detailed studies on potassium.

Variation in many leaf traits is related to variation in light exposure. In this study some traits were identified as being convergent and probably having adaptive value in relation to the natural variation in the environment. Some leaf traits were identified as probably resulting as trade-offs from selective pressures on other traits. The adaptive value of traits is best measured in terms of an optimization of the cost/benefit balance for the plant's leaf. However, more ecophysiological data on leaf characteristics of rain forest plants are needed to be able to fully understand the adaptive significance of the variation found.

Acknowledgments

We thank the staff of the Los Tuxtlas Biological Station for their continuous support. S. Sinaca assisted during field work, and J. Vazquez and M. Gutierrez with the anatomical analyses. We thank H. Lambers, E. van der Maarel and M.J.A. Werger for their comments on an earlier draft of the manuscript. The Laboratorio de Evertebrados, Facultad de Ciencias, Universidad Nacional Autónoma de México provided the equipment for the anatomical analyses. The IBM Scientific Center of Mexico and the Department of Experimental Plant Ecology of the Catholic University of Nijmegen, the Netherlands, kindly provided computer facilities. We were supported by grant nr. W84-204 of the Netherlands Foundation for the Advancement of Tropical Research (WOTRO).

References

Ashton, P.S. 1978. Crown characteristics of tropical trees. In: P.B. Tomlinson and M.H. Zimmermann (eds), Tropical Trees as Living Systems, pp. 591–615. Cambridge University Press, Cambridge.
Björkman, O. 1981. Responses to different quantum flux densities. In: O.L. Lange, P.S. Nobel, C.B. Osmond and H. Ziegler (eds), Physiological Plant Ecology I. Responses to the Physical Environment, pp. 57–107. Springer, Heidelberg, New York.
Blackman, G.E. and Wilson, G.L. 1951. Physiological and ecological studies in the analysis of plant environment. VII. An analysis of the differential effects of light intensity on the net assimilation rate, leaf area ratio and the relative growth rate of different species. Ann. Bot. 15: 374–408.
Boardman, N.K. 1977. Comparative photosynthesis of sun and shade plants. Ann. Rev. Plant Physiol. 28: 355–377.
Bongers, F. and Popma, J. 1988. Trees and gaps in a Mexican tropical rain forest; species differentiation in relation to gap-associated environmental heterogeneity. Ph.D. Thesis, Utrecht.
Bongers, F.J., Popma, J., Meave del Castillo, J. and Carabias, J. 1988a. Structure and floristic composition of the lowland rain forest of Los Tuxtlas, Mexico. Vegetatio 74: 55–80.
Bongers, F., Popma, J. and Iriarte-Vivar, S. 1988b. Response of *Cordia megalantha* Blake seedlings to gap environments in Mexican tropical rain forest. Functional Ecology 2: 379–390.
Chazdon, R. and Fetcher, N. 1984. Photosynthetic light environments in a lowland tropical rain forest in Costa Rica. J. Ecol. 72: 553–564.
Chiariello, N. 1984. Leaf energy balance in the wet lowland tropics. In: E. Medina, H.A. Mooney and C. Vazquez-Yanes (eds), Physiological Ecology of Plants in the Wet Tropics, pp. 85–98. Dr W. Junk Publishers, The Hague.
Evans, G.C. 1972. The analysis of plant growth. University of California Press, Berkeley and Los Angeles. 734 pp.
Fetcher, N., Strain, B.R. and Oberbauer, S.F. 1983. Effects of light regime on the growth, leaf morphology, and water relations of seedlings of two species of tropical trees. Oecologia 58: 314–319.
Field, C. and Mooney, H.A. 1986. The photosynthesis-Nitrogen relationship in wild plants. In: T.J. Givnish (ed.), On the Economy of Plant Form and Function, pp. 25–55. Cambridge University Press, Cambridge.
Fisher, J.B. 1986. Sun and shade effects on the leaf of *Guarea* (Meliaceae): plasticity of a branch analogue. Bot. Gaz. 147: 84–89.
Givnish, T.J. and Vermeij, G.J. 1976. Sizes and shapes of liane leaves. Am. Nat. 100: 743–778.
Givnish, T.J. 1987. Comparative studies of leaf form: Assessing the relative roles of selective pressures and phylogenetic constraints. New Phytologist 106 (suppl.): 131–160.
Jackson, L.W.R. 1967. Effect of shade on leaf structure of deciduous tree species. Ecology 48: 498–499.
Hirose, T. and Werger, M.J.A. 1987. Nitrogen use efficiency in instantaneous and daily photosynthesis of leaves in the canopy of a *Solidago altissima* stand. Physiologia Plantarum 70: 215–222.
McClendon, J.H. and McMillen, G.G. 1982. The control of leaf morphology and the tolerance of shade by woody plants. Botanical Gazette 143: 79–83.
Medina, E. 1984. Nutrient balance and physiological processes at the leaf level. In: E. Medina, Mooney, H.A. and Vazquez-Yanes, C. (eds), Physiological Ecology of Plants in the Wet Tropics, pp. 134–154. Dr W. Junk Publishers, The Hague.
Mooney, H.A., Ferrar, P.J. and Slatyer, R.O. 1978. Photosynthetic capacity and carbon allocation patterns in diverse growth forms of Eucalyptus. Oecologia 36: 103–111.
Napp-Zinn, K. 1984. Handbuch der Pflanzenanatomie. VIII Anatomie des Blattes, 2. Blattanatomie der Angiospermen, B. Experimentelle und ökologische Anatomie des Angiospermenblattes (part 1). Borntraeger, Berlin Stuttgart. 520 pp.

Oberbauer, S.F. and Strain, B.R. 1986. Effects of canopy position and irradiance on the leaf physiology and morphology of *Pentaclethra macroloba* (Mimosaceae). Amer. J. Bot. 73: 409–416.
Parkhurst, D.F. and Loucks, O.L. 1972. Optimal leaf size in relation to environment. J. Ecol. 60: 505–537.
Popma, J. and Bongers, F. 1988. The effect of canopy gaps on growth and morphology of seedlings of rain forest species. Oecologia 75: 625–632.
Popma, J., Bongers, F. and Meave del Cartillo, J. 1988. Patterns in the vertical structure of the lowland tropical rain forest of Los Tuxtlas, Mexico. Vegetatio 74: 81–91.
Roth, I. 1984. Stratification of tropical forests as seen in leaf structure. Dr W. Junk Publishers, The Hague. 522 pp.
Small, E. 1972. Photosynthetic rates in relation to nitrogen recycling as an adaptation to nutrient deficiency in peat bog plants. Can. J. Bot. 50: 2227–2233.
Sokal, R.R. and Rohlf, F.J. 1981. Biometry. 2nd edition. Freeman, San Francisco. 859 pp.
Taylor, S.E. 1975. Optimal leaf form. In: D.M. Gates and R.B. Schmerl (eds), Perspectives in Biophysical Ecology, pp. 73–86. Springer, Berlin, New York.
Vogel, S. 1968. 'Sun leaves' and 'shade leaves': differences in convective heat dissipation. Ecology 49: 1203–1204.
Wylie, R.B. 1951. Principles of foliar organization shown by sun-shade leaves from ten species of deciduous dicotyledonous trees. Am.J. Bot. 38: 355–361.

THE FUNCTIONAL SIGNIFICANCE OF SHORT STATURE IN MONTANE VEGETATION

JOHN GRACE
Department of Forestry and Natural Resources, University of Edinburgh, Edinburgh, EH9 3JU, UK

Abstract

The role of structure in the aerodynamic properties of vegetation is reviewed in relation to the transition from forest to dwarf shrubs at the treeline. Short vegetation above the treeline displays higher aerodynamic resistances whilst the temperature of meristems is highly correlated to the radiation flux. Forests, on the other hand, display a low aerodynamic resistance and the temperature of meristems is here correlated to air temperature. Stature therefore determines the coupling between climate and endoclimate, and short stature ensures a favourable endoclimate in a cold climate.

Introduction

Vegetational structure is an all-embracing term which means rather different things to different people. Few authors have attempted to define structure as applied to vegetation; and when they have, it has been from the viewpoint of descriptive plant ecology (see Barkman 1979 and this volume).

My interest has been in identifying those attributes of structure in wild vegetation which determine fundamental biophysical properties of the individuals and the stand, especially the exchanges of heat and mass between vegetation and the atmosphere. The most simple of all, and the one I want to elaborate upon in this paper, is vegetational height. The significance of height was realised long ago when Raunkiaer (1934) proposed his system of life forms. He observed that tall life forms (phanerophytes) were absent from cold places, but did not say why. Dwarf shrubs have life cycles and physiological responses which do not differ much in general from those of trees, and so their success in areas which are too cold for trees to grow may therefore stem from some functional consequence of dwarfness *per se*. Tall vegetation is aerodynamically rough and so the air from well above the vegetational surface is frequently swept down and mixed with the air close to the plants (Geiger 1966; Monteith 1973; Fritschen *et al.* 1985). The temperature of plant parts in these circumstances is likely to be coupled to that of the atmosphere as a whole. In contrast, short vegetation is aerodynamically smooth: the turbulent eddies are scaled to the vegetation height and so mixing with the atmosphere as a whole is much less efficient. Consequently, on sunny days the climate at the plant surface is likely to differ appreciably from that of the air above, being warmer. On cold nights with clear skies (radiation flux negative) we might expect plant surfaces to be colder than the atmosphere as a whole.

Recently, it has been possible to test these ideas at a natural treeline in the Cairngorm Mountains of Scotland. At this site, *Pinus sylvestris* L. forms the treeline with well-developed *krummholz* at 600 m a.s.l. and isolated individuals as seedlings or damaged saplings extending to about 800 m (Watt and Jones 1948; Pears 1968;

Table 1. Main climatological variables during a continuous record 27 May – 30 June 1985, as means of 15-minute means.

Altitude (m)	Vegetation	Height (m)	Net radiation (W m^{-2})	Air temperature (°C)	Water vapor pressure (kPa)	Wind speed (m s^{-1})
450	*Pinus sylvestris* L. forest	15–18	195	12.8	0.75	4.3
600	*Pinus sylvestris* krummholz	1–4	181	10.3	0.98	5.8
650	Dwarf shrubs, especially *Arctostaphylos uva-ursi* L.	0.1	176	8.7	1.12	8.4
850	Dwarf shrubs, especially *Loiseleuria procumbens* L. Desv.	0.1	152	9.1	1.9	9.4

Miller and Cummins 1982). Details of the site and techniques are given in Wilson *et al.* (1987).

Quantitative relationships

We used a well-known set of relationships to estimate the aerodynamic resistance (r_a) to heat transfer from tall or short vegetation to the atmosphere above the local boundary layer (Grace 1981).

$$r_a = [\ln (z - d) / z_0]^2 / (k^2 u)$$
$$z_0 = 0.1\, h$$
$$d = 0.7\, h$$

where h is the height of the vegetation, k is von Karman's constant (0.41), z is the height at which the wind speed was measured, z_0 is the roughness length and d is the zero plane displacement. This relationshop applies strictly to large uniform stands of vegetation, although the treeline site of the present study is very uneven. The relationship between vegetational height and aerodynamic roughness was set at $z_0 = 0.1\, h$ on the basis of a survey of roughness length of wild vegetation by Garrett (1977). His data permit on estimation of z_0 from vegetation of known height and silouette area per land area.

Using the calculated r_a, it is possible to estimate the temperature of plant surfaces by solving the energy balance. This was done using an iterative procedure described in detail by Grace (1983).

Measurements

Four stations were established on an altitudinal gradient (Table 1). Microclimatological instruments were installed at each station, taking measurements of air temperature, net all-wave radiation, humidity and wind speed every 10 sec and aver-

The functional significance of short stature in montane vegetation

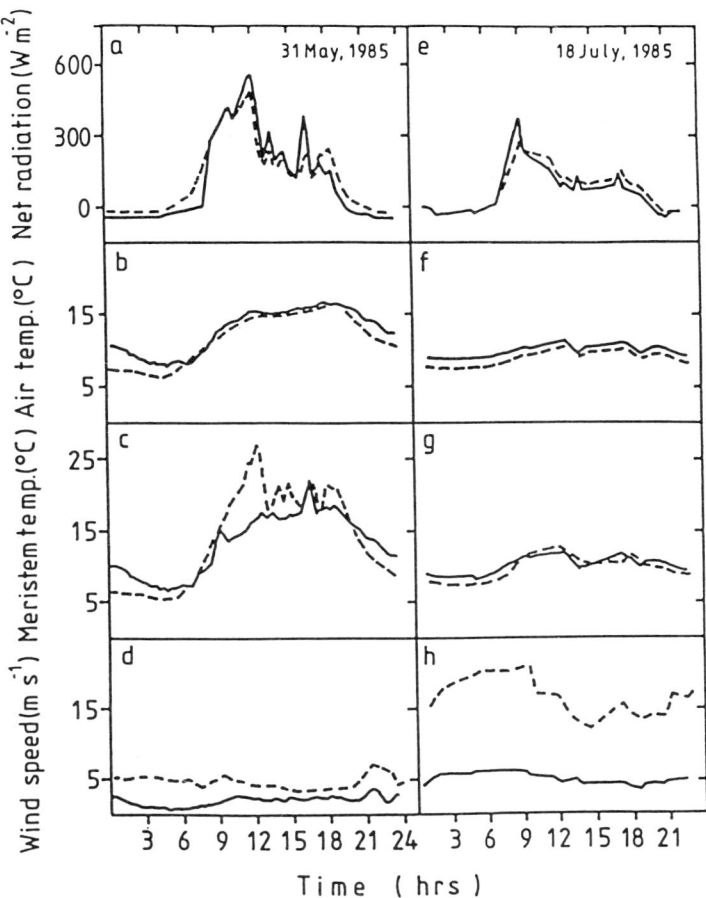

Fig. 1. Diurnal trends on a sunny day with light wind (a–d) and a dull day with strong wind (e–h). Each window displays data from two altitudes: (———), forest at 450 m; (---), dwarf shrubs at 650 m. From Wilson *et al.* (1987).

aging them over 15-min intervals using a data logger (CR21, Campbell Scientific Instruments, Utah, USA).

Aerodynamic resistances to heat transfer from apical meristems to the atmosphere were measured with a special heated probe, and the temperatures of apical meristems were sensed by inserting 0.1 mm-diameter copper-constantan thermocouples into holes bored axially into the buds or growing shoots of *Pinus*. In the dwarf shrubs *Arctostaphylos uva-ursi* (L.) Spreng. and *Loiseleuria procumbens* (L.) Desv. the terminal region of the shoot is too fine and so the thermocouple lead was coiled around the shoot and the junction placed in contact with the terminal area.

Results

To demonstrate typical inter-relationships between variables on a diurnal basis,

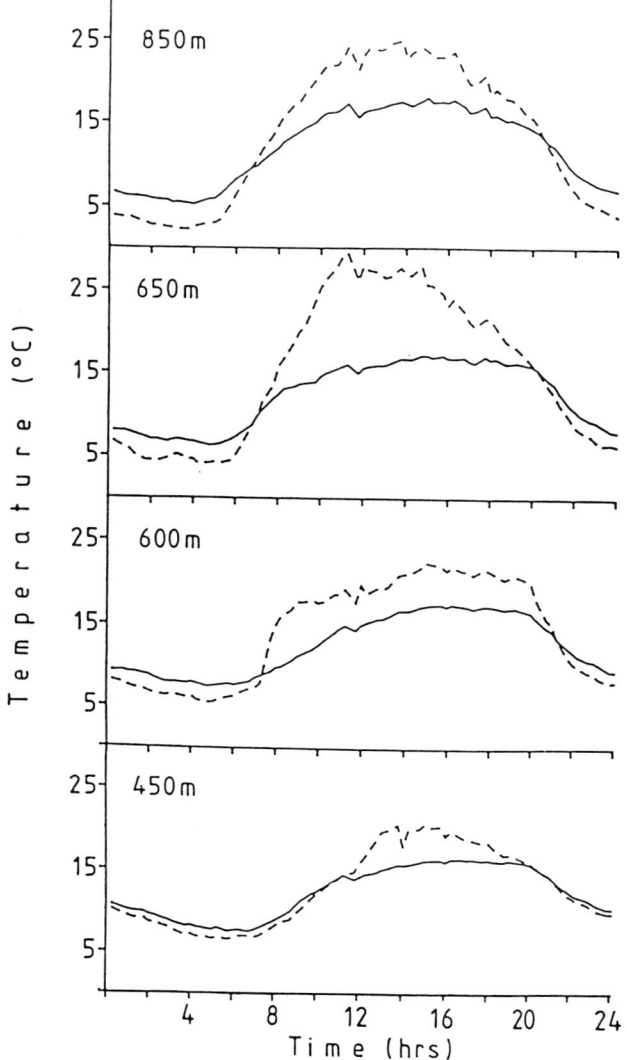

Fig. 2. Mean diurnal trends of air temperature (———) and meristem temperature (---) for the period 27 May to 30 June 1985. From Wilson *et al.* (1987).

results from two contrasting days are presented (Fig. 1). On calm days with sunshine, like 31 May 1985, differences between meristem and air temperatures were generally large. In the case of short vegetation, meristems could be as much as 15°C warmer than ambient air, but those of the tall vegetation were never more than 7°C warmer than air (Fig. 1). The meristem temperatures of short vegetation show a diurnal trend which parallels the march of net radiation. Those of tall vegetation make a diurnal trend which is like that of air temperature.

On clear nights the short vegetation was usually a few degrees colder than ambient air, whereas tall vegetation was scarcely at all colder than the air.

The functional significance of short stature in montane vegetation

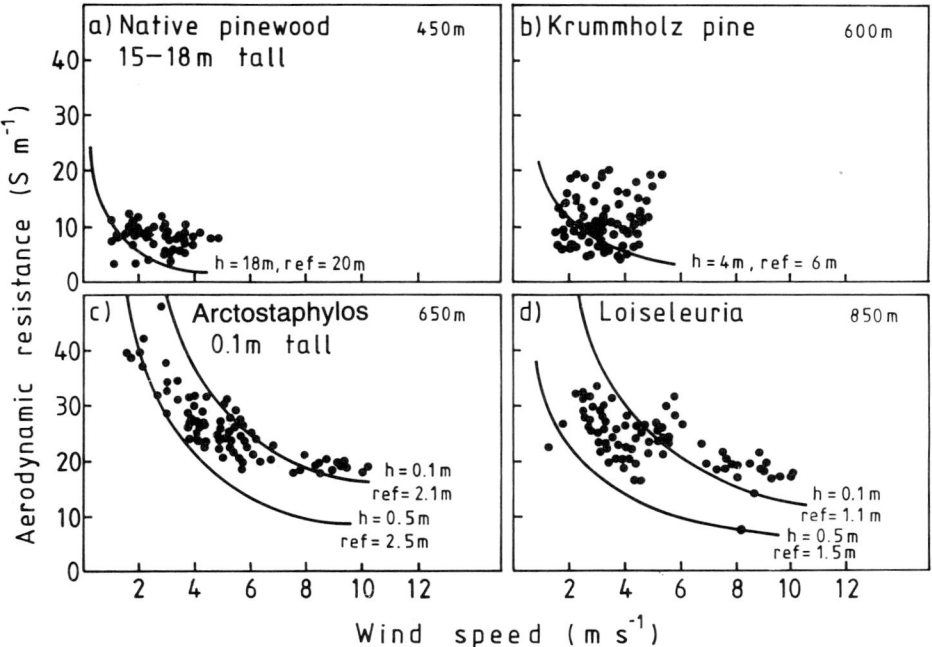

Fig. 3. Aerodynamic resistances for heat transfer from meristems to a reference height above the local boundary layer, plotted against wind speed at the reference height. Points are measured values, lines are calculated. From Wilson *et al.* (1987).

The situation on dull, windy days like 18 July 1985 was quite different. Meristem temperatures were nearly the same as air temperatures in all cases.

Over longer periods the differences between tall and short vegetation, which have just been described for 27 May, are still apparent (Fig. 2). To illustrate this, the period 27 May to 30 June 1985 is reported. The period includes budbreak and early extension growth. Over this part of Scotland June 1985 was colder and rather less sunny than average for the time of year, but not greatly so. The mean diurnal curve shows that in the short vegetation at 650 m the meristems were 11°C warmer than ambient air at noon whilst the corresponding difference for the tall vegetation at 450 m was less than two degrees (Fig. 2). Just before dawn, the short vegetation was measurably colder than ambient air whilst the tall vegetation was not (differences less than 0.5°C cannot be deemed significant because of limitation in the precision of the measurement and the statistical sampling errors).

Aerodynamic resistances measured with the special sensors were similar in magnitude to those calculated from the equations (Fig. 3). The solid lines give the values calculated when the reference height z is set to the height of the air temperature and wind speed sensors. Two lines are shown where there is some variation in the vegetational height, h.

Finally, the relationship between net all-wave radiation, temperature and transpiration rate may be calculated (Fig. 4). The calculation has been performed as-

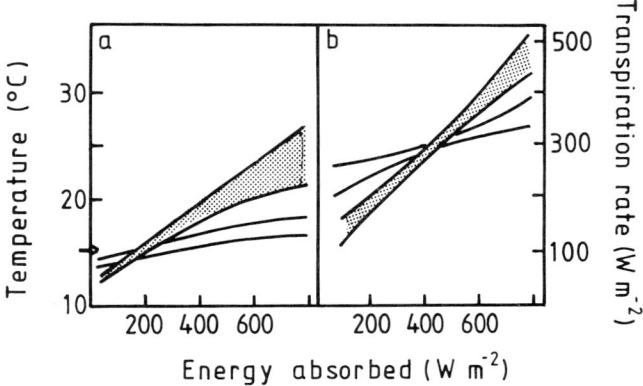

Fig. 4. Calculated meristem temperatures (a) and transpiration rates (b) when air temperature is 15° C. The shaded portions refer to chamaephytes (r_a = 20–50 s m^{-1}) the open portions are trees (r_a = 3–10 s m^{-1}). Stomatal resistance is set to 50 s m^{-1}.

suming an air temperature of 15°C and a water vapor pressure of 0.8 kPa. It has further been assumed that the effective stomatal resistance to water loss of the terminal region of shoot is 50 s m^{-1}. Thus, tall and short vegetation may be compared under conditions where everything other than height (and therefore r_a) is the same. The result shows that not only does short vegetation become appreciably warmer as the radiation flux increases, but it also uses more water at the highest values of radiation (Fig. 4).

Discussion

Simultaneous measurements of surface-to-air temperature differentials of tall versus short vegetation were also made by Körner and Cochrane (1983) in the Snowy Mountains of south-eastern Australia. They found leaves of dwarf shrubs to be 21°C warmer than air in bright sunshine, whereas trees of *Eucalyptus pauciflora* were only 7°C warmer than air. Very high leaf temperatures in dwarf vegetation have also been reported by Stoutjesdijk 1970, Cernusca 1976, Larcher and Wagner 1976, Körner and Diemer 1987. A survey of leaf temperatures of forests suggest that they never differ much from air temperatures (Yamaoka 1958; Tranquillini 1963; Vanderwaal and Holbo 1984).

All the data suggest that tall and short vegetation types may respond to their atmospheric environments in quite different ways. Short vegetation is decoupled from air temperature, develops a more extreme microclimate and its temperature responds especially to solar radiation and wind speed. Other structural attributes of the vegetation may also have an effect on coupling. The distribution of gaps, the grouping of leaves around stems and buds, and the size and shape of leaves are perhaps the most important factors after that of stature (see Grace 1981).

At high altitudes and latitudes, the growth of woody perennials is likely to depend crucially on temperature, and so those variables that influence temperature may also be expected to influence growth. One source of information about the effect of temperature on woody perennials is dendrochronology: near the treeline the correlation between annual radial wood increment and temperature of the current or previous year becomes very strong (Hustich 1947; Mikola 1962; Schweingruber *et al.* 1979; Norton 1984). The regenerative capacity of trees near the treeline is best in periods of unusually warm summers (Mikola 1962, Kullman 1981). At a physiological level it seems that the sites of temperature perception are the meristematic regions: for example, Juntilla (1986) showed the number of needle primordia to be included in the resting bud to be determined by temperature acting over the entire growing season preceeding the bud's formation. Moreover, budbreak in the spring is also under control of temperatures, and so is the elongation of the shoot and needles (Juntilla 1986). But temperature may act at several points in the life cycle. The summer temperature is said to influence the development and thickness of cuticle in conifers, so that low temperatures in the summer cause an excessive cuticular transpiration in the early spring and may lead to extensive desiccation (Lange and Schulze 1966; Tranquillini 1963).

The close coupling between air and tissue temperatures in trees presumably is the main reason why the altitudinal and latitudinal limit to forest is everywhere roughly co-incident with the 10°C isotherm of summer mean air temperature (Daubenmire 1954; Mikola 1962; Wardle 1965; Grace 1977). In contrast, the altitudinal zonation of *krummholz* and dwarf shrub vegetation is less predictable and much more related to topography and microtopography, which influence wind speed and radiation balance, and consequently, tissue temperatures (Watt and Jones 1948; Nägeli 1971).

To what extent may short stature be regarded as being a trait of adaptive value? The advantages of dwarfism would seem to include not only an elevated tissue temperature but also: (a) avoidance of winter desiccation by being underneath snow, (b) avoidance of wind damage, (c) completion of the life cycle within a shorter period, (d) less structural tissue requirement. An obvious disadvantage is poor competitive ability in the presence of trees (even if shade can be tolerated, the benefit of elevated temperatures is lost when radiation is reduced). A less obvious disadvantage, suggested by the energy balance calculations (Fig. 4), is that at high radiation loads dwarf vegetation may require very high rates of water loss to prevent temperatures from becoming so high as to be lethal. Measurements of tissue temperatures of rosette alpine plants emphasize this point (Larcher 1980). There have been a few studies demonstrating that dwarfism in tree species near the treeline is a genetic trait (Clausen 1965; Grant and Mitton 1977). In some conifers, such as *Pinus cembra* L. and *Juniperus communis* L., dwarf forms from high altitudes are used in horticulture, and in these forms the dwarfism persists in cultivation. Similarly, other studies on herbaceous plants and dwarf shrubs have shown by means of transplant experiments that the shorter or more prostrate high altitude forms are indeed genetic dwarfs (Clausen *et al.* 1940; Grant and Hunter 1962). Nevertheless, some component of short stature in *krummholz* is clearly not genetic, but simply the consequence of natural pruning of that part of the tree which undergoes winter desiccation or loss of limbs by the action of snow and ice (Fanta 1981).

The adaptive value of short stature cannot yet be measured, as this would require a much better understanding of the relationship between biomechanical properties of woody plants and their reproductive success. However, the current data from the Cairngorms suggest that as far as the thermal benefit is concerned, the effect of short stature is to increase the daytime temperatures of the plant to such an extent that its mean tissue temperatures are similar to those prevailing in tall vegetation growing two hundred metres lower down the mountain.

References

Barkman, J.J. 1979. The investigation of vegetation texture and structure. In: M.J.A. Werger (ed.), The Study of Vegetation, pp. 123–160. Dr W. Junk Publ., The Hague, Boston, London.
Cernusca, A. 1976. Bestandesstruktur, Bioklima und Energiehaushalt von alpinen Zwergstrauchbeständen. Oecol. Plant. 11: 71–102.
Clausen, J. 1965. Population studies of alpine and subalpine races of conifers and willows in the California High Sierra Nevada. Evolution 19: 56–68.
Clausen, J, Keck, DD. and Hesey, W.M. 1940. Experimental studies on the nature of species 1. The effect of varied environments on western North American plants. Publs. Carnegie Instn. 520: 1–452.
Daubenmire, R. 1954. Alpine timberlines in the Americas and their interpretation. Butler Univ. Bot. Stud. 2: 119–136.
Fanta, J. 1981. *Fagus sylvatica* L. und das Aceri-Fagetum an der Alpinen Waldgrenze in Mitteleuropäischen Gebirgen. Vegetatio 44: 13–24.
Fritschen, L.J., Gay, L. and Simpson, J. 1985. Eddy diffusivity and instrument resolution in relation to plant height. In: B.A. Hutchison and B.B. Hicks (eds), The Forest-Atmosphere Interaction, pp. 583–590. D. Reidel Publishing Company, Dordrecht, Boston, Lancaster.
Garrett, J.R. 1977. Aerodynamic roughness and mean monthly surface stress over Australia. CSIRO Australia, Division of Atmospheric Physics Technical Paper 29: 1–19.
Geiger, R. 1966. The climate near the ground. Harvard University Press, Cambridge, Massachusetts.
Grace, J. 1977. Plant response to wind. Academic Press, London.
Grace, J. 1981. Some effects of wind on plants. In: J. Grace, E.D. Ford and P.G. Jarvis (eds), Plants and their Atmospheric Environment, pp. 125–130, Blackwell Scientific Publications, Oxford.
Grace, J. 1983. Plant-atmosphere interactions. Chapman and Hall, London.
Grant, M.C. and Mitton, J.B. 1977. Genetic differentiation among growth forms and Englemann spruce and subalpine fir at tree line. Arctic and Alpine Research 9: 259–263.
Grant, S.A. and Hunter, R.F. 1962. Ecotypic differentiation of *Calluna vulgaris* (L.) in relation to altitude. New Phytologist 61: 44–55.
Hustich, I. 1947. Climate fluctuation and vegetation growth in northern Finland during 1890–1939. Nature 160: 478–479.
Juntilla, O. 1986. Effects of temperature on shoot growth in northern provenances of *Pinus sylvestris* L. Tree Physiology 1: 185–192.
Körner, Ch. and Cochrane, P. 1983. Influence of plant physiognomy on leaf temperature on clear midsummer days in the Snowy Mountains, south-eastern Australia. Acta Oecol./Oecol. Plant. 4: 117–124.
Körner, Ch. and Diemer, M. 1987. *In situ* photosynthetic responses to light, temperature and carbon dioxide in herbaceous plants from low and high altitude. Functional Ecology 1: 179–194.
Kullman, L. 1981. Recent tree-limit dynamics of Scots pine (*Pinus sylvestris* L.) in the southern Swedish Scandes. Wahlenbergia 8: 3–67.
Lange, O.L. and Schulze, E.-D. 1966. Untersuchungen über die Dickenentwicklung der kutikularen Zellwandschichten bei der Fichtennadel. Forstwissenschaftliches Centralblatt 85: 27–38.
Larcher, W. 1980. Physiological Plant Ecology. 2nd Edn. Springer-Verlag, Berlin.
Larcher, W. and Wagner, J. 1976. Temperaturgrenzen der CO_2-Aufnahme und Temperaturresistenz der Blätter von Gebirgspflanzen im vegetationsaktiven Zustand. Oecol. Plant. 11: 361–374.
Mikola, P. 1962. Temperature and tree growth near the northern timber line. In: T.T. Kozlowski (ed.), Tree Growth, pp. 265–274, Ronald Press, New York.
Miller, G.R. and Cummins, R.T. 1982. Regeneration of Scots pine *Pinus sylvestris* at a natural treeline in the Cairngorm Mountains, Scotland. Holarctic Ecology 5: 27–34.

Monteith, J.L. 1973. Principles of environmental physics. Arnold, London.
Nägeli, W. 1971. Der Wind als Standortsfaktor bei Anforstungen in der subalpinen Stufe (Stillbergalp im Dischmatal, Kanton Graubünden. Mitteilungen 47, Schweizerische Anstalt für das Förstliche Versuchswesen, 147 pp.
Norton, D.A. 1984. Tree-growth-climate relationships in subalpine Nothofagus forests, South Island, New Zealand. New Zealand J. Bot. 22: 471–481.
Pears, N.V. 1968. The natural altitudinal limit of forest in the Scottish Grampians. Oikos 19: 17–80.
Raunkiaer, C. 1934. The life forms of plants and statistical plant geography. (Translation of 1909 original). Oxford University Press, Oxford.
Schweingruber, F.H., Bräker, O.U. and Schär, E. 1979. Dendroclimatic studies on conifers from central Europe and Great Britain. Boreas 8: 427–452.
Stoutjesdijk, Ph. 1970. A note on vegetation temperatures above the timberline in southern Norway. Acta Bot. Neerl. 19: 918–925.
Tranquillini, W. 1963. Climate and water relations of plants in the subalpine region. In: A.J. Rutter and F.H. Whitehead (eds), The Water Relations of Plants. British Ecological Symposium 3, pp. 153–166, Blackwell Scientific Publications, Oxford.
Vanderwaal, J.A. and Holbo, H.R. 1984. Needle-air temperature differences of Douglas-fir seedlings and relation to microclimate. Forest Science 30: 635–644.
Wardle, P. 1965. A comparison of alpine timberlines in New Zealand and North America. New Zealand J. Bot. 3: 113–135.
Watt, A.S. and Jones, E.W. 1948. The ecology of the Cairngorms 1. The environment and altitudinal zonation of the vegetation. J. Ecol. 36: 283–304.
Wilson, C., Grace, J., Allen, S. and Slack, F. 1987. Temperature and stature: a study of temperatures in montane vegetation. Functional Ecology 1: 405–413.
Yamaoka, Y. 1958. The total transpiration from a forest. Transactions of the American Geophysical Union 39: 266–272.

PLANT GROWTH-FORM STRATEGIES AND VEGETATION TYPES IN ARID ENVIRONMENTS

AVI SHMIDA[1] and TONY L. BURGESS[2]
[1]Department of Botany, The Hebrew University, Jerusalem, Israel;
[2]Department of Ecology and Evolutionary Biology, University of Arizona, Tucson, Arizona

Abstract

In arid and semi-arid regions patterns of growth-form dominance are usually reasonable expressions of soil moisture regimes. The total biomass and height of the vegetation are closely coupled with total annual rainfall, but the relationship is not linear. Many apparent exceptions to this general trend can be understood as adaptations to the irregular and patchy distributions of water which is typical of more arid habitats. Water resources should be viewed in terms of their reliability and quality. Relatively subtle changes in soil or disturbance regime can cause major shifts in growth-form dominance. In arid regions temperature affects vegetation structure primarily through its action on soil moisture, but the relatively open canopy structure of woody plants appears to be an adaptation to maximize convective cooling. Advances in physiological ecology plant demography, and comparative studies have made it easier to interpret the relationships between growth-forms and their habitats, but there are significant gaps in our knowledge of underlying mechanisms, and several anomalies require explanation.

Introduction

It is well-known that plant growth-forms are major features differentiating plant communities along a moisture gradient from mesic aboreal formations to xeric true deserts (Adamson 1939; Cain 1950; Raunkiaer 1934; Schimper 1903; Walter 1973; Whittaker 1975). This paper offers a global perspective with emphasis on the distribution patterns of the lower growth-forms (annuals, hemicryptophytes, chamaephytes) which become especially common towards the drier part of the moisture gradient. Two related questions are addressed: Can the relative dominance of various growth-forms be predicted from environmental characteristics, and what is the adaptive significance of each growth-form? The concept of growth-form is used here in a non-restrictive sense; it includes morphological and functional aspects that are elsewhere covered by the concepts of life-form and plant strategy type.

From an ecological point of view an arid environment can be defined as an ecosystem in which water is the most limiting resource for primary production (Noy-Meir 1973; Orians and Solbrig 1977; Walter 1973). The biological effect of resource scarcity in deserts becomes apparent in a transect from arid to humid ecosystems (Shmida et al. 1986). At both the community and species level biomass production and diversity increase in a non-linear way in response to greater rainfall (Fig. 1). There is a trend in plants as well as in animals to become smaller (lower biomass, less surface area) and to have a smaller surface/volume ratio as conditions become more arid. There are two reasons for this trend. A desert organism has to invest more on maintenance (Fig. 2) and the cost per unit of energy

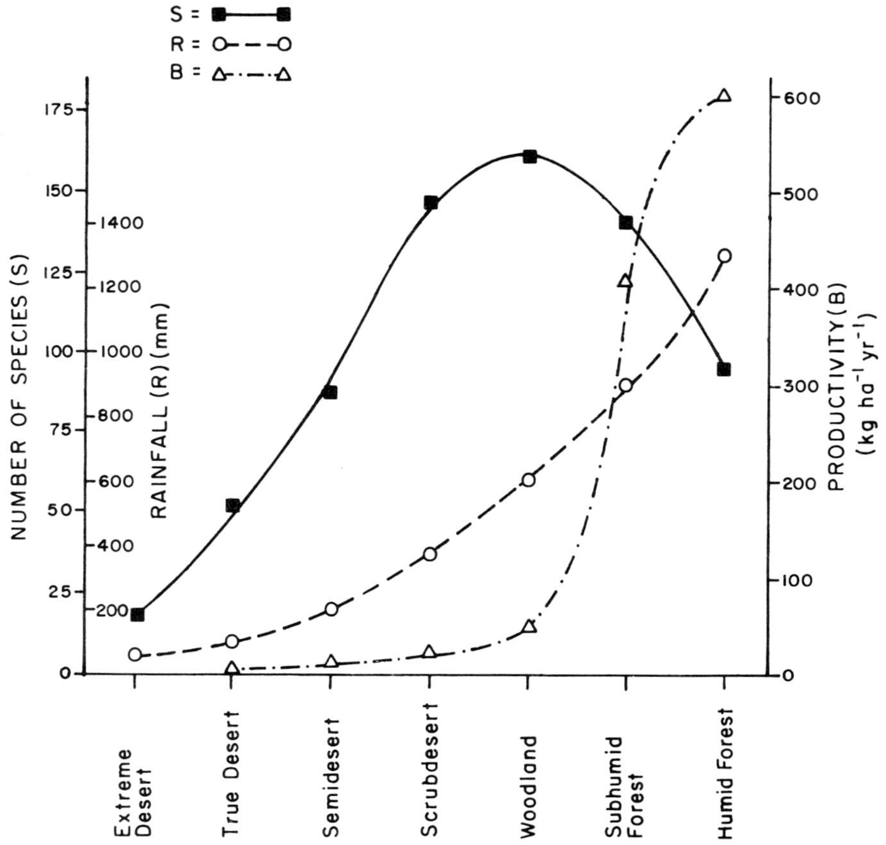

Fig. 1. Ecological trends in species diversity, biomass production and rainfall (as resource availability) from extreme desert to humid climates in Israel. R – precipitation, S – species diversity, B – biomass production (after Shmida *et al.* 1986).

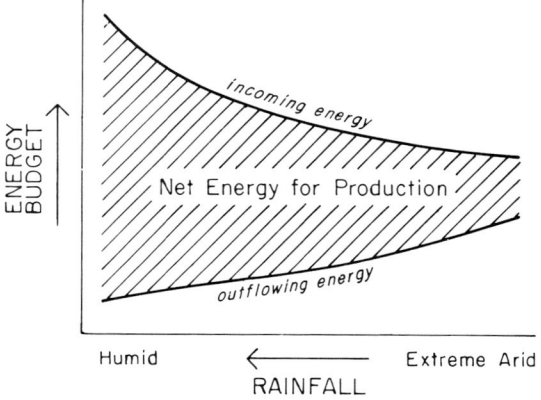

income is higher for a desert organism because its resources are more spatially and temporally scattered.

Furthermore, in most deserts the effects of water deprivation are aggravated by extreme temperatures and low humidity, which impose additional costs in maintaining favorable water and energy balances. Periods when resources are scarce are generally long, and their duration is very unpredictable.

Two sets of terminologies are used in this paper to avoid confusion between climate and vegetation systems. 'Contracted desert', 'diffuse desert', 'steppe', etc., refer to vegetation formations, defined in a separate section below. The following terms are used to classify climate, following Shmida (1985):

Semi-desert – mean annual precipitation less than 120 mm.

Extreme desert – mean annual precipitation less than 70 (to 50) mm.

These specifications are necessary because there is an enormous number of desert definitions in the literature (see Goodall and Perry 1979; McGinnies 1979; McGinnies *et al.* 1968 for a review). Mean annual precipitation is a simple criterion, and often is the only available climatological statistic. Le Houerou (1970) has stated that these simple data show an adequate correspondence with indices developed from more complex calculations such as those of Thornthwaite (1948) and Koppen (1954).

Vegetation types in water-limited environments

As most formations are identified mainly by the tallest or dominant growth-form, there is an inherent correlation between the distributions of growth-forms and vegetation types. Much of the confusion in comparisons of vegetation among different continents stems from inconsistencies in the use of terms such as steppe, thornscrub, savanna, etc. To avoid such confusion relevant plant communities are defined below, based on Whittaker (1975) and Shmida (1981, 1985, Shmida and Whittaker 1979 (Figs 3 and 4).

Contracted desert (= annual desert) – A formation in which perennial plants are restricted to sites receiving additional precipitation runoff from surrounding areas (Monod 1954). In years of above-normal rainfall, annual plants can grow on the interfluves and slopes between drainages.

Diffuse desert – a formation in which perennial plants are dispersed, yet cover less than 10% of the ground (Shmida 1985). The vegetation is generally dominated by sparse dwarf shrubs (chamaephytes) or occasionally by larger shrubs. In the rainy season a flush of annuals covers much of the ground. The total plant coverage, even in the best years is less than 50%.

Steppe (= chamaephyte semidesert) – a formation dominated by dwarf-shrubs (chamaephytes) which cover more than 10% (usually 10–30%) of the ground (Zohary 1962). Herbaceous perennials may grow among the suffrutescent shrubs but they are not dominant. This term is equivalent to subshrub and dwarfshrub com-

←

Fig. 2. Energy budget trends for organisms, from humid to arid ecosystems. As the incoming energy decreases the expended energy increases, hence the net energy gain decreases in more arid environments.

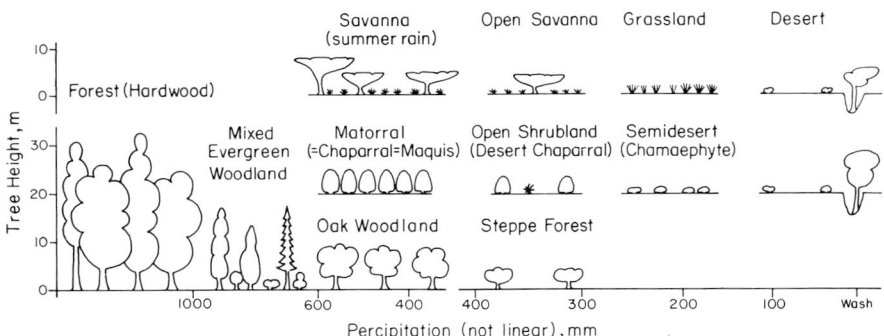

Fig. 3. Schematic diagram representing growth-form changes in vegetation along a gradient of decreasing precipitation.

munities recognized in other studies (Daubenmire 1978; Werger 1978, 1976, Zohary 1973).

(Desert) Grassland – A semi-desert formation dominated by perennial grasses. We include most of the Australian spinifex hummock communities in this category (Bradle 19817.

Arboreal plants (trees and shrubs above 1 m tall) are often absent from the formations defined above. However, even in contracted deserts trees and shrubs may occur, but they are restricted to the wadis (= arroyos, washes). If the trees in the drainages form a more or less continuous band, the vegetation is sometimes referred to as a riparian woodland or gallery forest (McKinley and Brown 1982; Zohary 1973).

Shrubland – A formation dominated by shrubs between 0.7 and 2 m tall, which cover more than 30% of the ground (Shmida 1981). If the shrubs are more widely spaced, composing only 3 to 30% of the cover, the formation is defined as 'open shrubland', equivalent to the 'desert chaparral' used by Barbour and Major (1977). The shrubs may be evergreen or deciduous.

If a shrubland is composed mostly of broad-leafed evergreens, the vegetation can correspond to Mediterranean Matorral (= chaparral, maquis, mulga, etc.). If the shrubs are leptophyll evergreens the vegetation is usually referred to as heath (Specht 1979). If the shrubs are completely or partially drought deciduous the appropriate terminology is rather vague; thornscrub, scrub, thickets, and other local terms have been used (Beard 1955). If the shrubs are cold deciduous the vegetation can be called shibeliek (Turril 1930) or montane scrubland (Brown 1982).

Woodland – An open arboreal formation in which the trees and sometimes the shrubs are taller than 2 m, and cover between 20 and 80% of the ground (Shmida 1981). Woodlands that are traditionally associated with temperate climates are composed of broadleafed trees (*e.g.*, oak) or conifers. Woodlands in areas of savanna climates can be broadleafed and fine-compound leafed (Werger 1983).

If the trees are sparse, with less than 20% cover, it can be called 'open wood-

→

Fig. 4. Vegetation catena from an arboreal temperate formation to contracted desert. A – Mediterranean catena. B – subtropical catena.

Plant growth-form strategies and vegetation types

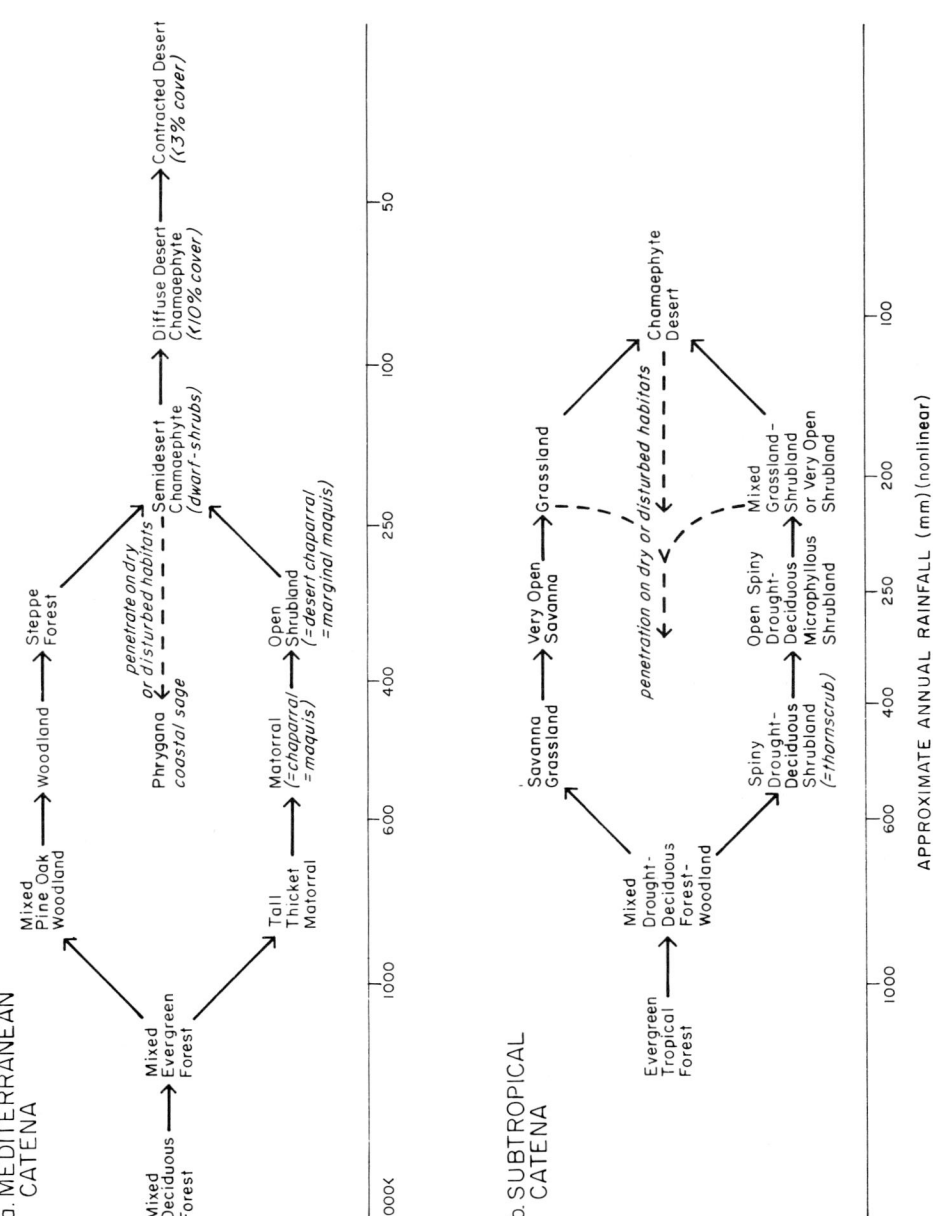

land'. This formation is sometimes called 'steppe forest' in Asia (Walter 1971, 1973; Zohary 1973) or 'savanna' in California (Griffin 1977). It is difficult to derive a definition based on cover which can effectively distinguish savanna from woodland unless biogeographical affinities are incorporated (Werger 1983; Zohary 1973). Originally savanna denoted tropical grassland (Hills 1965; Werger 1983), but the term has become associated with grasslands which have a very open overstorey of trees. Commonly savanna trees are said to be drought deciduous as in extensive areas in Africa, but evergreen or nearly evergreen trees dominate most Australian savannas (Beadle 1981) and can also be important in equivalent vegetation of the Neotropics (Cole 1960; Sarmiento and Monasterio 1983). The transition from a closed aboreal canopy to a treeless grassland or steppe may occur in a broad ecocline or an abrupt ecotone.

Succulent growth-forms

Shreve (1951) and Werger (1978, 1986) pointed out that succulents contribute significantly to the diversity of growth-forms characteristic of deserts, especially those in subtropical and tropical climates. The presence of succulent species in a community alters the aspect of the vegetation and indicates a different resource spectrum from sites in which succulents are rare or absent. Traditionally the general label 'succulent' has included a wide variety of forms, ranging from leafless, thick-stemmed cactoids to caudical or rosette plants with fleshy leaves and/or fleshy stembase. These forms occur in wide range of sizes with inherent life history implications. Thus we can distinguish among rosette trees and rosette shrubs, cactoid shrubs and cactoid chamaephytes, etc. Most of the conclusions below apply equally to both succulent and non-succulent members of a given growth-form category, but there are differences in resource utilization which are pointed out. We have dealt with the rather exceptional conditions favoring dominance and diversity of succulent plants in a separate paper (Burgess and Shmida 1988) and therefore give them peripheral attention here.

The major trend in growth-form dominance along a mesic-xeric gradient

Figure 3 illustrates a generalized transect along a gradient of decreasing precipitation proceeding from closed forest to contracted desert. The tallest growth-form shrinks from tall trees to annuals. This replacement series represent an adaptive response to differences in the most limiting resources. There is a shift from communities structured by competition for light to those in which competition for soil moisture predominates. But the transition from tall closed canopies to the open vegetation of truly arid regions is not smooth, as evidenced by the complex mosaic of communities which frequently develops in semi-arid climates. Decreasing plant height reflects both a reduced commitment of resources to building and maintaining above-ground structural biomass, and a shift of available resources from growth into maintenance (Figs 1 and 2). A relatively lower investment of resources in above-ground biomass allows a relatively higher allocation to roots. The smaller shoot biomass also decreases water loss, because the potential transpiring surface is diminished.

Table 1. Growth-form spectra of desert floras in different arid regions. Numbers are percentages of species with a given growth-form. Data from Bowers and Turner 1986; Eig 1931; Leistner 1967; Noy-Meir 1970; Ozenda 1977; Paulsen 1912, and unpublished records.

Growth-form	Australia	South Africa	Central Asia	Middle East	Sahara	Arizona (Sonoran)	Sonora	Calif. (desert)	Great basin
Trees (over 2 m)	12.1%	2.7%	1.0%	3.9%	5.2%	2.3%	4.0%	2.8%	2.7%
Trees 2–4 m	7.8		1.0	1.6	2.2	2.0	3.0	1.4	2.2
Trees 4–8 m	3.4	2.7		1.9	1.8		2.0	1.4	0.5
Trees over 8 m	0.9			0.3	1.2	0.3			
Shrubs 0.6–2 m	22.6	8.3	5.5	5.6	7.6	10.4	17.4	4.7	9.7
Chamaephytes	16.1	12.2	28.4	24.2	26.7	16.4	30.9	26.0	28.1
True Chamaephytes	8.0			2.5	3.7	6.0	3.5	9.8	11.4
Suffrutescent Cham.	6.7			16.0	15.1	5.7	17.4	11.5	7.6
Nano-Chamaephytes	1.4			6.4	7.9	4.7	10.0	4.7	9.2
Hemicryptophytes (total)	16.7	20.5	30.4	12.7	15.6	14.8	21.4	10.9	29.2
Hemicryptophyte Grasses	7.9	14.0				6.4		3.1	7.0
Geophytes	1.6	11.4	9.0	4.6	0.7	1.7	1.5	1.8	2.7
Biennials	0.2	ca. 2.3				0.3		0.3	1.0
Annuals	27.4	30.4	25.0	38.9	34.8	49.7	24.9	46.1	
Vines (climbing)	0.5	?		0.5	0.9	3.7		0.6	
Parasites	1.3	?	0.7	0.5	0.7	0.7		0.3	1.1
Species total	552	444	514	604	765	298	201	360	185

Table 1 presents growth-form spectra from dry areas in different floristic regions. The total flora of a given region is broken down into percentages of species with the same growth-form. Because each region has more than one vegetation type and a range of climates which may include both extreme desert and semi-desert types, the table conveys vary generalized impressions of the floras. For example the Central Asian data include many Russian steppe species, whereas the Sahara flora represents a higher proportion of true desert plants, and the South African data represent an area that is transitional between desert grassland and savanna.

Several trends are apparent.

1. Hemicryptophytes are much more abundant in the Central Asian and Great Basin deserts (30.4% and 29.2% respectively). This includes the perennial grasses which dominate vast areas of the Russian steppes.

2. Geophytes are generally rare in deserts except in the South African Kalahari where they comprise 11.4% of the flora.

3. Overall, there is a low proportion of trees and shrubs. Most grow in riparian habitats within the deserts, but are included in floristic lists. Australia is exceptional, with a much greater richness in large woody plants than other regions.

4. Annuals are the most abundant growth-form in almost all deserts. Australia, Central Asia and Sonora have the lowest proportion of annual plants, and Arizona and California, the highest. Note that the annual percentages in Table 1 are much lower than the annual percentages in Fig. 5. The difference is due to the different sizes of areas sampled. The floristic data of Table 1 are on a regional scale, whereas data in Fig. 5 are from 0.1 hectare Whittaker diversity plots (Shmida 1984). The beta-diversity of the perennials is much lower than the annuals at small scales, hence smaller samples in desert environments almost always have a higher percentage of annuals (50–80%) in comparison with larger samples (30–46%).

In moving from taller to lower vegetation four important boundaries are evident.

The tree border

This important landscape demarcation between forests and treeless regions occurs in temperate climates generally around the 350–400 mm annual precipitation isohyet. In subtropical areas with summer or biseasonal rainfall (*e.g.*, the Kalahari or Australian mallee) the tree border may reach the 150 or 200 mm isohyet on sandy soils (Leistner and Werger 1973; Werger 1978, Noy-Meir, pers. comm.). In tropical latitudes the tree border can be seen at rainfalls as high as 1300 mm, but these mesic demarcations always reflect particular edaphic conditions or disturbance regimes (Menaut and Cesar 1979; Cole 1982; Werger 1983).

The tree border often corresponds with the hydrological line of yearly drainage to underground water. Beyond this boundary trees are usually confined to sites where water is concentrated under aerobic soil conditions, as in desert washes and oases.

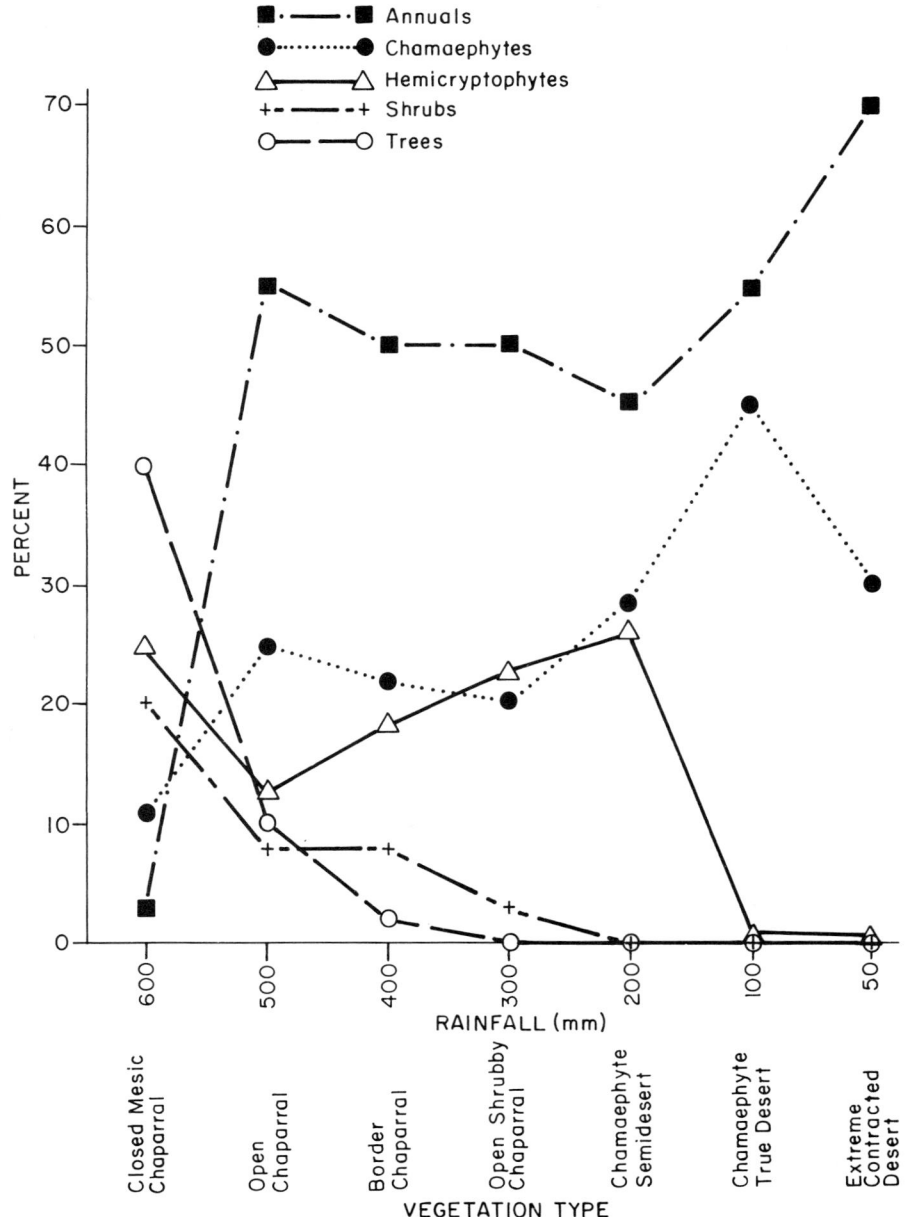

Fig. 5. Growth-form spectra in selected California vegetation types arrayed along a mesic-xeric gradient. Only the main classes of growth-forms are shown.

The shrub border

In Mediterranean climates the proportion of the shrub growth-form (nanophanerophyte sensu Raunkiaer 1934) decreases and chamaephytes become dominant at annual rainfalls between 200 and 300 mm. This transition is not abrupt and is heavily influenced by edaphic conditions. Open shrub communities can still be found in areas with less than 150 mm of annual rain, but they are usually restricted to rocky habitats (Werger 1978, 1986) often on polar-facing slopes or sandy soils (see the ecogeographical rule of Boyko 1947). *Larrea** in New World deserts is an apparent exception, for this shrub is a common dominant in true desert habitats in North America.

The grassland border

In climates with warm-season rainfall there is an analogous transition from perennial grassland to an open vegetation of shrubs and chamaephytes. On both their arid and mesic borders grasslands usually are replaced by woody plants, but in some regions increasing aridity is correlated with a change from perennial to annual grass dominance (Penning de Vries and Djitèye 1982; Walter 1973). For reasons discussed in a following section, intrinsic differences between woody plants and grasses give rise to dynamic behavior affecting the boundaries between grassland and other vegetation types. On grassland margins edaphic patterns, disturbance, rainfall, and temperature regimes interact to influence relative dominance of grasses, making these boundaries unstable and relatively diffuse on a regional scale.

The chamaephyte border (contracted vs. diffuse desert)

When annual rainfall is under 20 mm chamaephytes are confined to sites where soil receives supplemental moisture. Monod (1954) coined the term 'diffuse vegetation' for habitats in which chamaephytes are generally dispersed, in contrast to contracted vegetation in which the perennials are restricted to washes. In the contracted desert, interfluvial slopes and hammadas support annual vegetation, but vegetative growth occurs only sporadically in 'good' rainy years (Shmida *et al.* 1986; Walter 1973). Even though desert annuals generally are shorter and have less biomass than chamaephytes, it would be an oversimplification to assert that the annual growth-form is more adapted to withstand dry conditions. Annuals represent an alternative life history in which stress is imposed on embryos rather than on developed vegetative organs (see below).

The chamaephyte border corresponds generally to the transition from true desert to extreme desert climates (Shmida 1985), but the demarcation is heavily influenced by local conditions (Evenari *et al.* 1971). Runon habitats receiving water from the surrounding terrain, such as minor washes or crevices in rock outcrops, can support chamaephytes even under 50 mm of rainfall. In contrast runoff habitats losing precipitation to other sites (*e.g.*, marly slopes, badlands) may lack perennial vegetation at rainfalls of 200 mm (Yair and Shachak 1982). Slope aspect, inclination, and stone cover (desert pavement) are also very important determinants of vegetative cover (Lee 1988; Shmida 1972). Usually the chamaephyte border is difficult to delineate on a regional scale but it can be defined on a basis of mesoscale geomorphology (Danin 1970).

---

* Throughout this paper nomenclature follows that in the papers cited.

The 'major trend' in growth-form replacement described above can be readily explained. In approaching arid conditions the advantage of height in competing for light disappears, and trees and shrubs give way to lower growth-forms. However, if the main selective factor favors water conservation during drought, we should expect hemicryptophytes and geophytes to be better adapted to aridity than chamaephytes. Chamaephytes retain metabolically active stems after the growing period, whereas hemicryptophytes shed most of their living canopy, giving the chamaephytes an inherently higher rate of transpiration and respiration in the dry season. Table 1 and Fig. 5 show a contrary pattern. In many regions there are relatively few perennial herbs in true and extreme desert climates, and chamaephytes dominate in terms of both biomass and species richness. Accounting for the paucity of hemicryptophytes in more arid regions requires a consideration of another selective factor affecting life history strategies – the temporal and spatial variation of soil moisture.

Chamaephytes versus hemicryptophyte grasses

Chamaephytes and hemicryptophyte grasses dominate vast areas of semi-desert and desert regions (Rikli 1943–1948; Schimper 1903; Walter 1973). Sometimes the two growth-forms occur in mixed stands as in semi-deserts of the Fertile Crescent in the Middle East (Zohary 1973), parts of the Russian steppes (Walter 1973), and some tropical savannas (Sarmiento and Monasterio 1983). Chamaephytes totally dominate large areas in the Middle-East, the Sahara, the South African Karoo and parts of South America but in other semi-arid climates grass covers the landscape – e.g., the Pampas of South America, the Great Plains of North America, the Kalahari and part of the Sahel of Africa.

What ecological factors affect the development of grasslands and steppes? Many local studies have been conducted to reveal the mechanisms favoring one of these growth forms over the other (Buffington and Herbel 1965; Ellison 1960; Humphrey 1958), and many factors have been implicated including geomorphology, substrate, rainfall patterns, temperature, and various regimes of disturbance (Estes et al. 1982; Leistner and Werger 1973; Penning de Vries and Djitèye 1982; Werger 1978). With the possible exception of frequent disturbance by burning or grazing, no single factor can universally predict whether grassland, shrubland or steppe will form on a given site in transitional semi-desert climates. In some cases a slight shift in some environmental variable can have effects on the community structure which are disproportionate to the amount by which the variable is changed (Breman 1982).

Intrinsic differences between woody growth-forms and hemicryptophytes have logical consequences which can help clarify the complex vegetation patterns often observed in semi-desert climates.

1. Chamaephytes must divert a higher proportion of their resources to maintaining meristematic buds and conductive tissue in their more massive shoot and root systems, even during periods when photosynthesis is not possible. In active growth, additional resources must be committed to making non-productive tissue for support, storage, and protection (e.g., suberized bark).

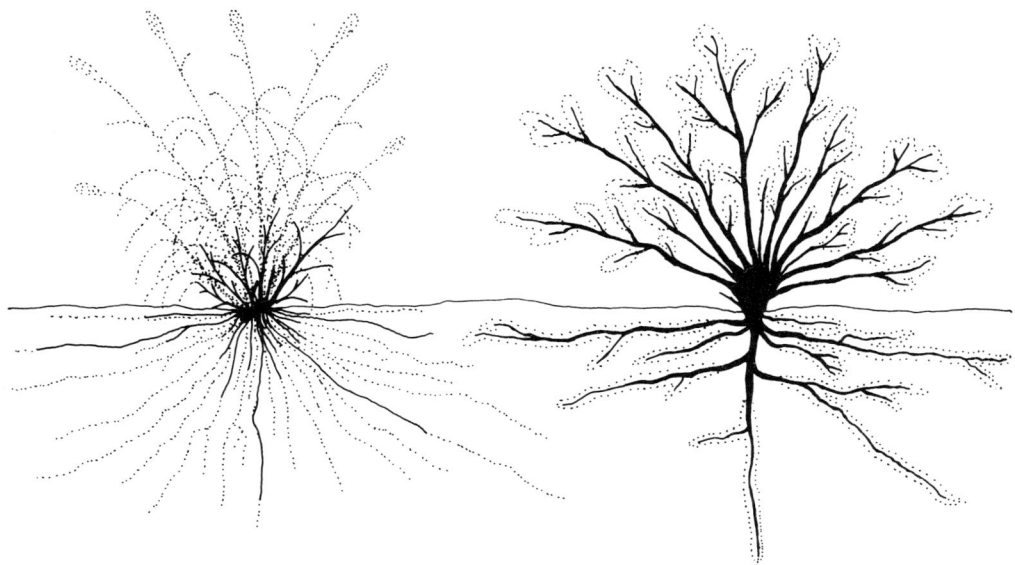

Fig. 6. Schematic diagram of annual change in biomass in hemicryptophyte and chamaephyte plants. Persistent biomass is outlined with solid lines. Dotted lines indicate extent of short-lived biomass.

2. By definition hemicryptophytes have very reduced perennial shoot organs. Grasses, the most important hemicryophytes of semi-desert climates, also appear to have substantial turnover rates of root biomass (Parton *et al.* 1978; Redmann and Reekie 1982). In grasses, resources which in chamaephytes would be sequestered for structural survival and maintenance, can be put into photosynthetic tissue and absorptive roots, allowing a more intensive exploitation of light and soil moisture. Under favorable conditions grasses can generally grow faster than chamaephytes. Annual flux of biomass in these two life forms is shown schematically in Fig. 6. The more open root system composed of a few large perennial roots is typical of arid-adapted chamaephytes (Cannon 1911; Evenari *et al.* 1971). These major woody roots may become quite long, and their branching patterns and depth are usually variable, even within a species. Most grass root systems have an extensive development of adventitious roots, resulting in a relatively dense network of small roots centered near the crown of the plant (Cable 1975; Noble 1981; Penning de Vries 1982).

3. The shoot architecture shows differences analogous to those described for root systems. Chamaephytes have a system of indeterminate branches formed from terminal shoot growth. Grass stems usually cease terminal growth after flowering. Although some species have extensive above-ground branching, most grasses add to their canopy by sprouting new stems from the crown near the soil surface. This has important consequences for the responses of grasses and chamaephytes to grazing and burning. Removal of large proportions of the shoot biomass will not impair regeneration of a grass canopy by basal sprouting, whereas the potential growth of most chamaephytes will be severely reduced. Four primary ecological factors determine the dominance relations between

chamaephytes and hemicryptophyte grasses (henceforth referred to simply as 'grasses').

Soil moisture regime

Because grasses loose much of their biomass during stressfull seasons their growth and height are closely coupled with soil moisture conditions of the current or most recent growing season (Cable 1975; Webb et al. 1978). For most grasses to reproduce, soil moisture must be available for a sufficient period to allow the plant to generate enough leaves and roots to effectively utilize light and soil moisture. In chamaephytes the basic structure of the canopy and root system is already in place when soil moisture becomes available. Only leaves and small absorbing roots need be grown to bring the plant to full productive capacity.

Figure 7 shows a typical frequency distribution of discrete rainfall events in a true desert climate (Button and Ben-Asher 1983; Shanan et al. 1967). About 50% of the total rainfall comes from events of less than 4 mm, and 74% is from storms delivering less than 8 mm. This means that more than half of the total rainfall moistens only the uppermost layers of the soil and evaporates very rapidly after the storms. Generally a 5 mm storm penetrates about 5 mm into the soil in the common desert soil types. Most of this soil moisture has evaporated after only 5–10 sunny days (Kadnor, pers. comm.). Thus about half of the potential water resource is unavailable to the plant unless it can be used rapidly (Noy-Meir 1973; Shreve 1934).

Woody perennials are better adapted to exploit this transient resource than most grasses. This is especially apparent in those species with long shoot-short shoot architecture. Many arid-adapted woody plants initially respond to rainfall by producing fascicled leaves from axillary buds or short shoots dispersed along branches (*e.g.*, *Fouquieria, Zygophyllum, Lycium*). By greatly reducing internode length the plant is able to grow this type of leaf more rapidly. If growth is possible over extended periods, branches elongate in the long shoot mode (Westman 1981).

When soil moisture is available for longer periods due to a higher frequency and intensity of storms, the faster growth of grasses gives them a competitive advantage. The dense fibrous root system can extract water and nutrients from a given volume of soil more efficiently than the coarser root system of chamaephytes, and the higher surface/volume ratio of roots with smaller diameters seems to allow a more thorough utilization of soil moisture for a given amount of root biomass (Penning de Vries 1982).

If seedlings of grasses and chamaephytes colonize the same open site the regime of soil moisture has a major influence on which plant will eventually dominate. If the soil moisture is adequate for a sufficient length of time the grasses will overtop the chamaephytes and deprive them of light. Their more rapid growth and denser root system also tends to exhaust soil moisture and nutrients before the chamaephytes can get them. On the other hand, if moisture is only sporadically available, the rapid response of the chamaephytes allows them to grow in short bursts, and use resources which are too limited to produce a competitively efficient grass canopy.

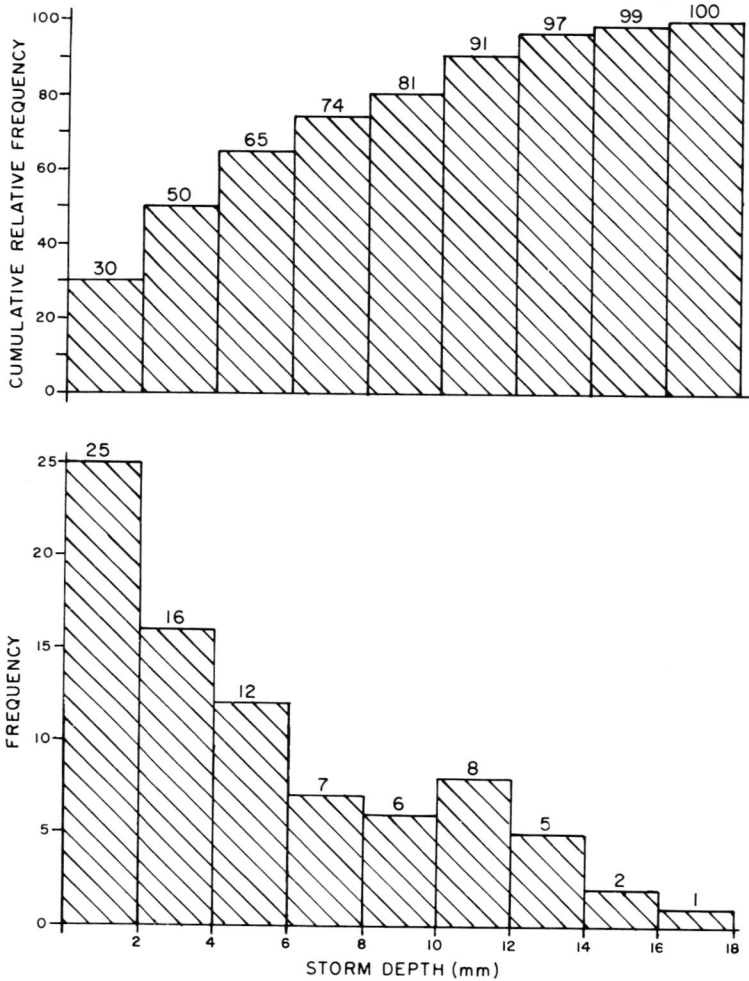

Fig. 7. Frequency distribution of precipitation amounts delivered by single storms in Avdat, Israel. Mean annual rainfall is 87 mm (after 20).

If there is regular annual alternation between drought and waterlogging, grasses almost always dominate (Cole 1982). These 'hyperseasonal' conditions (Sarmiento and Monasterio 1983) typically occur on low-lying clay soils in semi-arid climates with highly seasonal rainfall. In such circumstances improving soil drainage and aeration can convert grassland into shrubland or woodland (Tinley 1982). Apparently few woody plants can tolerate regular shifts from anaerobic to desiccated soils, and the mechanisms used by grasses to survive in hyperseasonal environments are not well-understood.

Compelling evidence for the advantage of a chamaephyte strategy in true deserts is the fact that some of the most arid-adapted perennial grasses have a growth-form functionally equivalent to a chamaephyte (Gillett 1941; Werger

1978). Convergent evolution has given rise to unrelated species that have relatively long-lived, branched shoots which remain functional during drought, with axillary buds which can initiate growth very soon after rain (Lazarides 1970; Nobel 1980). Examples include *Hilaria rigida* and *Muhlenbergia porteri* in North America, *Stipagrostis sabulicola*, *S. amabilis* and *Panicum turgidum* from Africa, and various species of *Triodia* and *Plectrachne* in Australia.

Temperature

Temperature can be an important factor in determining relative dominance of grasses and chamaephytes, but the relationship is not simple. Because most metabolic activities proceed more rapidly at higher temperatures, we should expect photosynthesis to be greater under warm conditions. But for most plants with normal Calvin cycle, or C_3 photosynthesis, the net gain of carbon decreases at higher temperatures due to photorespiration, a metabolic artifact of photosynthetic enzymes (Lorimer 1981; Smith 1976). C_4 plants generally have higher rates of carbon fixation under warm conditions, higher levels of light saturation, and lower carbon dioxide compensation points than C_3 plants (Hatch and Osmond 1976; Laetsch 1974). Thus under high temperatures and high light intensities, C_4 plants can fix more carbon than most C_3 plants. Available evidence suggests that C_4 metabolism is usually less efficient than C_3 photosynthesis under cool temperatures or low light levels (Ehleringer 1978; Ehleringer and Bjorkman 1977).

Laetsch (1974) proposed that the C_4 syndrome can function as a 'trap' for carbon dioxide within the leaf, reducing its diffusion from the leaf under conditions of stress when stomatal opening is minimal. This can keep leaves metabolically active for longer periods during drought and enhance the ability of the leaf to respond rapidly to rain. Sala et al. (1981) investigated the water balance of *Bouteloua gracilis*, a dominant C_4 grass of North American semi-arid plains. They found that leaf conductance and hence possible carbon intake is closely related to leaf water potential, which at dawn is correlated with the wettest area of the soil which the *Bouteloua* roots can reach. As the soil dries, leaf conductance decreases, but the leaves don't become dormant. Within less than one day after a simulated rainfall of 5 mm *Bouteloua gracilis* leaf water potential and conductance were restored to normal levels (Sala and Lauenroth 1982). The C_4 grass responds in a manner similar to a desert chamaephyte. If C_4 metabolism enables a plant to have a rapid response to rain, why isn't it more prevalent in deserts?

If the presence of soil moisture is correlated with cool conditions C_3 plants should dominate, and when rains are associated with warm temperatures C_4 plants should be favored (Ehleringer 1978). In semi-desert climates, but not in true deserts (see below) this relationship is generally observed, with C_3 shrubs and chamaephytes dominating in Mediterranean climates and extensive stands of C_4 grasses in subtropical areas (Werger and Ellis 1981). Where soil moisture is available in winter and summer the two types of plants can coexist by separating their major growth periods (Cable 1969; Kemp and Williams 1980). In areas with biseasonal rainfall, years with dry summers will favor the establishment and growth of C_3 plants, and the reverse situation promotes C_4 species. When species pools of both types are present, short term climatic fluctuations can induce major changes in the vegetation (Neilson 1986).

It is tempting to attribute the ecological differences between chamaephytes and grasses to the fact that many desert chamaephytes and shrubs have C_3 photosynthesis (Szarek 1979) and most grasses of semi-desert climates have C_4 metabolism (Brown 1977; Werger and Ellis 1981). These correlations work well in some situations, but they are not universal. C_4 photosynthesis is known from many chamaephytes (Downton 1975; Shomer-Ilan et al. 1981; Winter 1981), and there are arid grasslands dominated by C_3 grasses (Teeri and Stowe 1976). The ability of the C_3 grass *Bromus tectorum* (Waller and Lewis 1979) to develop its root system under cold conditions allows it to deplete soil moisture in early spring before other plants can use it (Harris 1967). This seems to be an important factor promoting its invasion of perennial grassland and steppe in the western United States. Rapid growth is the critical property of this grass, and for cold, C_3 metabolism is favored. C_3 and C_4 photosynthesis seem adaptations to temperature. They are not closely correlated with life form, and their relative merits are not consistent with respect to the contrasting strategies of rapid photosynthetic response versus fast vegetative growth.

Soil structure

Often grasses dominate on very heavy fine-textured or on very sandy soils, and chamaephytes tend to occupy well-drained and shallow soils on rocky slopes. An example is the shift from grasslands of basalt soils to chamaephyte steppe on very shallow limestone soils in the Middle East and east Africa or the shift from grasslands on heavy loam in dry riverbeds as well as on deep sand to chamaephyte steppe on limestone in the Kalahari (Leistner and Werger 1973). Heavier loams and clays promote the hyperseasonal moisture conditions favoring grasses, but other mechanisms may also be involved. If soil moisture is homogeneously distributed and fairly close to the surface, the fibrous root system of grasses can extract it efficiently. If moisture is concentrated or more permanent at specific microsites, as for example below rocks in the soil profile (Evenari et al. 1971), or at some depth below the crown, a system of long woody roots is desirable. Such long-lived roots concentrate absorption in the favorable sites and serve as dependable conduits between the rest of the plant and patchy resources which are difficult to locate (Shmida 1982). When loam or clay-rich horizons are overlain by a veneer of sand which minimizes runoff, grasses often dominate even under relatively arid conditions (Tinley 1982). If the sand layer is removed increased runoff can significantly reduce and redistribute available moisture, allowing more drought-tolerant chamaephytes to replace the grasses.

Walker et al. (1981) have used simple models to simulate dynamics of grass and woody elements of savanna ecosystems. Their models partition soil moisture into topsoil and subsoil components, giving woody plants exclusive access to subsoil moisture, and allowing both growth-forms to compete for topsoil water. By allowing feedback between grass cover, infiltration rate, and soil water, dynamics can be generated which conform to observations of natural systems. The exclusive access to some portion of the soil moisture is a key component allowing woody plants to persist. Soil structure which generates a very patchy distribution of soil moisture, either spatially or temporally, seems to favor chamaephytes and shrubs over grasses. In the case of hyperseasonal conditions, anaerobic soil moisture may be largely unavailable to roots of most woody plants. During the wet season, shrubs would be forced to depend on shallower soil horizons where competition with grass roots is intense. If this scenario is correct, soils promoting hyperseasonal conditions may actually reduce the patchiness of soil moisture as perceived by roots, and favor grasses which exploit topsoil more thoroughly.

Disturbance

The biomass of grasses can be replaced more rapidly than chamaephyte stems or roots. This makes grasses well-adapted to above-ground disturbance, and they can usually regenerate very rapidly after burning and sometimes after grazing. Fire and heavy grazing have had major impacts in most arid and semi-arid regions, their severity being closely related to the level of human occupation. How do these kinds of disturbance effect the relative importance of chamaephytes and perennial grasses? The trend is relatively simple for fire but less clear for grazing.

There are no fires in diffuse and contracted desert. The canopy is too open and fuel too scattered to propagate a fire. In semi-deserts with a continuous vegetative cover fires develop very easily. When burning is frequent grasses which can regenerate their canopy between fires are favored, but woody species with lignotubers or extensive underground organs can also persist if the burn is not too hot (Parsons 1981; Sarmiento and Monasterio 1983; Werger 1983).

Cutting, grazing and mowing have been frequently imposed upon steppes and grasslands over the last century (Le Houerou 1970; Naveh and Dan 1973; Walker *et al.* 1981). The studies which have been conducted to analyze and predict vegetation dynamics under various disturbance regimes have mostly failed to show a simple pattern. From a theoretical point of view it appears that grassland will overcome chamaephyte vegetation if disturbance frequency is close to the regeneration rate of the grasses, which is usually faster than that of chamaephytes (White 1977). But grazing is selective, and most stock prefer herbaceous to woody stems.

Many vast areas of 'nice' grassland have converted to chamaephyte vegetation over the past century. According to Zohary (1973) much of the fertile crescent in the Middle East used to be grassland, but heavy grazing has eliminated most perennial grasses from this region. The situation in South Africa and North America is similar. Among the better documented studies are those of Acocks (1953), Buffington and Herbal (1965), Hastings and Turner (1965), York and Dick-Peddie (1969) and Werger (1980), all showing extensive replacement of grassland by communities of shrubs and chamaephytes. Once the invasion of grassland by shrubs is underway cessation of grazing may not halt the process (Hennessy *et al.* 1983; Walker *et al.* 1981). We can speculate that some differences in the resistance of grasslands to grazing are due to the kinds of herbivores they have coevolved with. High levels of ungulate herbivory are an integral part of African grasslands (McNaughton 1983, 1985), but it was relatively unimportant in pristine Neotropical grasslands (McNaughton *et al.* 1982).

In summary, frequent burning usually drives the system to a grassland, but there is no simple pattern relating grazing regime with the relative importance of chamaephytes and grasses.

Unstable sand

Compared with the complex situations we have examined above, the sand dune habitat has a relatively straightforward pattern of the favored life form – perennial grasses. Sand dunes are widespread along shores, especially in true desert climates (Brown 1968; Shmida 1985), and huge areas of the Sahara, Arabia, Central Asia, southern Africa and Australia are covered with sand fields (McKee 1979). When the sand forms moving dunes with crested ridges the vegetation cover is very sparse (0%–3% coverage) and it is often dominated completely by perennial grasses. Those non-graminoid species which are common on sand dunes often show grass-like characteristics of rapid stem growth and adventitious root formation (Bowers 1986). In very unstable sand no plants occupy the deflating sides of the dunes (Shmida 1972; Zohary 1962), but grasses can grow on the leeward areas.

In dunes the grass growth-form finds a suitable 'niche'. The plant can grow quickly after sudden burial by sand. There is almost no investment necessary for support tissue because the sand holds the buried stems up. Good examples of ecological convergence occur among disjunct sand dunes on different continents, each harboring similar-looking but unrelated grasses, as *Zygochloa paradoxa* in Australia, *Stipagrostis scoparia* in the Middle East, and *Swallenia alexandrae* in California (Bowers 1986; Danin 1983; Purdie 1984; Shmida 1985).

Open canopy structure in woody desert plants

A very open, diffuse branching architecture is typical of woody desert plants, whereas in montane and alpine ecosystems chamaephytes have a tendency to form very dense, low mounds. In the Middle East there are good examples of structural differentiation even within a species. For example *Noaea spinosa* has a sparse, tall ecotype in the desert and a cushion-like form on ledges in the subalpine belt of Mt. Hermon (Shmida 1972, 1977). Trees and shrubs of the thorny microphyll scrub, as in the Sonoran region of North America and the Sudanian region of Africa, also can have an open canopy and diffuse leaf arrangement (Shreve 1951; Van der Meulen and Werger 1984; Whittaker and Niering 1964; Zohary 1973).

What is the advantage of an open structure? Light is not usually limiting in the upper strata of these communities, but water frequently is. The diffuse canopies in the subtropical deserts and semi-deserts promote maximum convective heat exchange between the leaves and the air. Models developed by Parkhurst and Loucks (1972) and Givnish (1979) indicate that in environments with high light intensity, transpiration will be minimized by reducing leaf size, thereby coupling leaf and air temperatures more closely. An open canopy should further enhance convective cooling. In contrast a cushion-like form tends to behave more like a single large leaf, which can be much warmer than the surrounding air. As usual in biology, there are exceptions. In extreme desert climates of the Sahara, *Fredo-*

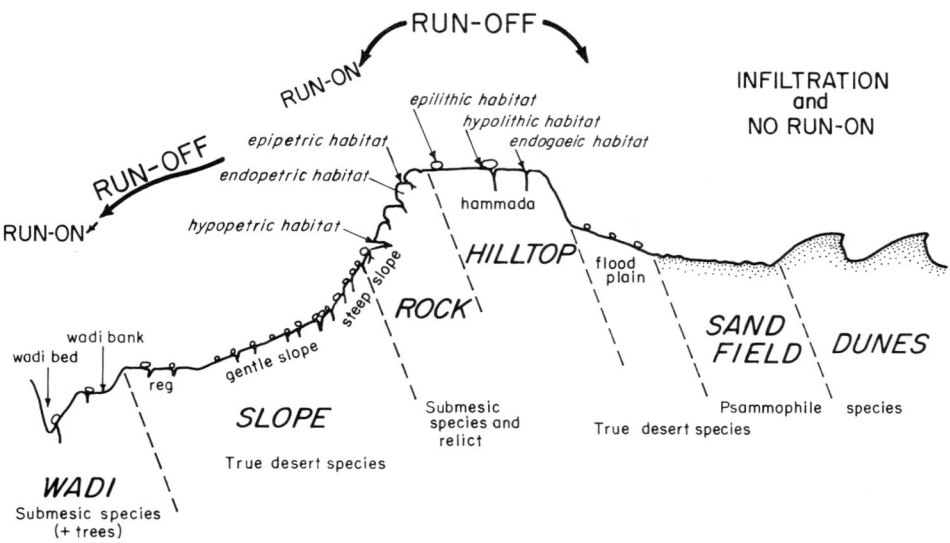

Fig. 8. Schematic representation of the major desert habitats as defined by geomorphology.

lia artioides and *Limoniastrum feei* form very dense, stone-like mounds which differ greatly from the expected canopy structure (Ozenda 1977).

Low energy and high energy strategies

Arid environments can be generally characterized by a scarcity of water and often by an unfavorable balance of minerals, but these resources are highly variable and unpredictable (Evenari *et al.* 1971; McDonald 1956; Noy-Meir 1973). This heterogeneity exists at both spatial and temporal scales and at different levels. The availability of these vital resources changes greatly through time and space, among years as well as within a single year. Water and nutrients are abundant after rain, but their temporal duration and quantity are unpredictable. Spatial heterogeneity in deserts results from the following causes:

a. Geological and geomorphological heterogeneity (Fig. 8).
b. Redistribution of rainfall water by runoff from some areas and concentration in others.
c. Rainfall patchiness, which in part is predictable (*e.g.*, orographic) and in part erratic and unpredictable (*e.g.*, storm cells (Sharon 1972, 1978).
d. Biologically induced patchiness (*e.g.*, special microclimatic and microedaphic habitats around plants, detritus heaps and animal burrows (Charley and McGarity 1964; Evenari *et al.* 1971).

Spatial variation in concentration and availability of water has more important consequences in arid environments than in more productive biomes. Diverse habitats with very different water regimes are created (Fig. 8). The pattern is influenced by relief, rainfall intensity, and the presence of surfaces which impede

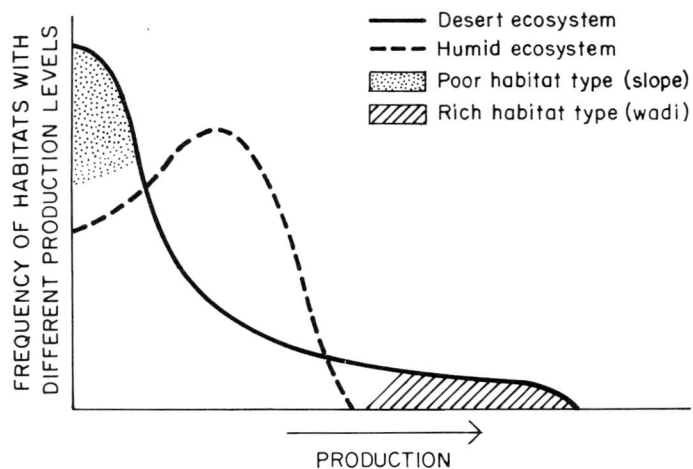

Fig. 9. Frequency distribution of habitat types ordered by their relative primary production. Desert and humid ecosystems are compared.

infiltration (rocks, crust-forming soils) (Musick 1975; Yair and Shachak 1982; Werger 1986). In deserts where rainfall is usually less than about 100 mm and the surface produces runoff, there is a drastic ecological contrast between widespread dry, poor habitats and very favorable sites with water accumulation (*e.g.*, wadis and bases of outcrops) which occupy only a small proportion of the area (Evenari *et al.* 1971; Hekman and Berkas 1981; Lee 1988).

Desert habitats may differ greatly and it is not surprising that there are diverse growth-forms adapted to different environments. Highly contrasting sites can partially compensate for temporal variability and uncertainty, because in dry years more mesic habitats provide a refuge for many plants and animals. In some desert areas with strong spatial differentiation, biological fluctuations are smaller and total productivity is higher than in deserts with a similar climate but less ecological differentiation (Fig. 9). The high proportion of runoff sites gives the tiny habitats predictably moist, favorable conditions.

In true deserts the extreme unpredictability of resources should favor dominance of species with 'r-type' strategies – short lived, fast growing, extensively dispersing organisms (Orians and Solbrig 1977; Pianka 1973). But desert growth-forms adapted to employ other strategies are very common. Westoby (1980) claimed that in desert environments each of the various growth-forms uses water and other resources differently in space and time. Noy-Meir (1973) reviewed the different life history strategies in deserts and found that there are some studies which document 'high-performance' organisms with good competitive ability and fast growth. Other studies emphasize 'low-performances' species showing small net biomass gains and low competitive ability. How can we resolve this apparent contradiction? How can slow-growing cactoids and chamaephytic Chenopodiaceae coexist with annuals and fast-growing chamaephytes?

The diverse strategies of desert plants can be ordered along an axis (Fig. 10) oriented according to what we term a 'low energy strategy' and a 'high energy

Plant growth-form strategies and vegetation types 231

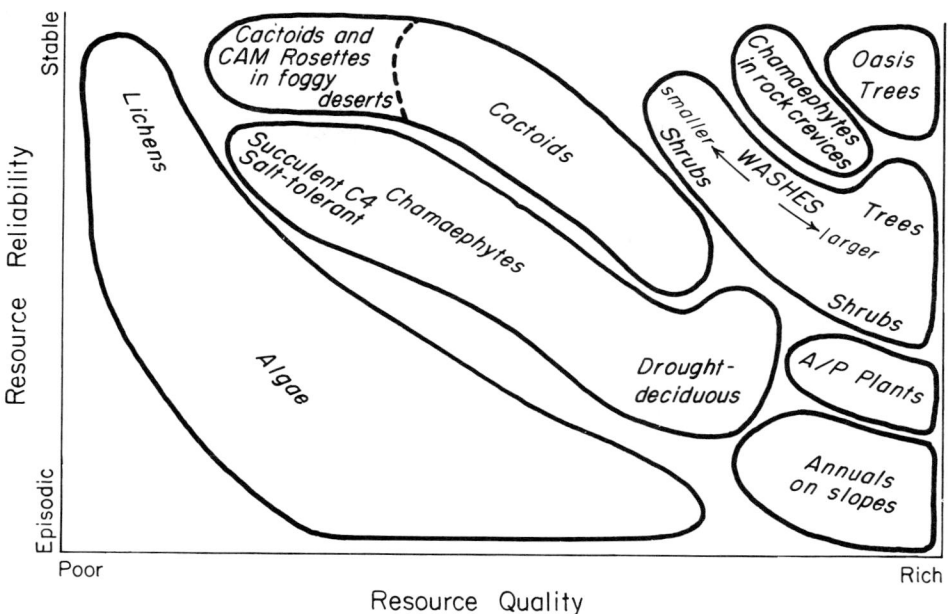

Fig. 10. Plant growth-forms/strategies plotted against resource quality and resource reliability.

strategy' (Shmida *et al.* 1986). We define a low energy strategy as one using poor resources with low rewards, hence production is also poor (slow growth). A high energy strategy can be recognized by its high rate of resource income, a high metabolic rate, and a rapid growth rate. Often but not always these attributes are coupled, making it easy to segregate different growth-form strategies along this niche axis. Succulents with Crassulacean Acid Metabolism (CAM) are an exception, because they show high rates of resource income yet have relatively slow net growth rates (Burgess and Shmida 1988; Kluge and Ting 1978). We emphasize that the high energy strategy and low energy strategy are mainly differentiated by their relative efficiency – their income per unit time – and not their total production. In distinguishing between the two strategies it does not matter how many days per year the plant is active, but rather how much production occurs on an active day (*e.g.*, poikilohydric ferns have a high energy strategy, see below). At one terminus of the axis are organisms which are adapted to habitats with resources which are unpredictable, episodic, and have high energetic rewards (habitats are defined here on a temporal as well as spatial basis, *e.g.*, hamadas in 'good' rainy years). At the other extreme are growth-forms which utilize the more temporally predictable habitats with low energetic rewards. We postulate that segregating the different growth-form types described above according to their resource exploitation strategies can explain a major part of the diversity of growth-forms observed in deserts.

The different types of resources and habitats in deserts can be characterized according to two main attributes: richness (the concentration and abundance of the resources, the ease of getting and using them, their quality, and energetic reward) and reliability (the frequency and predictability of the resource). Reliable rich resources can be found in deserts only in oases which are very limited in area. Away from the oases, resources which are rich and easy to use are usually episodic, that is short-lived and unreliable (*e.g.*, free water and topsoil moisture just after a rain). Resources which last longer or are more reliable either have low concentrations (dew, fog), or are difficult to find (soil moisture in deep horizons or under scattered stones), or have low quality (saline water).

We can distinguish two contrasting modes of adaptation in desert organisms in response to this gradient of resources: adaptation for monopolizing high-quality episodic resources versus adaptation for more thorough use of poor resources.

Poor resources

Attributes necessary for utilization of poor but relatively predictable water resources include high physiological tolerance to drought and temperature extremes, efficient (even if slow) processing of resources, special mechanisms for obtaining and using particular kinds of unusual resources, and low population turnover with increased individual longevity.

Some types of desert plants have evolved mainly in this direction: true perennial xerophytes, CAM succulents, lichens using dew. These have been called drought-resistant or arido-active plants (Evenari 1981). A general tendency of all organisms which utilize poor resources in deserts is to be adapted to reduce water loss. This can be achieved by physiological properties such as efficient stomatal responses and nocturnal activity (CAM plants). Structural adaptations are also involved, for example special cuticular and epidermal structures, and specialized storage organs. These structural adaptations have an energetic cost, and thus reduce potential net production (Fig. 2). Low amounts and poor quality of resources restrict metabolic activity in time and/or in rate, imposing low productivity which is typical for arido-active desert organisms.

Rich resources

To make good use of episodic but rich resources in deserts, a different set of attributes is needed: rapid (even if inefficient) use of resources, high reproductive rates, and a capability of changing quickly from dormant to active stages in the life cycle in response to water availability. Examples of adaptation in this opportunistic direction are desert annuals and geophytes. A further distinction should be made between resources which are episodic in time but dependable in space, and resources which are also unpredictable in space (Ellner and Shmida 1981). If the moisture occurs in the same place after every rain, annuals established there should not disperse, and dormant stages of perennials must be capable of surviving long droughts. Utilization of rich resource patches which shift irregularly in space is possible only if a plant produces large amounts of widely dispersed seeds (Shmida and Ellner 1984).

As a rough illustration of these ideas, we have arrayed some types of desert plants according to the richness and reliability of the resources they use (Fig. 10). The diagram indicates that a population may persist by using only the best habitats during dry periods and dispersing into a wider range of poorer sites in wetter times (Mott 1973).

Adaptations for opportunistic use of episodic resources can also be useful in habitats subjected to surface disturbance (dunes, water-courses). These sites may be colonized by organisms with well-developed dispersal mechanisms. Other species may survive by tolerating surface disturbances and regenerating from remaining viable organs.

The annuals, 'skimming the cream'

Annual plants are adapted to go through dry periods as a packaged embryo which is the most desiccation-resistant morphological unit of angiosperms. Before extensive human disturbance favored their spread, the richest annual communities were centered in two ecosystems.

1. *The Mediterranean region.* The open chaparral, phrygana and annual pastures have the most diverse assemblages of annuals in the world (Shmida 1981, 1986). In Mediterranean climates annuals pass through the dry summer as seeds. It is a predictable system in which the annuals can successfully reproduce almost every winter (Shmida and Ellner 1984).

2. *Desert environments.* In true and extreme desert climates the development of extensive stands of annuals growing on slopes (which may account for as much as 90% of the landscape) is rare. This happens every five to seven years in the Negev of Israel (Evenari *et al.* 1971) but intervals may be much longer. Seeds of annuals must not only remain viable after long droughts but also must detect large rainfall events and germinate only when the soil water supply is adequate for growth and reproduction (Ellner and Shmida 1981; Venable and Lawlor 1980). For accomplishing this they have ingenious 'rain guage' mechanisms which have been studied in detail for a few species (Evenari *et al.* 1971; Mott 1974). In areas with biseasonal rainfall seeds also are able to suppress germination if temperatures are unfavorable for vegetative growth (MacMahon and Wagner 1985; Mulroy and Rundel 1977; Went 1949).

The notion that the annual growth-form is the dominant type in true deserts should not be uncritically accepted, because most of the time they are hidden in the soil seed bank. We can always find chamaephyte growth-forms, but the annual growth-form is restricted to temporally limited habitats.

Raunkiaer (1934) was among the first to observe that 'A therophyte climate is in the regions of the subtropical zone with winter rain'. He also noted that 'therophytes are distributed much more easily by cultivation than other life-forms'. The annual growth-form probably evolved in semi-arid and arid environments with a long dry season and an episodic but rich resource. This selected for a life history strategy which was a good preadaptation for the weedy habitats which many annuals have invaded worldwide (Baker 1974; Baker and Stebbing 1965).

Summer annuals, winter annuals and A/P plants

Annuals have a successful strategy in a region with regular drought. In Mediterranean climates the dry and moist periods are very predictable. In the northern

Middle East and northern Sahara there are absolutely no summer rains and the winter storms are unpredictable. In such regimes winter annuals are quite common and summer annuals are absent.

In deserts of southwestern North America summer rains increase in frequency from California to northeastern Mexico (Huning 1978; MacMahon and Wagner 1985; Mitchell 1976; Shreve 1951). In Arizona and New Mexico both winter and summer storms create a biseasonal rainfall regime, and both winter and summer annuals coexist (Freas and Kemp 1983; MacMahon and Schimpf 1981; Orians and Solbrig 1977). In these areas there are more species of winter annuals than obligate summer annuals (Bowers and Turner 1986; Kemp 1983; Shreve 1951). Also common in these regions are short-lived perennials (*e.g.*, *Machaeranthera pinnatifida*) and annuals which may survive over more than one rainy season, which can be termed A/P plants (*e.g.*, *Baileya multiradiata*).

To reduce confusion, we offer the following tentative definitions as a basis for further investigations:

1. A/P life history – 10 to 50% of those individuals alive at the end of their first rainy season (*i.e.*, the one in which they germinated) will survive until a second rainy season, and 0 to 10% of the initial cohort survive until a third rainy season.

2. P/A life history – roughly equivalent to the 'pauciennial' life history described by Grubb (1986) – of those individuals alive at the end of their first rainy season, 70 to 95% are present at the end of the second rainy season and 20 to 50% survive until the end of their third rainy season.

There is a continuum of life history schedules between A/P and P/A types, but the distinction is conceptually useful and has a real basis. A/P plants are more typical of deserts, whereas P/A plants are more often found in weedy habitats and open grasslands of more humid climates (Schenkeveld and Verkaar 1984).

The higher diversity of winter annuals in comparison with obligate summer annuals, and the advantage of the A/P life history can be explained by comparing winter- and summer-rainfall climates. In winter rainfall regimes (*e.g.*, Jericho in the Middle East and Death Valley in California) the occurrence of rain in at least some winters and drought every summer is predictable (Aschman 1973; Atlas of Israel 1970; Bailey 1966), hence a good strategy is to be an obligate annual. It does not pay to leave vegetative organs with energetic and transpiration costs if drought is predictable for at least four to five months, which is normal for Mediterranean climates (Shmida 1981).

In contrast, summer rainfall climates frequently have more than one period of rain within a year, or the rain is more scattered over several months (Flohn 1969; Minckley and Brown 1982; Neilson 1986; Rumney 1968; Sellers and Hill 1974). In such environments the probability of a short dry period is much higher than in winter-rain deserts. In these situations obligate, 'big-bang' annuals (sensu Schaffer and Gadgil 1975) which die at the onset of dry conditions, may produce fewer seeds than plants which can survive a short drought and produce a second crop of seeds in the next rainy period. The optimal life history strategy may be to become an A/P plant – a therophyte which after flowering diverts some energy in vegetative organs in order to survive poor conditions and rapidly initiate growth if another rain occurs. Some examples of A/P plants in North American deserts are *Erioneuron pulchellum*, *Verbesina encelioides* (Shreve 1951),

Aphanostephus ramosissimus, and *Abronia angustifolia* (Royce and Cunningham 1982). In some A/P species there is a marked tendency to shift from an A/P strategy in biseasonal rainfall areas to a strict annual strategy in winter rainfall regions. For example, *Erioneuron pulchellum* behaves as an annual in Death Valley.

In the Old World the transition from winter to summer or biseasonal rainfall regimes occurs over a very long geographical gradient (Shmida 1985), and in the middle, the extreme Central Saharan desert has almost no rain at all (5 mm average, which actually comes from a single storm every 20–50 years, Ozenda 1977). Along this gradient obligate annuals are totally dominant in the northern margins of the Sahara and Middle East deserts, comprising up to 97% of the total flora. The A/P plants begin to appear when winter annual rainfall drops below 50 mm and there is a low probability of warm-season rain in late spring or autumn. The same shift from obligate annuals to A/P plants can be observed both on the floristic level and within a species. Examples are *Diplotaxis harra*, *Reseda verrucosa*, and *Trichodesma africana*, all obligate annuals in the northern Negev and A/P plants farther south around Eilat and the Sinai peninsula.

Ferns and resurrection plants

While the 'classic' studies include lichens, ferns, and club mosses (*Selaginella*) among the poikilohydric xerophytes (Gaff 1981; Walter and Stadelmann 1974), we prefer to separate them along the 'high – low energy axis'. The lichens have an extreme low energy strategy (poor resource and very slow net annual growth) (Kappen et al. 1980; Rogers 1971) whereas the ferns and clubmosses have a high energy strategy type (Eickmeier 1983; Nobel 1978). Through much of the year the latter plants may be dormant but they can resume full photosynthetic function within a few hours after rain. There is still much that is poorly understood about plants with this 'risky' strategy (Ellner 1987; Hatch and Osmond 1976).

Desert species survive on reliable poor resources, episodic rich resources, or a combination of the two. But some resources are not used at all in deserts because they are insufficient to maintain any populations, or they are so erratic that no population is able to synchronize with their availability, or catastrophic surface disturbances have destroyed populations which could exploit them. Though water is the limiting resource to desert plants, a considerable proportion of the rain falling on deserts is not used by plants. Instead it evaporates from areas where it has not been sufficient to induce germination, or it seeps into unvegetated mobile dunes and flood channels.

Acknowledgments

We are very grateful to Raymond Turner, Michael Rosenzweig, James Aaronson, and Gadi Polak for discussions and encouragement. Many people have helped with field work contributing to this analysis, including Robert Thorne, Reid Moran, Richard Felger, Mitchell Beauchamp, Michael Barbour, and Jack Major. A.S. expresses his thanks to his teachers M. Zohary, R. Thorne, H. Ellenberg and

Z. Sachs. T.R. received much conceptual help from A. Gibson and J. O'Leary. We are also very indebted to Raymond Turner, whose support made this paper possible.

References

Acocks, J.P.H. 1953. Veld types of South Africa. Mem. Bot. Surv. S. Afr. 28: 1–192.
Adamson, R.S. 1939. The classification of life-forms of plants. The Botanical Review 5: 546–561.
Aschman, H. 1973. Distribution and peculiarity of Mediterranean ecosystems. In: F. Di Castri and H.A. Mooney (eds), Mediterranean-type ecosystems; Origin and structure, pp. 11–19. Springer Verlag.
Atlas of Israel. 1970. Surveys of Israel Ministry of Labor. Jerusalem and Elsevier Pub. Co., Amsterdam.
Baily, H.P. 1966. The climate of southern California. California Natural History Guide 17, University of California Press, Berkeley.
Baker, H.G. 1974. The evolution of weeds. Ann. Rev. Ecol. Syst. 5: 1–25.
Baker, H.G. and Stebbins, G.L. (eds) 1965. The genetics of colonizing species. Academic Press, New York. 588 pp.
Beadle, N.C.W. 1981. The vegetation of Australia. Gustav Fischer Verlag, Stuttgart & New York. 690 pp.
Beard, J.S. 1955. The classification of tropical American vegetation types. Ecology 36: 89–100.
Bowers, J.E. 1986. Seasons of the wind: sand dunes of the southwest. Northland Press.
Bowers, J.E. and Turner, R.M. 1986. A revised vascular flora of Tumamoc Hill, Tucson, Arizona. Madrono 32: 225–252.
Boyko, H. 1947. On the role of plants as quantitative climate indicators and the geoecological law of distribution. J. Ecol. 35: 1–27, 138–157.
Breman, H. 1982. La productivité des herbes pérennes et des arbres. In: F.W.T. Penning de Vries and M.A. Djitèye (eds), La Productivité des Paturages Sahéliens, pp. 284–389. Centre for Agricultural Publishing and Documentation, Wageningen.
Brown, D.E. 1982. Great Basin montane scrubland. Desert Plants 4: 83–84.
Brown, W.V. 1977. The Kranz syndrome and its subtypes in grass systematics. Memoirs of the Torrey Botanical Club 23(1): 1–97.
Buffington, L.C. and Herbel, C.H. 1965. Vegetational changes on a semidesert grassland range from 1858 to 1963. Ecol. Mono 35: 139–164.
Burgess, T.L. and Shmida, A. In press. Succulent growth-forms in arid environments. In: E.E. Whitehead, C.F. Hutchinson, B.N. Timmermann and R.G. Verity (eds), Arid Lands: Today and Tomorrow. Proceedings of an international research and development conference. Office of Arid Lands Studies, University of Arizona, Tucson.
Button, B.J. and Ben-Asher, J. 1983. Intensity-duration relationships of desert precipitation of Avdat, Israel. J. Arid Envir. 6: 1–12.
Cable, D.R. 1969. Competition in the semidesert grass-shrub type as influenced by root systems, growth habits, and soil moisture extraction. Ecology 50: 27–38.
Cable, D.R. 1975. Influence of precipitation on perennial grass production in the semidesert Southwest. Ecology 56: 981–986.
Cain, S.A. 1950. Life-forms and phytoclimate. The Botanical Review 16: 1–32.
Cannon, W.A. 1911. The root habits of desert plants. Publication No. 131, Carnegie Institution of Washington, D.C. 96 pp.
Charley, J.L. and McGarity, J.W. 1964. High soil nitrate levels in patterned saltbush communities. Nature 201(4926): 1351–1352.
Cole, M.M. 1960. Cerrado, caatinga and pantanal: the distribution and origin of the savanna vegetation of Brazil. Geogr. J. 126: 168–179.
Cole, M.M. 1982. The influence of soils, geomorphology and geology on the distribution of plant communities in savanna ecosystems. In: B.J. Huntley and B.H. Walker (eds), Ecology of Tropical Savannas, pp. 145–174. Springer-Verlag, New York.
Danin, A. 1970. A phytosociological-ecological study of the Northern Negev of Israel. Ph.D. thesis. Dept. of Botany, The Hebrew University, Jerusalem.

Danin, A. 1983. Desert vegetation of Israel and Sinai. Cana Publ. House, Jerusalem, Israel. 148 pp.
Daubenmire, R. 1978. Plant geography, with special reference to North America. Academic Press, New York. 338 pp.
Downton, W.J.S. 1975. The occurrence of C_4 photosynthesis among plants. Photosynthetica 9: 96–105
Ehleringer, J. 1978. Implications of quantum yield differences on the distribution of C_3 and C_4 grasses. Oecologia (Berlin) 31: 255–267.
Ehleringer, J. and Bjorkman, O. 1977. Quantum yields for CO_2 uptake in C_3 and C_4 plants: dependence on temperature, CO_2, and O_2 concentration. Plant Physiology 59: 86–90.
Eickmeier, W.G. 1983. Photosynthetic recovery of the resurrection plant *Selaginella lepidophylla* (Book. and Grev.) spring: effects of prior dessication damage. Oecologia (Berlin) 58: 115–120.
Eig, A. 1931. Les éléments et les groupes phytogéographiques auxiliares dans la flore Palaestinienne. Fedde Report. Spec. Nov. regn. veget. Beihf. 63, Vol. II. Dahlem-Berlin, 120 pp.
Ellison, L. 1960. Influence of grazing on plant succession of rangelands. The Botanical Review 26: 1–78.
Ellner, S. 1987. Alternate plant life-history strategies and coexistence in randomly varying environments. Vegetatio 69: 199–208.
Ellner, S. and Shmida, A. 1981. Why are adaptations for long-range seed dispersal rare in desert plants? Oecologia 51: 133–144.
Estes, J.R., Tynl, R.J. and Brunken, J.N. 1982. Grasses and grasslands-systematics and ecology. University of Oklahoma Press, Norman. 312 pp.
Evenari, M. 1981. Synthesis. In: Goodall, D.W. and Perry, R.A. (eds), Arid Land Ecosystems: Structure, Functioning and Management. Vol. 2, pp. 555–599. Cambridge University Press, Cambridge.
Evenari, M., Shanan, L. and Tadmor, N. 1971. The Negev: the challenge of a desert. Harvard University Press, Cambridge, MA. 245 pp.
Flohn, H. 1969. Climate and weather. World Univ. Library, Weidenfield and Nicolson. 252 pp.
Freas, K.E. and Kemp, P.R. 1983. Some relationships between environmental reliability and seed dormancy in desert annual plants. J. Ecol. 71: 211–218.
Gaff, D.F. 1981. The biology of resurrection plants. In; J.S. Pate and A.J. McComb (eds), The Biology of Australian Plants, pp. 114–146. University of Western Australia Press, Nedlands, Western Australia.
Gillett, J.B. 1941. The plant formations of Western British Somaliland and the Harar Province of Abyssinia. Bulletin of Miscellaneous Information No. 2, Royal Botanic Gardens, Kew, 37–75.
Givnish, J.J. 1979. On the adaptive significance of leaf form. In: O.T. Solbrig, P.H. Raven, S. Jain and G.B. Johnson (eds), Topics in Plant Population Biology, pp 375–407. Columbia University Press, New York.
Goodall, D.W. and Perry, R.A. (eds) 1979. Arid land ecosystems: structure, functioning and management. Vol. 1. Cambridge University Press, Cambridge. 381 pp.
Griffin, J.R. 1977. Oak woodland. In: M. Barbour and J. Major (eds), Terrestrial Vegetation of California, pp. 383–416. Wiley, New York.
Grubb, P.J. 1986. Problems posed by sparse and patchily distributed species in species-rich plant communities. In: J. Diamond and T.J. Case (eds), Community Ecology, pp. 207–225. Harper and Row, New York.
Harris, G.A. 1967. Some competitive relationships between *Agropyron spicatum* and *Bromus tectorum*. Ecol. Mon. 37: 89–111.
Hatch, M.D. and Osmond, C.B. 1976. Compartmentation and transport in C_4 photosynthesis. In: C.B. Stocking and V. Heber (eds), Transport in Plants III: Intracellular Interactions and Transport Processes, pp. 144–184. Springer-Verlag, New York.
Hekman, L.H. and Berkas, W.R. 1981. Modeling desert runoff. In: D.D. Evans and J.L. Thames (eds), Water in Desert Ecosystems, pp. 244–264. US/IBP Series 11. Dowden and Hutchinson, Stroudsburg, Pennsylvania.
Hennessy, J.T., Gibbens, R.P., Tromble, J.M. and Cardenas, M. 1983. Vegetation changes from 1935 to 1980 in mesquite dunelands and former grasslands of southern New Mexico. J. range Managem. 36: 370–374.
Hills, T.L. 1965. Savannas: a review of a major research problem in tropical geography. Canadian Geographer 9: 216–228.
Humphrey, R.R. 1958. The desert grassland, a history of vegetational change and an analysis of causes. Bot. Rev. 24: 193–252.

Huning, J.R. 1978. A characterization of the climate of the California desert. Desert Planning Staff, Bureau of Land Management, Riverside, California. 220 pp.
Kappen, L., Lange, O.L., Schulze, E.D., Buschbom, U. and Evenari, M. 1980. Ecophysiological investigations on Lichens of the Negev Desert VII. The influence of the habitat exposure on dew inhibition and photosynthetic productivity. Flora 169: 216–229.
Kemp, P.R. 1983. Phenological patterns of Chihuahuan desert (New Mexico, USA) plants in relation to the timing of water availability. J. Ecol. 71: 427–436.
Kemp, P.R. and Williams, G.J. III. 1980. A physiological basis for niche separation between *Agropyron smithii* (C_3) and *Bouteloua gracilis* (C_4). Ecology 61: 846–858.
Kluge, M. and Ting, I.P. 1978. Crassulacean acid metabolism, analysis of an ecological adaptation. Ecological studies, Vol. 30. Springer, New York. 211 pp.
Koppen, W. 1954. Classification of climates and the world patterns. In: G.T. Trewartha (ed.), An Introduction to Climate (3rd ed.). pp. 225–226. McGraw-Hill, New York. 402 pp.
Laetsch, W.M. 1974. The C_4 syndrome: a structural analysis. Ann. Rev. Plant Physiol. 25: 27–52.
Lazarides, M. 1970. The grasses of Central Australia. Australian National University Press, Canberra. 282 pp.
Le Houerou, H.N. 1970. North Africa: Past, present, future. In: H.E. Dregne (ed.), Arid Lands in Transition, pp. 227–278. American Association for the Advancement of Science, Washington, D.C.
Lee, M. In press. The development of a distributed computer simulation model of a reconstructed ancient water-harvesting system. In: E.E. Whitehead, C.F. Hutchinson, B.N. Timmermann and R.G. Verity (eds), Arid Lands: Today and Tomorrow. Proceedings of an international research and development conference. Office of Arid Lands Studies, University of Arizona, Tucson.
Leistner, O.A. 1967. The plant ecology of the southern Kalahari. Mem. Bot. Surv. S. Afr. 38: 1–172.
Leistner, O.A. and Werger, M.J.A. 1973. Southern Kalahari phytosociology. Vegetatio 28: 353–399.
Lorimer, G.H. 1981. The carboxylation and oxygenation of ribulose 1,5-bisphosphate: the primary events in photosynthesis and photorespiration. Ann. Rev. Plant Physiol. 32: 349–383.
MacMahon, J.A. and Schimpf, D.J. 1981. Water as a factor in the biology of North American desert plants. In: D.D. Evans and J.L. Thames (eds), Water in Desert Ecosystems, pp. 114–171. Dowden & Hutchinson, Stroudsburg Pennsylvania.
MacMahon, J.A. and Wagner, F.H. 1985. The Mojave, Sonoran, and Chihuahuan Deserts of North America. In: M. Evenari, I. Noy-Meir and D.W. Goodall (eds), Hot Deserts and Arid Shrublands, pp. 105–202. Elsevier, Amsterdam.
McDonald, J.E. 1956. Variability of precipitation in an arid region. Technical report of the meteorology and climatology of arid regions, No. 1, University of Arizona Institute of Atmospheric Physics. 56 pp.
McGinnies, W.G. 1979. Arid-land ecosystems – common features throughout the world. In: R.A. Perry and D.W. Goodall (eds), Arid Land Ecosystems: Structure, Functioning and Management, Vol. 1, pp. 299–316. Cambridge University Press, Cambridge.
McGinnies, W.G., Goldman, B.J. and Paylore, P. (eds) 1968. Deserts of the world: an appraisal of research into their physical and biological environments. University of Arizona Press, Tucson, Arizona. 788 pp.
McKee, E.D. 1979. A study of global and seas. Geological Survey Professional Paper 1052, U.S. Government Printing Office, Washington, D.C. 429 pp.
McNaughton, S.J. 1983. Serengeti grassland ecology: the role of composite environmental factors and contingency in community organization. Ecol. Mon. 53: 291–320.
McNaughton, S.J. 1985. Ecology of a grazing ecosystem: The Serengeti. Ecol. Mon. 55: 259–294.
McNaughton, S.J., Coughenour, M.B. and Wallace, L.L. 1982. Interactive processes in grassland ecosystems. In: J.R. Estes, Tyrl, R.J. and Brunken, J.N. (eds), Grasses and Grasslands, Systematics and Ecology, pp. 167–193. University of Oklahoma Press, Norman, Oklahoma.
Menaut, J.C. and Cesar, J. 1979. Structure and primary productivity of Lamto Savannas, Ivory Coast. Ecology 60: 1197–1210.
Minckley, W.L. and Brown, D.E. 1982. Wetlands. Desert Plants 4: 223–287.
Mitchell, V.L. 1976. The regionalization of climate in the western United States. J. Appl. Meter. 15: 920–927.
Monod, T. 1954. Modes contracte et diffuse de la végétation saharienne. In: J.L. Cloudsley-Thompson (ed.), Biology of Deserts, pp. 35–37. Travistock House, London.
Mott, J.J. 1973. Temporal and spatial distribution of an annual flora in an arid region of western Australia. Tropical Grasslands 7: 89–97.

Mott, J.J. 1974. Mechanisms controlling dormancy in the arid zone grass *Aristida contorta*: I. Physiology and mechanisms of dormancy. Austr. J. Bot. 22: 635–645.
Mulroy, T.W. and Rundel, P.W. 1977. Annual plants: adaptations to desert environments. BioScience 27: 109–114.
Musick, H.B. 1975. Barrenness of desert pavement in Yuma County, Arizona. J. Arizona Acad. Sci. 10: 24–28.
Naveh, Z. and Dan, J. 1973. The human degradation of Mediterranean landscape in Israel. In: F. Di Castri and H. Mooney (eds), Mediterranean-type Ecosystems; Origin and Structure, pp. 378–390. Springer-Verlag, Berlin.
Neilson, R.P. 1986. High-resolution climatic analysis and Southwest biogeography. Science 232: 27–34.
Nobel, P.S. 1978. Microhabitat, water relations, and photosynthesis of a desert fern, *Notholaena parryi*. Oecologia (Berlin) 31: 293–309.
Nobel, P.S. 1980. Water vapor conductance and CO_2 uptake for leaves of a C_4 desert grass, *Hilaria rigida*. Ecology 61: 252–258.
Nobel, P.S. 1981. Spacing and transpiration of various sized clumps of a desert grass, *Hilaria rigida*. J. Ecol. 69: 735–742.
Noy-Meir, I. 1970. Vegetation of Central Australia. Ph.D. Dissertation. CSIRO, Canberra.
Noy-Meir, I. 1973. Desert ecosystems: environment and producers. Ann. Rev. Ecol. Syst. 4: 25–51.
Orians, G.H. and Solbrig, O.T. 1977. A cost-income model of leaves and roots with special reference to arid and semiarid areas. Amer. Nat. 111: 677–690.
Ozenda, P. 1977. Flore du Sahara. Centre National de la Recherche Scientifique, Paris. 622 pp.
Parkhurst, D.F. and Loucks, O.L. 1972. Optimal leaf size in relation to environment. J. Ecol. 60: 505–537.
Parsons, R.F. 1981. Eucalyptus scrubs and shrublands. In: R.H. Groves (ed.), Australian Vegetation, pp. 227–252. Cambridge University Press, New York. 449 pp.
Parton, W.J., Singh, J.S. and Coleman, D.C. 1978. A model of production and turnover of roots in shortgrass prairie. J. Appl. Ecol. 15: 515–542.
Paulsen, O. 1912. The second Danish Pamir expedition. Studies on the vegetation of the Transcarpian woodland. Arbejder fra den botanisk Havens i. Kobenhaven 90: 1–279.
Penning de Vries, F.W.T. 1982. La production potentielle des paturages naturels. In: F.W.T. Penning de Vries and M.A. Djitèye (eds), La Productivité des Paturages Sahéliens, pp. 165–181. Centre for Agricultural Publishing and Documentation, Wageningen.
Penning de Vries, F.W.T. and Djitèye, M.A. (eds) 1982. La productivité des paturages sahéliens. Centre for Agricultural Publishing and Documentation, Wageningen. 525 pp.
Pianka, E.R. 1973. Evolutionary ecology. Harper & Row, New York.
Purdie, R. 1984. Land systems of the Simpson Desert region. Natural Resources Series No. 2, Division of Water and Land Resources, Institute of Biological Resources, Commonwealth Scientific and Industrial Research Organization, Australia. 71 pp.
Raunkiaer, C. 1934. The life form of plants and statistical plant geography. Clarendon Press, Oxford. 632 pp.
Redmann, R.E. and Reekie, E.G. 1982. Carbon balance in grasses. In: J.R. Estes, Tyrl, R.J. and Brunken, J.N. (eds), Grasses and Grasslands, Systematics and Ecology, pp. 196–231. University of Oklahoma Press, Norman.
Rikli, M. 1943–1948. Das Pflanzenkleid der Mittelmeerländer. 3 Vols. Hans Huber, Berlin. 418 pp.
Rogers, R.W. 1971. Distribution of the lichen *Chondropsis semiviridis* in relation to its heat and drought resistance. New Phytol. 70: 1069–1077.
Royce, C.L. and Cunningham, G.L. 1982. The ecology of *Abronia angustifolia* (Nyctaginaceae): 1. Phenology and perennation. Southwestern Naturalist 27: 413–424.
Rumney, G.R. 1968. Climatology and the world's climates. Macmillan, New York. 653 pp.
Sala, O.E., Lauenroth, W.K., Parton, W.J. and Trlica, M.J. 1981. Water status of soil and vegetation in a shortgrass steppe. Oecologia (Berlin) 48: 327–331.
Sala, O.E. and Lauenroth, W.K. 1982. Small rainfall events: an ecological role in semiarid regions. Oecologia (Berlin) 53: 301–304.
Sarmiento, G. and Monasterio, M. 1983. Life forms and phenology. In: F. Bourliere (ed.), Tropical Savannas. Ecosystems of the World, Vol. 13, pp. 79–108. Elsevier, New York.
Schaffer, W.M. and Gadgil, M.D. 1975. Selection for optimal life history in plants. In: M.L. Cody and

J.M. Diamond (eds), Ecology and Evolution of Communities, pp. 142–157. Harvard University Press, Cambridge.

Schenkeveld, A.J. and Verkaar, H.J. 1984. On the ecology of short-lived forbs in chalk grasslands. Ph.D. Thesis. Utrecht. 180 pp.

Schimper, A.F.W. 1903. Plant geography upon a physiological basis. Oxford University Press. 1143 pp.

Sellers, W.D. and Hill, R.H. (eds) 1974. Arizona climate 1932–1972. University of Arizona Press, Tucson, Arizona. 616 pp.

Shanan, L., Evenari, M. and Tadnor, N.H. 1967. Rainfall patterns in the Central Negev desert. Israel Exploration Journal 17: 163–184.

Sharon, D. 1972. The spottiness of rainfall in a desert area. J. Hydrol. 17: 161–175.

Sharon, D. 1978. Rainfall fields in Israel and Jordan and the effect of cloud seeding on them. J. Appl. Meteor. 17: 40–48.

Shmida, A. 1972. The vegetation of Gebel Maghara, North Sinai. M.Sc. thesis. Dept. of Botany, The Hebrew University of Jerusalem (summary in English).

Shmida, A. 1977. A quantitative analysis of the Tragacanthic vegetation of Mt. Hermon and its relation to environmental factors. Ph.D. Dissertation, The Hebrew University of Jerusalem. (Hebrew, English summary). 170 pp.

Shmida, A. 1981. Mediterranean vegetation of Israel and California, similarities and differences. Israel J. Bot. 30: 105–123.

Shmida, A. 1982. Life forms of rocky Mediterranean plants in Israel. Rotem No. 2, 4–10. (Hebrew, Summary in English).

Shmida, A. 1984. Whittaker's plant diversity sampling method. Israel J. Bot. 33: 41–46.

Shmida, A. 1985. Biogeography of the desert flora. In: M. Evenari, I. Noy-Meir and D.W. Goodall (eds), Hot Deserts and Arid Shrublands Ecosystems of the world, Vol. 12A, pp. 23–77. Elsevier, Amsterdam.

Shmida, A. 1986. The Mediterranean annual plants – Richness and evolution of the annual flora of the Mediterranean basin. Rotem 18: 57–69.

Shmida, A. and Whittaker, R.H. 1979. Convergent evolution of arid regions in the New and Old World. In: R. Tuxen (ed.), Werden und Vergehen von Pflanzengesellschaften, pp. 437–450. Cramer, Vaduz.

Shmida, A. and Ellner, S. 1984. Coexistence of plant species with similar niches. Vegetatio 58: 29–55.

Shmida, A. Evenari, M. and Noy-Meir, I. 1986. Hot desert ecosystems: an integrated view. In: M. Evenari, I. Noy-Meir and D.W. Goodall (eds), Hot Deserts and Arid Shrublands. Ecosystems of the World, Vol. 12B, pp. 379–387. Elsevier, Amsterdam.

Shomer-Ilan, A., Nissenbaum, A. and Waisel, Y. 1981. Photosynthetic pathways and the ecological distribution of the Chenopodiaceae in Israel. Oecologia (Berlin) 48: 244–248.

Shreve, F. 1934. Rainfall, runoff and soil moisture under desert conditions. Ann. Ass. Amer. Geogr. 24: 131–156.

Shreve, F. 1951. Vegetation of the Sonoran desert. Carnegie Institution of Washington Publ. 591, Washington, D.C. 192 pp.

Smith, B.N. 1976. Evolution of C_4 photosynthesis in response to changes in carbon and oxygen concentrations in the atmosphere through time. Biosystems 8: 24–32.

Specht, R.L. 1979. Heathlands and related shrubslands of the world, a descriptive study. Ecosystems of the World, Vol. 9A&B. Elsevier, Amsterdam. 497 pp.

Szarek, S.R. 1979. Primary production in four North American deserts: indices of efficiency. J. Arid Envir 2: 187–209.

Teeri, J.A. and Stowe, L.G. 1976. Climatic patterns and the distribution of C_4 grasses in North America. Oecologia (Berlin) 23: 1–12.

Thornthwaite, C.W. 1948. An approach toward a rational classification of climate. Geographical Review 38: 55–94.

Tinley, K.L. 1982. The influence of soil moisture balance on ecosystem patterns in Southern Africa. In: B.J. Huntley and B.H. Walker (eds), Ecology of Tropical Savannas, pp. 175–192. Springer, New York.

Turril, W.B. 1930. The plant life of the Balkan peninsula. A phytogeographical study. Oxford University Press. 490 pp.

Van der Meulen, F. and Werger, M.J.A. 1984. Crown characteristics, leaf size and light throughfall of some savanna trees in southern Africa. S. Afr. J. Bot. 3: 208–218.

Venable, D.L. and Lawlor, L. 1980. Delayed germination and dispersal in desert annuals: escape in

space and time. Oecologia (Berlin) 46: 272–282.
Walker, B.H., Ludwig, D., Holling, C.S. and Peterman, R.M. 1981. Stability of semi-arid savanna grazing systems. J. Ecol. 69: 473–498.
Waller, S.S. and Lewis, J.K. 1979. Occurrence of C_3 and C_4 photosynthetic pathways in North American grasses. J. Range Managem. 32: 12–28.
Walter, H. 1971. Ecology of tropical and subtropical vegetation. Oliver and Boyd, Edinburgh. 539 pp.
Walter, H. 1973. Vegetation of the earth in relation to climate and the ecophysiological conditions. Springer, New York. 237 pp.
Walter, H. and Stadelmann, E. 1974. A new approach to the water relations of desert plants. In: G.W. Brown, Jr. (ed.), Desert Biology, Vol. 2, pp. 213–310. Academic Press, New York.
Webb, W., Szarek, S., Lauenroth, W., Kinerson, R. and Smith, M. 1978. Primary productivity and water use in native forest, grassland, and desert ecosytems. Ecology 59: 1123–1247.
Went, F.W. 1949. Ecology of desert plants. II: The effect of rain and temperature on germination and growth. Ecology 30: 1–13.
Werger, M.J.A. 1978. The Karoo-Namib region. In: M.J.A. Werger (ed.), Biogeography and Ecology of Southern Africa, pp. 231–299. Dr W. Junk Publishers, The Hague.
Werger, M.J.A. 1980. A phytosociological study of the Upper Orange River valley. Mem. Bot. Surv. S. Afr. 46: 1–92.
Werger, M.J.A. 1983. Tropical grasslands, savannas, woodlands: natural and manmade. In: W. Holzner, M.J.A. Werger and I. Ikusima (eds), Man's Impact on Vegetation, pp. 107–137. Dr W. Junk Publishers, The Hague.
Werger, M.J.A. 1986. The Karoo and southern Kalahari. In: M. Evenari, I. Noy-Meir and D.W. Goodall (eds), Hot Deserts and Arid Shrublands. Ecosystems of the World Vol. 12B, pp. 283–359. Elsevier, Amsterdam.
Werger, M.J.A. and Coetzee, B.J. 1978. The Sudano-Zambesian region. In: M.J.A. Werger (ed.), Biogeography and Ecology of Southern Africa, pp. 301–462. Dr W. Junk Publishers, The Hague.
Werger, M.J.A. and Ellis, R.P. 1981. Photosynthetic pathways in the arid regions of South Africa. Flora 171: 64–75.
Whittaker, R.H. and Niering, W.A. 1964. Vegetation of the Santa Catalina Mountains, Arizona. I. Ecological classification and distribution of species. J. Arizona Acad. Sci. 3: 9–34.
Winter, K. 1981. C_4 plants of high biomass in arid regions of Asia – occurrence of C_4 photosynthesis in Chenopodiaceae and Polygonaceae from the Middle East and U.S.S.R. Oecologia (Berlin) 48: 100–106.
Yair, A. and Shachak, M. 1982. A case study of energy, water and soil flow chain in an arid ecosystem. Oecologia (Berlin) 54: 389–397.
York, J.C. and Dick-Peddie, W.A. 1969. Vegetation changes in southern New Mexico during the past hundred years. In: W.G. McGinnies and B.J. Goldman (eds), Arid Lands in Perspective, pp. 157–166. University of Arizona Press, Tucson, Arizona.
Zohary, M. 1962. Plant life of Palestine – Israel and Jordan. The Ronald Press, New York.
Zohary, M. 1973. Geobotanical foundation of the Middle East. Gustav Fischer Verlag, Stuttgart. 737 pp.

THE IMPORTANCE OF PREDATION AND SMALL SCALE DISTURBANCE TO TWO WOODLAND HERB SPECIES

D.F. WHIGHAM and J. O'NEILL
Smithsonian Environmental Research Center, Box 28, Edgewater, MD 21037, USA

Abstract

The ecology of woodland herbaceous species is reviewed. Many factors affect woodland herbs but there have been few studies of their impacts on long-term patterns. Results from a 10-year study of two species in a deciduous forest in Maryland are presented. Populations of *Cynoglossum virginianum* were influenced by their proximity to tree gap disturbances. Populations of *Tipularia discolor* were primarily influenced by the activities of leaf and corm predators.

Introduction

Information and concepts about the population and community ecology of plants has expanded greatly in recent years (Dirzo and Sarukhan 1984; Givnish 1986; Grime 1979; Harper 1977; Solbrig 1980; Solbrig *et al.* 1979; White 1985). A number of investigations have focused on herbaceous and woody species in forests and various aspects of their ecology have been studied (Brewer 1980; Cook 1985; Davison and Forman 1982; Good and Good 1972; Hough 1965; Kawano 1985; Kawano *et al.* 1982; Nakagoshi 1985; Nakashizuka and Numata 1982; Newell *et al.* 1981; Pitelka and Ashmun 1986; Shorina and Smirnova 1985; Solbrig *et al.* 1980; Tamm 1972; Traczyk and Traczyk 1977; Werger and Van Laar 1985). A brief summary of our current understanding of the ecology of woodland species follows:

Life history strategies

Kawano and his colleagues have published a series of articles and provide a thorough analysis of life history strategies and biomass allocation patterns of herbs in temperate forests. Kawano (1985) has recently summarized his work and suggests that life history characteristics of the groups that he has identified are primarily influenced by variations in yearly environmental factors associated with temperate forests. Others (Bratton 1976; Shorina and Smirnova 1985) have also recognized temporally distinct life history patterns and Bratton suggested that woodland herbs form guilds along microtopographic gradients. Distinct patterns of convergence and a biomechanical basis for the formation of guilds has been described by Givnish (1982, 1986).

Distribution

The distribution of herbs and seedlings of woody species is influenced by many factors. The most important appear to be microtopographic gradients (Bratton 1976), the distribution of safe sites (Thompson 1980), and animals which act as seed predators, seed dispersal agents, and herbivores (Culver and Beattie 1978; Hough 1965; Kawano *et al.* 1982; Newell *et al.* 1981; Sork 1984).

Tree gaps

Tree gaps have been shown to be important in many ecosystems (Pickett and White 1985) and it has been proposed that they play an important role in controlling the diversity of understory herbs and seedlings of woody species (Ehrenfeld 1980; Maguire and Forman 1983; Runkle 1984). Recent work, however, has demonstrated that their role is complex and contradictory results have been obtained. Brewer (1980) found that long-term fire disturbance had a greater impact on herb diversity than tree gap disturbances in a Michigan forest. Collins and Pickett (1987) also found that the experimental creation of gaps had little effect on herb cover or species richness. They concluded that the lack of understory response was due to the relatively small size of their gaps (< 150 m^2) and/or that plant responses may take a longer time than the period of their study. Davison and Forman (1982) provide some support for this conclusion. Over a 30-year period, the Hutcheson Memorial Forest in New Jersey had undergone a significant change due to an increase in the frequency of gap creation (*i.e.*, more area of the forest was disturbed) caused by drought and gypsy moth defoliation. We will demonstrate in this paper, however, that the population structure of an understory herb species can change dramatically soon after a tree gap is created.

Primary production

Biomass production of understory herbs varies spatially (Kawano 1985) and temporally. Reasons for spatial variations are not clear although microtopographic and resource heterogeneity are undoubtedly important (Bratton 1976). Yearly differences in aboveground biomass have been shown to vary with weather conditions (Rogers 1983) but the long-term impacts of those variations are unknown.

Age and size class distribution

Most populations of understory species are relatively stable over long periods of time (Falinski 1986; Inghe and Tamm 1985) even though annual mortality and natality rates can be high (Newell *et al.* 1981). Most species have Type I and II Deevey survivorship curves with high mortality of seedlings and decreasing mortality with increasing age or size (Kawano *et al.* 1982). Development from seedling to reproductive (sexual or asexual) stages is usually a slow process (Shorina and Smirnova 1985).

Competition

There is little evidence for inter- or intraspecific competition (Bazzaz and Bliss 1971; Rogers 1985) even though many species are clonal and can form dense stands (Cook 1985; Pitelka and Ashmun 1985; Sobey and Barkhouse 1977; Whitford 1949). It has been suggested that understory plants can avoid competition by asexually spreading over the forest floor or by increasing rates of ramet mortality and natality in order to maintain optimum population density for given levels of environmental resources (Pitelka and Ashmun 1985). Herbs do, however, appear to influence reproduction of tree seedlings (Maguire and Forman 1983).

Reproduction

The majority of woodland species are perennial or pseudo-annual (Kawano 1985; Salisbury 1942) and reproduce by asexual propagation. Only a few species have been shown to rely completely on sexual reproduction (Kawano et al. 1982; Muller 1980). Sexual reproduction often has a high energy cost and many individuals often die or decrease in size following flowering (Inghe and Tamm 1985; Kawano 1985).

Physiology

Woodland species have evolved a variety of adaptations to the range of microclimatic conditions that occur in the forest. Most physiological characteristics seem to be related to phenological patterns of the overstory vegetation (Kawano et al. 1978) and leaf geometry and morphology has been shown to be important (Givnish 1986). Many clonal species also show a range in physiological integration between ramets (Pitelka and Ashmun 1985).

This introduction is intended to demonstrate that woodland species have evolved many life history characteristics and that we are beginning to compile enough information to draw broad generalities about their ecological implications. It is also clear that the establishment, growth, and maintenance of forest understory plants are influenced by many biotic and abiotic variables and that those variables act over a wide range of spatial and temporal scales. Given all of this complexity, can we predict the long-term fate of any woodland population or community and do the long-term changes in population structure alter ecosystem function? We don't believe that an answer to this question is possible without much more information on the degree of and cause of changes that occur in populations of woodland species. We will use two examples to demonstrate that populations of long-lived woodland species can change significantly in response to herbivory and to small scale disturbance (tree gaps). Our objective is to use these two examples to suggest that we need much more information on factors that are responsible for seedling recruitment and factors that control the rate at which individuals initiate and maintain successful sexual or asexual reproduction.

Methods

The data that we present were compiled as part of a long-term study of plants in a deciduous forest on the Inner Coastal Plain of Maryland. The forest is located on the property of the Smithsonian Environmental Research Center which is near Annapolis, Maryland. The forest is dominated by a variety of hardwood species and the herb and shrub strata are well-developed (Whigham 1984). Only selected results from the research are presented here as other detailed papers are in preparation.

In 1976 we identified 34 distinct groups (hereafter referred to as populations) of *Tipularia discolor* L. (*Orchidaceae*). It is a winter-green perennial that only rarely reproduces asexually. The annual pattern of biomass and nutrient allocation and pollination ecology have been described (Whigham 1984; Whigham and McWethy 1980) for plants in this forest. For purposes of this paper, it is important to note that each plant normally produces only one leaf and one underground corm per year. The locations of the populations are permanently marked with wire flags. The locations of all individuals within each population are marked with aluminum tags. The number of each plant is written on the tag which is placed on the ground within a few centimeters of the plant. The tags are held in place with aluminum nails that are pushed into the ground. All individuals were monitored yearly and the leaf area measured. When appropriate, we also estimated the percent of the leaf area removed by predators and counted the number of flowers and fruits of sexually reproductive plants. When leaves were not present at the time of the annual census, we determined whether or not the underground corms were present and characterized their status. When the plants were missing completely, we attempted to identify the source of mortality.

Cynoglossum virginianum L. (*Boraginaceae*) is a long-lived summer-green perennial that does not reproduce asexually. Each plant produces a basal rosette of leaves each year and the belowground stem can be quite large. In 1978, we located three populations of *Cynoglossum* and have monitored all individuals yearly. Procedures for marking the plants were similar to those described for *Tipularia*. The number of leaves was counted and percent leaf predation estimated. The number of flowers and fruits produced by sexually reproductive plants were determined and we also counted the number of leaves on each inflorescence. In this paper, we will present data from two populations: one in an area where there have not been any canopy disturbances since the start of the study; the other population is located within an area of disturbance created by the death of a canopy tree in 1980.

Results

Tipularia discolor

Percent leaf predation is variable from year to year (Fig. 1) but is so high that few plants go for more than two or three years without having their leaves eaten. White-tailed deer are the primary predators and they almost always remove 100% of the leaf tissue. The main impact of leaf predation is to inhibit the plant from increasing in size and we have demonstrated experimentally that either 50% or 100% annual leaf predation results in a significant decrease in plant size and a delay in the frequency of flowering (Whigham unpublished).

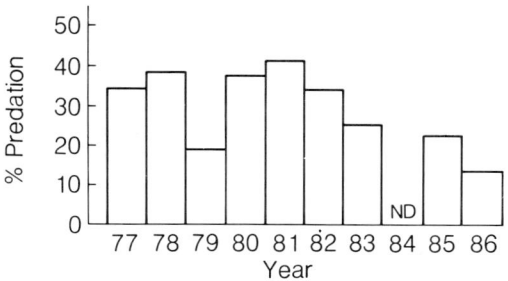

Fig. 1. Percent of all marked plants in populations of *Tipularia discolor* that suffered some amount of leaf predation on an annual basis for a 10-year period. ND indicates that data were not taken.

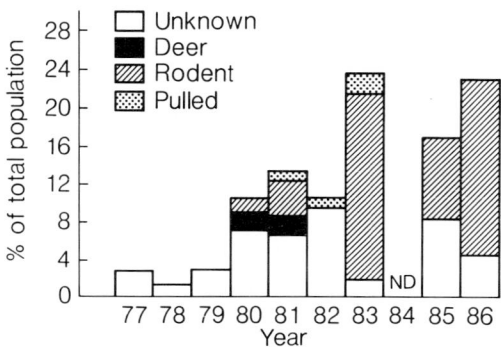

Fig. 2. Mortality of *Tipularia* plants. Sources of mortality are identified by the legend. Deer kill plants by stepping on them or by pulling them out of the ground. Small mammals (rodents) completely eat underground corms of the plants.

Complete predation of plants also occurs and over the 10 years of this study, 12 of the 34 populations have been completely eaten. We have identified two types of mortality associated with deer (stepping on plants and pulling small plants from the soil) and small mammals consume underground corms. Small mammals have accounted for most of the mortality since 1982 (Fig. 2). Mortality due to small mammals has been high in the 1980's. 1982 was a mast year for nut bearing trees in the forest and there was a resultant increase in the density of small mammals (James Lynch, unpublished data). This suggests that the dynamics of *Tipularia* populations may be indirectly controlled by mast cycles of forest trees.

The impact of both types of predation has been to reduce the number of plants being monitored from 159 to 55 over the 10-year period. The size distribution of the populations has also changed over that time as the larger populations have been destroyed or converted into smaller groupings (Table 1).

Cynoglossum virginianum

This species relies completely on sexual reproduction for propagation and suffers very little leaf predation. Only a small number of established plants in the popula-

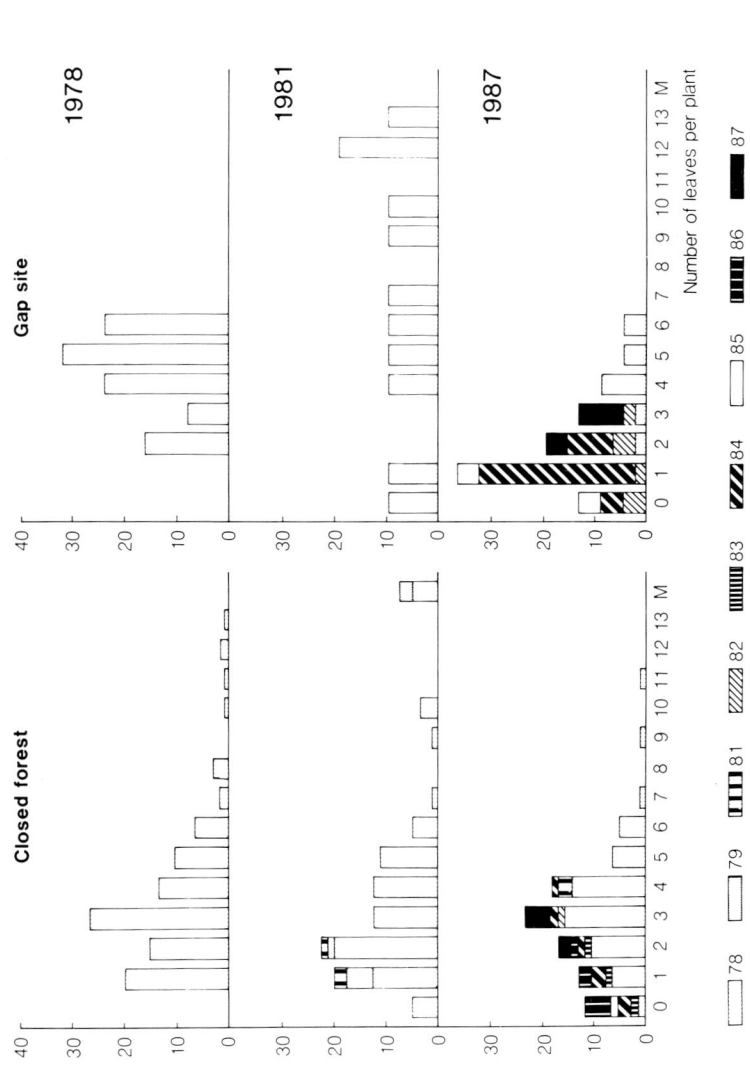

Fig. 3. Changes in the size class distribution of plants in two populations of Cynoglossum virginianum in an area where the canopy has remained intact (Closed) and in an area where a tree gap (Gap) was formed in 1980. Years when individuals were recruited into the population are indicated on the legend. M indicates mortality and 0 represents plants that did not produce any aboveground tissue but which were still alive. The horizontal axis is the number of leaves per plant. Data are given for the first year of the study (1978), the first year after the canopy disturbance (1981), and the most recent census year (1987).

Table 1. Eleven year changes in the number of individuals in monitored populations of *Tipularia discolor*. Values across the top of the table are numbers of plants per population. Values in the Table are numbers of populations. Data are only given for the first and most recent years of the study.

Number of plants per population	1	2	3	4	5	6	7	8	9	10	11	...	16
1976	9	5	5	2	2	2	4	3	0	1	0	...	1
1987	9	4	3	3	0	1	0	1	1	0	0	...	0

tions have died since 1978 (Fig. 3). There have been few changes in the size class distribution of the population in the area where the canopy has remained intact but plants in the tree gap area have undergone large changes. Most individuals increased in size after the creation of the tree gap (Fig. 3) and many flowered for approximately three years after which time the 'window-of-opportunity' had passed. The size-class structure of the population then returned to pre-gap form with the exception that the number of seedlings and juveniles had increased. The overall impact of this disturbance had been a three-fold increase in the size of that population, from 14 plants in 1978 to 46 in 1987. During the same time period, the population in the undisturbed area declined from 108 to 77 individuals.

Discussion

Rogers (1985) and Inghe and Tamm (1985) have demonstrated that climatological variables can affect species of woodland plants in forests. Others (Falinski 1986; Kawano *et al.* 1982) have demonstrated the degree of spatial variation that can occur within forests. In this study, we have focused on changes that can occur in established populations of two long-lived species. Leaf and corm predation were the most important factors that affected *Tipularia*. Frequent removal of leaf tissue results in a decrease in plant size and a decrease in the frequency of flowering (Whigham unpublished). Unfortunately, we still know little about the factors that are required for successful establishment of juveniles and there has not been any successful recruitment of juveniles into the populations that we have been monitoring. In contrast, *Cynoglossum* was not affected by predation. It appears to be a species which can persist in the understory for long periods of time without any reproductive activity. When a canopy disturbance occurs near an established plant, it is able to increase size rapidly and begin to reproduce sexually. The period of sexual activity lasted for only a few years in the one population that we have been studying but the size of the disturbance is probably the factor that determines how long that period lasts. Recruitment of juveniles into populations occurs primarily during disturbance events.

The impact of disturbances can be large. In the case of *Tipularia*, entire populations were lost due to corm predation and the total number of plants being studied has decreased by more than 65%. There has not been any seedlings recruitment and reproduction by asexual branching is very slow. In *Cynoglossum*, one population increased in size by a factor of three while the other had a net loss of several individuals even though recruitment of seedlings had occurred. These results demonstrate the importance of understanding the factors that are responsible for the establishment of new individuals in populations and for the increase in size that enables individuals to either reproduce sexually or asexually. For *Tipularia*, the size of individual plants appears to be primarily controlled by the frequency of leaf predation which tends to cause plants to be smaller. Reproduction in *Cynoglossum* is clearly limited by light and it takes advantage of short-term increases in light availability associated with tree gaps. Similar results have been shown for *Ranunculus repens* L. in a floodplain forest in Poland (Falinski 1986).

These results also demonstrate the importance of knowing more about the establishment requirements of species in the understory (Cook 1979) as seedling recruitment may only occur infrequently in ecosystems with well-established vegetation (Tamm 1972). Similar problems can be identified for herbaceous species in other types of ecosystems (Wells 1981; Willems 1982; Zhang 1983) and especially for rare and endangered species (Harvey 1985). As indicated, over the 11 years of the *Tipularia* research, no new individuals have successfully become established from seed. For *Cynoglossum*, we know that seedlings can become established in both the intact forest and in disturbed areas. Populations are, however, only likely to expand in disturbed areas as mortality has been greater than seedling recruitment in the population in the intact forest.

Acknowledgments

We recognize the financial support of the Smithsonian Environmental Sciences Program and the Nederlandse Organisatie voor Zuiver-Wetenschappelijk Onderzoek (Grant B 84-272). The logistical support of the Department of Plant Ecology, University of Utrecht – particularly Jos Verhoeven – is appreciated, especially during the writing of the manuscript. We thank Jo Willems for his comments on the manuscript.

References

Bazzaz, F.A. and Bliss, L.C. 1971. Net primary production of herbs in a central Illinois deciduous forest. Bull. Torrey Bot. Club 98: 90–94.
Bierzychudek P. 1982. Life histories and demography of shade-tolerant temperate forest herbs: A review. New Phytol. 90: 757–776.
Bratton, S.P. 1976. Resource division in an understory herb community: response to temporal and microtopographic gradients. Am. Nat. 110: 679–693.
Brewer, R. 1980. A half-century of changes in the herb layer of a climax deciduous forest in Michigan. J. Ecol. 68: 823–832.
Collins, B.S. and Pickett, S.T.A. 1987. Influence of canopy opening on the environment and herb layer in a northern hardwood forest. Vegetatio 70: 3–10.
Cook, R.E. 1979. Patterns of juvenile mortality and recruitment in plants. In: O.T. Solbrig, S. Jain, G.B. Johnson and P.H. Raven (eds), Topics in Plant Population Biology, pp. 207–231. Columbia University Press, New York.
Cook, R.E. 1985. Growth and development in clonal plant populations. In; J.B.C. Jackson, L.W. Buss and R.E. Cook (eds), Population Biology and Evolution of Clonal Organisms, pp. 259–296. Yale University Press, New Haven.
Culver, D.C. and Beattie, A.J. 1978. Myrmecochory in *Viola*: Dynamics of seed-ant interactions in some West Virginia species. J. Ecol. 66: 53–72.

Davison, S.E. and Forman, R.T.T. 1982. Herb and shrub dynamics in a mature oak forest: a thirty-year study. Bull. Torrey Bot. Club 109: 64–75.
Dirzo, R. and Sarukhan, J. 1984. Perspectives on plant population ecology. Sinauer Associates Inc. Sunderlands. 478 pp.
Ehrenfeld, J.G. 1980. Understory response to canopy gaps of varying size in a mature oak forest. Bull. Torrey Bot. Club 107: 29–41.
Falinski, J.B. 1986. Vegetation dynamics in temperate lowland primeval forests. Ecological studies in Bialowieza forest. Dr W. Junk Publishers, Dordrecht. 537 pp.
Givnish, T.J. 1982. On the adaptive significance of leaf height in forest herbs. Am. Nat. 120: 353–381.
Givnish, T.J. 1986. Biomechanical constraints on crown geometry in forest herbs. In: T.J. Givnish (ed.), On the Economy of Plant Form and Function, pp. 525–579. Cambridge University Press, New York.
Good, N.F. and Good, R.E. 1972. Population dynamics of tree seedlings and saplings in a mature eastern hardwood forest. Bull. Torrey Bot. Club 99: 172–178.
Grime, J.P. 1979. Plant strategies and vegetation processes. John Wiley and Sons, New York. 222 pp.
Harper, J.L. 1977. Population biology of plants. Academic Press, New York. 892 pp.
Harvey, H.J. 1985. Population biology and the conservation of rare species. In: J. White (ed.), Studies of Plant Demography. A Festschrift for John L. Harper, pp. 111–122. Academic Press, New York.
Hough, A.F. 1965. A 20-year record of understory vegetational change in a virgin Pennsylvania forest. Ecology 46: 370–373.
Inghe, O. and Tamm, C.O. 1985. Survival and flowering of perennial herbs. IV. The behaviour of *Hepatica noblis* and *Sanicula europaea* on permanent plots during 1943–1981. Oikos 45: 400–420.
Kawano, S. 1985. Life history characteristics of temperate woodland plants in Japan. In: J. White (ed.), The Population Structure of Vegetation, pp. 515–549. Dr W. Junk Publishers, Dordrecht.
Kawano, S., Takasu, H. and Nagai, Y. 1978. The productive and reproductive biology of flowering plants. IV. Assimilative behaviour of some temperate woodland herbs. J. College Liberal Arts, Toyama Univ., Japan 11: 33–60.
Kawano, S., Kiratsuka, A. and Hayashi, K. 1982. Life history characteristics and survivorship of *Erythronium japonicum*. The productive and reproductive biology of flowering plants. V. Oikos 38: 129–149.
Maguire, D.A. and Forman, R.T.T. 1983. Herb cover effects on tree seedling patterns in a mature hemlock-hardwood forest. Ecology 64: 1367–1380.
Muller, R.N. 1980. The phenology, growth, and ecosystem dynamics of *Erythronium americanum* in the northern hardwood forest. Ecol. Monogr. 45: 1–20.
Nakagoshi, N. 1985. Buried viable seeds in temperate forests. In: J. White (ed.), The Population Structure of Vegetation, pp. 551–570. Dr W. Junk Publishers, Dordrecht.
Nakashizuka, T. and Numata, M. 1982. Regeneration process of climax beech forest I. Structure of a beech forest with the undergrowth of *Sasa*. Jap. J. Ecol. 32: 57–67.
Newell, S.J., Solbrig, O.T. and Kincaid, D.T. 1981. Studies on the population biology of the genus *Viola* III. The demography of *Viola blanda* and *Viola pallens*. J. Ecol. 69: 997–1016.
Pickett, S.T.A. and White, P.S. 1985. The ecology of natural disturbance and patch dynamics. Academic Press, New York. 472 pp.
Pitelka, L.F. and Ashmun, J.W. 1985. Physiology and integration of ramets in clonal plants. In: J.B.C. Jackson, L.W. Buss and R.E. Cook (eds), Population Biology and Evolution of Clonal Organisms, pp. 399–436. Yale University Press, New Haven.
Rogers, R.S. 1983. Annual variability in community organizations of forest herbs. Effect of an extremely warm and dry early spring. Ecology 64: 1086–1091.
Rogers, R.S. 1985. Local coexistence of deciduous-forest groundlayer species growing in different seasons. Ecology 66: 701–707.
Runkle, J.R. 1984. Development of woody vegetation in treefall gaps in a beech-sugar maple forest. Holarct. Ecol. 7: 157–164.
Salisbury, E.J. 1942. The reproductive capacity of plants. Bell, London. 244 pp.
Shorina, N.I. and Smirnova, O.V. 1985. The population biology of ephemeroids. In: J. White (ed.), The Population Structure of Vegetation, pp. 226–240. Dr W. Junk Publishers, Dordrecht.
Sobey, D.G. and Barkhouse, P. 1977. The structure and rate of growth of the rhizomes of some forest herbs and dwarf shrubs of the New Brunswick – Nova Scotia border region. Can. Field-Nat. 91: 377–383.

Solbrig, O.T. (ed.) 1980. Demography and evolution in plant populations. Blackwell Scientific, London. 222 pp.
Solbrig, O.T., Jain, S., Johnson, G.B. and Raven, P.H. (eds) 1979. Topics in plant population biology. Columbia University Press, New York. 589 pp.
Solbrig, O.T., Newell, S.J. and Kincaid, D.T. 1980. Studies on the population biology of the genue *Viola* I. The demography of *Viola sororia*. J. Ecol. 68: 521–546.
Sork, V.L. 1984. Examination of seed dispersal and survival in Red Oak, *Quercus rubra* (Fagaceae), using metal-tagged acorns. Ecology 63: 1020–1022.
Tamm, C.O. 1972. Survival and flowering of some perennial herbs. II. The behaviour of some orchids on permanent plots. Oikos 223: 23–28.
Thompson, J.N. 1980. Treefalls and colonization patterns of temperate forest herbs. Am. Midl. Nat. 104: 176–184.
Traczyk, T. and Traczyk, H. 1977. Structural characteristics of herb layer and its production in more important forest communities of Poland. Ekol. Pol. 25: 359–378.
Wells, T.C.E. 1981. Population ecology of terrestrial orchids. In: H. Synge (ed.), The Biological Aspects of Rare Plant Conservation, pp. 281–295. John Wiley and Sons, New York.
Werger, M.J.A. and van Laar, J.M. 1985. Seasonal changes in the structure of the herb layer of a deciduous woodland. Flora 176: 351–364.
Whigham, D.F. 1984. Biomass and nutrient allocation patterns of *Tipularia discolor* (Orchidaceae). Oikos 42: 303–313.
Whigham, D.F. and McWethy, M. 1980. Studies on the pollination ecology of *Tipularia discolor* (Orchidaceae). Am. J. Bot. 67: 550–555.
White, J. (ed.) 1985. The population structure of vegetation. Dr W. Junk Publishers, Dordrecht. 666 pp.
Whitford, W.G. 1949. Distribution of woodland plants in relation to succession and clonal growth. Ecology 30: 199–208.
Willems, J.H. 1982. Establishment and development of a population of *Orchis simia* Lamk. in the Netherlands, 1972–1981. New Phytol. 91: 757–765.
Zhang, L. 1983. Vegetational ecology and population biology of *Fritillaria meleagris* L. at the Kungsangen Nature Reserve, Eastern Sweden. Acta Phytogeogr. Suec. 73. Uppsala. 96 pp.

EFFECTS OF SPRUCE BUDWORM OUTBREAKS ON VEGETATION, STRUCTURE, AND SUCCESSION OF BALSAM FIR FORESTS ON CAPE BRETON ISLAND, CANADA

DAVID A. MACLEAN
Canadian Forestry Service – Maritimes, P.O. Box 4000, Fredericton, New Brunswick, Canada E3B 5P7

Abstract

Thirty 0.05-ha plots (five plots in each of six stands) were measured annually during a spruce budworm (*Choristoneura fumiferana* Clem.) outbreak. Balsam fir (*Abies balsamea* (L.) Mill.) mortality began after three to five years of severe current defoliation and increased from an average of 9% of the trees in 1977 to 82% in 1986. Mortality reduced live crown coverage in the stands from 77% in 1976 to less than 10% in 1986. In response to the opening of the stand, cover of shrub and some herbaceous species increased, while moss cover decreased. *Rubus idaeus* L. dominated several of the plots, increasing from 1% cover in 1979 to 78% in 1985. Most balsam fir regeneration adequately established before the budworm outbreak survived and, in 1985, there were 9 750 to 79 000 stems per hectare, or 67 to 100% stocking. Results of this study support the hypothesis of Baskerville (1975) that budworm outbreaks act as a cycling mechanism that allows advanced fir regeneration to succeed the fir overstory. Thus, budworm populations and fir forest may be viewed holistically as a cyclic ecological system that undergoes fluctuations in insect population level and tree mortality, but is essentially stable.

Introduction

The spruce budworm (*Choristoneura fumiferana* Clem.) is a forest insect that periodically undergoes population increases to epidemic levels in spruce (*Picea* sp.) and balsam fir (*Abies balsamea* (L.) Mill.) stands in eastern North America. Reduced tree growth results from budworm larvae feeding on the newly developing foliage. If severe defoliation persists for several years, trees will die. Effects of spruce budworm outbreaks on forest productivity and stability were reviewed in detail by MacLean (1984, 1985).

Spruce budworm outbreaks are the major forest insect problem in much of eastern North America because of their extent and severity. Three outbreaks have occurred in the 20th century, starting in about 1910, 1940, and 1970, and affecting areas of 10, 25, and 57 million hectares, respectively (Blais 1983). Outbreaks generally persist about 10 years. After the 1910 to 1920 outbreak, it was estimated that over 720 million m^3 of balsam fir and spruce timber, or 40 to 50% of the host tree volume, had been killed in eastern Canada (Swaine and Craighead 1924). Detailed studies of spruce budworm population dynamics have been conducted (Morris 1963; Royama 1984), but the cause of outbreaks is not known and biological control trials have been unsuccessful. Aerial spraying of chemical and biological insecticides is currently used in parts of Canada and the United States to prevent defoliation and keep trees alive (*e.g.*, Irving and Webb 1981).

Since 1978, I have studied effects of a severe spruce budworm outbreak on balsam fir forest on Cape Breton Island, Nova Scotia, Canada. The objectives of this paper are (1) to describe the changes in vegetation and stand structure that resulted from defoliation and (2) to test the hypothesis of Baskerville (1975) that budworm and fir forest together form a self-regulating system in which budworm outbreaks act as a cycling mechanism that allows fir regeneration to succeed the fir overstory.

Materials and methods

A severe spruce budworm outbreak began in 1974 on Cape Breton Island. During 1977, five 0.05-ha circular permanent sample plots were established in each of six stands. Five of the sampled stands (stands 1 to 5) were mature 55- to 80-year-old balsam fir; the sixth (stand 6) was a 30-year-old fir stand that had been operationally thinned to 2.4 × 2.4 m spacing in 1971. All trees in the sample plots were numbered, mapped, and measured for diameter at breast height, height, and crown dimensions. Each year, current and cumulative defoliation of each tree was estimated using binoculars. Trees were checked for mortality twice per year, by examining the cambium for discoloration and dryness. Plot establishment and measurement methodology were described in detail by MacLean (1979).

Ground vegetation was sampled biennially in each plot, using 12 randomly located 1-m^2 quadrats (four transects, with three quadrats per transect), giving a total of 60 quadrats per stand. Percentage cover by each plant species was estimated using a modified Braun-Blanquet cover scale (10 classes) and frequency of each species (the percentage of quadrats in which it was present) was calculated. Cumulative cover of shrub, herbaceous, fern, and moss species was calculated each sample year.

Amount of regeneration was determined biennially, by species and height class, in 12 randomly located 4-m^2 quadrats per plot. Regeneration was tallied in height classes of 0–5, 6–15, 16–30, 31–45, 46–60, and > 60 cm. In 1984, 10 balsam fir seedlings from each of these height classes were sampled in stands 2 and 3. Annual height growth of each seedling was measured to the nearest 0.1 cm. A disc about 1.0 cm thick was cut from the base of each seedling and annual ring widths were measured to the nearest 0.01 mm along four average radii on the disc, using a Holman DIGI-MIC ring measuring machine (Jordan and Ballance 1983). Mean annual height and radial increment were calculated for seedlings in each height class in each stand.

Results and discussion

Stand characteristics and history

Data on average stand structure for the five sample plots in each stand are presented in Table 1. All plots were nearly pure balsam fir, with 92 to 100% of the basal area consisting of fir. The five mature stands became established after budworm infestations that occurred between 1891 and 1896 and 1911 and 1915 (Hawboldt 1955); stand 6 originated after clear-cut harvesting in the early 1950s. Stand 3 was

Table 1. Summary of characteristics of six balsam fir stands sampled on the Cape Breton Highlands, Nova Scotia, Canada. Data are the average for five 0.05-ha plots in each stand.

Stand number	Age (years)	Density (stems/ha)	Mean dbh (cm)	Merchantable volume (m^3/ha)	Cumulative mortality (% stems)
1	80	1340	17	138	94
2	55–60	960	21	196	81
3	55–60	2660	14	181	65
4	55–60	1240	21	229	68
5	55–60	1000	20	161	41
6	25–30	1680	11	45	94

Note: Stands 4 and 5 were harvested in 1981 and 1980.

more densely stocked than the other four mature stands, with over 2600 stems/ha, and had smaller trees (Table 1). Stands 4 and 5 were harvested in 1981 and 1980, respectively, and thus only partial data on the effects of the budworm outbreak on these stands are available.

The first year of defoliation was 1974 in Stands 1 and 2, 1975 in 4 and 5, and 1976 in 3 and 6. Defoliation of current-year's foliage was moderate (26–75%) or severe (>75%) for five or six years in most stands. The budworm population subsided in 1981, by which time many trees had died. Average current defoliation between 1976 and 1980 ranged from 62% in stand 2 to 93% in stand 3. Thus, trees in these stands had lost about four or five age-classes of foliage.

Tree mortality

In response to defoliation, the proportion of dead trees in the stands increased from an average of 9% in 1977 to 82% in 1986 (Fig. 1). Tree mortality increased after three or four years of defoliation, with 60 to 80% of the stand dying in a two- to four-year period. Mortality was less in stand 3 than in the other stands, only about 15% of the stand dead after six years of defoliation, but thereafter, the pattern was similar. Trees in the immature, thinned stand 6 died over a relatively short period, increasing from 8% dead in 1979 to 88% dead in 1982 (Fig. 1). Budworm populations were very high in the vicinity of stand 6 from 1976 to 1978, with the result that insects fed on both current and older foliage (Piene and MacLean 1984). This probably resulted in the rapid increase in dead trees. Mortality patterns observed in this study were similar to those found by MacLean (1980) in reviewing studies of uncontrolled spruce budworm outbreaks in other regions and times.

The mortality caused by defoliation drastically altered the structure of these balsam fir stands. Crown coverage, calculated from the crown widths of all trees in the plots, declined from an average of 77% in 1976 to less than 10% in 1986. Mortality in the five plots sampled in each stand varied from 85 to 100% in stands 1 and 6, from 73 to 90% in stand 2, and from 36 to 90% in stand 3.

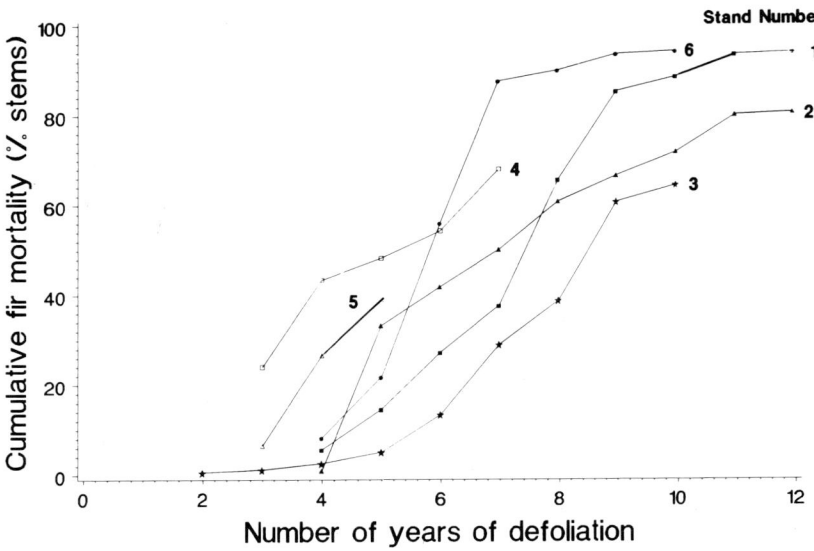

Fig. 1. The pattern of tree mortality in six balsam fir stands during a severe spruce budworm outbreak. Each stand was first sampled in 1977, except stand 6 which was first sampled in 1979. Stands 4 and 5 were harvested in 1981 and 1980.

Tree growth

Defoliation during the first four years of the outbreak reduced radial increment at breast height by 20 to 60% (MacLean 1979). Mean annual growth from 1977 to 1985 of balsam fir trees that survived was 0.6–0.7 mm/yr in stands 1 and 3, 1.14 mm/yr in stand 6, and 1.66 mm/yr in stand 2. Stand 2 was least defoliated and this was reflected in higher growth rates. In comparison, though, growth rates of white birch (*Betula papyrifera* Marsh.) averaged 3.99 mm/yr and yellow birch (*Betula alleghaniensis* Britton) 3.02 mm/yr in the sample plots.

Growth rate of surviving fir trees from 1977 to 1985 was directly related to cumulative defoliation. Mean annual radial increment declined from 2.32 mm/yr for trees with cumulative defoliation (considering all age-classes of foliage) of 0–10%, to 1.15 mm/yr for trees with 11–25% defoliation, to 0.70 mm/yr for trees with 26–50% defoliation, 0.64 mm/yr for 51–75% defoliation, 0.44 mm/yr for 76–90% defoliation, to 0.38 mm/yr for trees with greater than 90% cumulative defoliation.

Ground vegetation

Reduced crown cover associated with defoliation and tree mortality stimulated the growth of ground vegetation and caused major changes in its structure. Cover of shrubs and some herbaceous species generally increased, while moss cover decreased as the stand opened. Species composition varied considerably among plots in a stand, as a result of differences in drainage, stand density, and other factors. Stand averages would tend to mask some of the temporal differences occur-

Table 2. Changes in percentage cover of classes of ground vegetation from 1979 to 1985, which accompanied overstory tree mortality in three balsam fir plots.

Stand number	Plot number	Tree mortality (%)	Cumulative percentage cover of plants[1]									
			Ferns		Shrubs		Herbs		Total vascular plants		Mosses	
			1979	1985	1979	1985	1979	1985	1979	1985	1979	1985
1	5	100	19	36	2	86	53	32	74	154	31	6
2	3	90	24	50	4	19	65	105	92	174	12	3
3	1	87	22	40	0	1	52	159	74	201	67	6

[1]Cumulative cover is the sum of individual species cover within a vegetation class. It is being used as an indicator of changes in the relative abundance of different types of plants over time. Total vascular plant cover is the sum of fern + shrub + herb cover.

ring in plots; therefore, for three representative plots, percentage cover by vegetation class (ferns, shrubs, herbaceous, total vascular plants, mosses) in 1979 and 1985 is described in Table 2. Eighty-seven to 100% of the trees died in these plots. The total vascular plant cover doubled, from 74–92% in 1979 to 154–201% in 1985, and the moss cover decreased from 12–67% in 1979 to 3–6% in 1985. However, different vegetation strata and species were involved in the three plots.

In stand 1 plot 5, *Rubus idaeus* L. (classed as a shrubby species) increased from 1% cover in 1979, to 23% in 1981, to 46% in 1983, and to 78% cover in 1985, by which time it dominated the plot (Table 2). *Dryopteris austriaca* (Jacq.) Woynar, a fern, increased from 19 to 36% cover, but mosses and herbaceous species decreased (*e.g.*, *Oxalis montana* Raf., from 21 to 7%; *Clintonia borealis* (Ait.) Raf., from 17 to 4%).

In stand 2 plot 3, fern, shrub, and herbaceous species increased in abundance and moss species decreased. Species that increased by at least 10% cover from 1979 to 1985 included *Rubus idaeus* (from 1% to 13% cover), *Dryopteris austriaca* (from 21 to 34%), *Oxalis montana* (from 7 to 26%), *Cornus canadensis* L. (from 15 to 41%), and *Solidago macrophylla* Pursh (from 0 to 14%). *Clintonia borealis* declined from 15 to 5% cover.

Stand 3 plot 1 exemplifies a case of greatly increased herbaceous cover, a substantial decline in moss cover, and some increase in ferns (Table 2). Species which increased by at least 10% cover included *Dryopteris austriaca* (from 11% to 40% cover), *Oxalis montana* (from 20 to 47%), *Cornus canadensis* (from 9 to 57%), *Clintonia borealis* (from 7 to 19%), and *Aralia nudicaulis* L. (from 2 to 17%). *Pleurozium schreberi* (Brid.) Mitt., the dominant moss species, declined from 63% cover in 1979 to 5% in 1985, following death of the overstory trees. There were virtually no shrub species in this plot.

Regeneration

Advanced balsam fir regeneration was abundant in 1979 in the sample stands, ranging from 13 200 to 132 000 per hectare, or 65 to 98% stocking (Fig. 2A, B). Survival of regeneration during the budworm outbreak was quite high and in 1985 there were still 9 750 to 79 000 stems per hectare, or 67 to 100% stocking. Most mortality occurred in small (< 5 cm tall) regeneration which was not adequately established. In 1979, 39% of the fir regeneration was < 5 cm tall and 90% was < 15 cm tall. By 1985, however, only 2% was < 5 cm tall, with 31% 6–15 cm tall, 40% 16–30 cm, 14% 31–45 cm, and 16% > 45 cm. This indicates that there were few newly germinated (< 5 cm) fir seedlings in the stand in 1985 and that much of the fir regeneration had survived the budworm outbreak and grown.

In addition to the balsam fir regeneration, many seedlings of white birch, yellow birch, and pin cherry (*Prunus pensylvanica* L.f.) became established in several of the stands. In stand 1, 2% was white birch seedlings; the remainder was fir. Stand 2 had 26% white birch and 2% pin cherry seedlings, whereas stand 3 had 11% white birch seedlings. In the thinned immature stand 6, however, 52% of the sampled seedlings were hardwood species: 33% white birch, 12% yellow birch, and 7% pin cherry. It is possible that the hardwood component of the stand may increase somewhat as the dead overstory breaks up, but it would seem that the

Effects of spruce budworm outbreaks

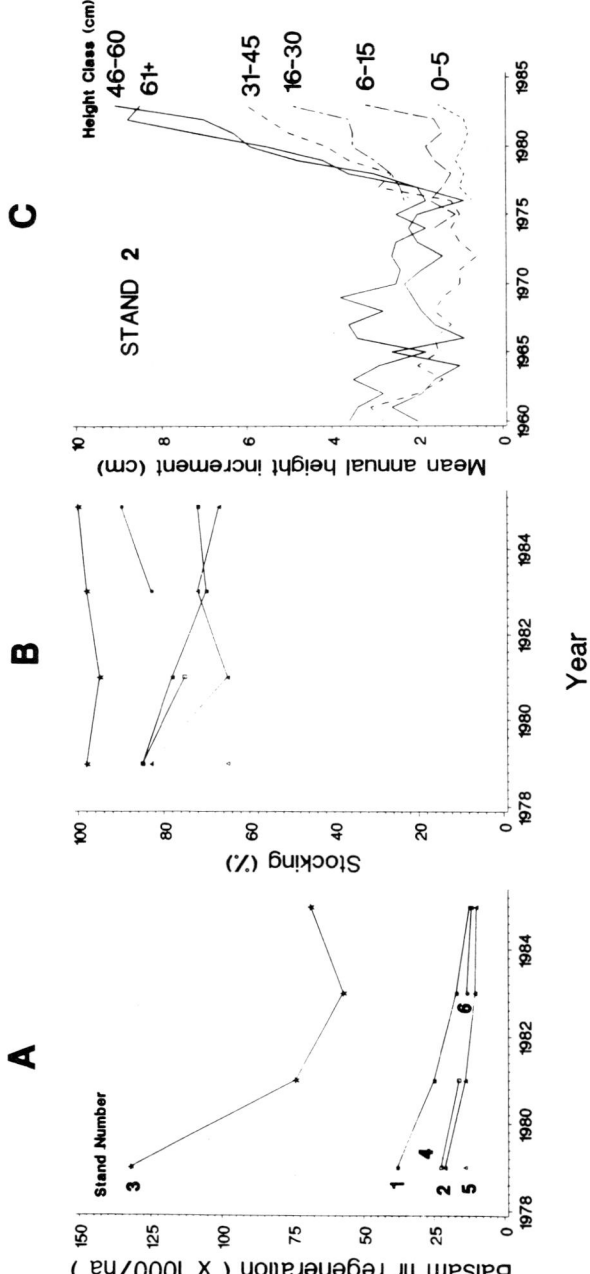

Fig. 2. Changes in balsam fir regeneration in six balsam fir stands during a severe spruce budworm outbreak. A. Balsam fir regeneration, B. Balsam fir stocking, and C. Mean annual height increment of regeneration in stand 2, sampled in six height classes.

mature fir stands will regenerate to primarily balsam fir forest. The immature fir stand will probably develop into a mixed fir-birch stand.

The fir regeneration that was sampled for analysis of radial and height increment clearly grew in response to the overstory defoliation and mortality, beginning two to three years after the first defoliation (Fig. 2C). Growth response was directly proportional to the regeneration height class, with the taller regeneration (> 45 cm) showing a three- to fivefold increase in height and radial growth.

Conclusions

A spruce budworm outbreak reduced the growth rate of balsam fir trees and, more significantly, caused the death of an average 82% of the trees in six sampled stands on Cape Breton Island. Opening of the stand stimulated the growth of ground vegetation, such that the total vascular plant cover doubled (1.80x – 2.27x) over a six-year period. Adequate fir regeneration survived to restock the stands and new fir stands will succeed those killed by budworm.

The pure fir stands examined in this study are only one of the stand types in this region, however, and spruce, mixed spruce-fir, fir-intolerant hardwoods, and other types also occur. The long-term fate of these stand types during budworm outbreaks is more complex and not well understood.

Results of this study support the hypothesis of Baskerville (1975) that spruce budworm and fir forest together form a self-regulating system. Further evidence exists in long-term (1955 to 1985) data on development of balsam fir stands in northwestern New Brunswick, Canada after a budworm outbreak. (M. Huot and D.A. MacLean, unpublished data). Although budworm outbreaks cause tree mortality and dramatically alter the structure and productivity of forest ecosystems over the short term, it appears that, over the long term, budworm outbreaks act as a cycling mechanism which allows advanced fir regeneration to succeed the fir overstory. Spruce budworm populations and balsam fir forests therefore form an interdependent, cyclic ecological system which possesses long-term successional stability.

References

Baskerville, G.L. 1975. Spruce budworm: super silviculturist. For. Chron. 51: 138–140.
Blais, J.R. 1983. Trends in the frequency, extent, and severity of spruce budworm outbreaks in eastern Canada. Can. J. For. Res. 13: 539–547.
Hawboldt, L.S. 1955. The spruce budworm. Nova Scotia Department Lands and Forests, Truro, N.S., Canada. Unpublished paper. 15 pp.
Irving, H.J. and Webb, F.E. 1981. Forest protection against spruce budworm in New Brunswick – an aerial applicator's perspective. Pulp Pap. Can. 82: 23–31.
Jordan, G.A. and Ballance, R.H. 1983. A microcomputer-based annual ring measurement system. For. Chron. 59: 21–25.
MacLean, D.A. 1979. Spruce budworm-caused balsam fir mortality on the Cape Breton Highlands 1974–1978. Can. For. Serv., Marit. For. Res. Cent., Fredericton, N.B., Canada. Inf. Rep. M-X-97.
MacLean, D.A. 1980. Vulnerability of fir-spruce stands during uncontrolled spruce budworm outbreaks: a review and discussion. For. Chron. 56: 213–221.

MacLean, D.A. 1984. Effects of spruce budworm outbreaks on the productivity and stability of balsam fir forests. For. Chron. 60: 273–279.
MacLean, D.A. 1985. Effects of spruce budworm outbreaks on forest growth and yield. In: C.J. Sanders, R.W. Stark, E.J. Mullins and J. Murphy (eds), Recent Advances in Spruce Budworms Research, pp. 148–175. Proc. CANUSA Spruce Budworms Research Symp., Bangor, ME Sept. 16–20, 1984. Can. For. Serv., Ottawa, Ont., Canada.
Morris, R.F. (ed.) 1963. The dynamics of epidemic spruce budworm populations. Mem. Entomol. Soc. Can. 31.
Piene, H. and MacLean, D.A. 1984. An evaluation of growth response of young, spaced balsam fir to three years of spruce budworm spraying with *Bacillus thuringiensis*. Can. J. For. Res. 14: 404–411.
Royama, T. 1984. Population dynamics of the spruce budworm, *Choristoneura fumiferana* (Clem.). Ecol. Monogr. 54: 429–462.
Swaine, J.M. and Craighead, F.C. 1924. Studies on the spruce budworm (*Cacoecia fumiferana* Clem.). Part 1. A general account of the outbreaks, injury and associated insects. Can. Dep. Agric., Ottawa, Ont., Canada. Tech. Bull. 37 (n.s.). pp. 3–27.

INSECT HERBIVORY AND VEGETATIONAL STRUCTURE

V.K. BROWN, A.C. GANGE and C.W.D. GIBSON
Imperial College at Silwood Park, Ascot, Berks SL5 7PY, U.K.

Abstract

We address the question 'What effects do insect herbivores have on the structure of natural plant communities?' The nature and density of insect herbivores associated with a five-year old plant community on a sandy, acid soil at Silwood Park and a calcareous (limestone) soil at Wytham, Oxford are compared. Manipulative field experiments, in which the natural levels of insect herbivory are reduced by the application of chemicals, are used to compare the effects of these consumers. Vegetation cover, species composition and above-ground structure are strongly influenced by insect herbivory. Furthermore, life-history groupings of plants respond in different ways. By this means, insect herbivory has a profound effect both on the rate and direction of plant succession.

Introduction

Over the last decade there has been an ever-growing interest in the ecology of plant/animal interactions and more especially the impact of herbivorous species on vegetation. Apart from classic studies on larger herbivores (mammals: Watt 1981; Belsky 1986; Gomez Sal *et al.* 1986; Gibson *et al.* 1987a, b; reptiles: Merton *et al.* 1976b; Gibson *et al.* 1983; birds: Patton and Frame 1981; Bazely and Jefferies 1986) there is now a substantial body of evidence highlighting the more subtle effects that invertebrate herbivores have on selected attributes of plant populations (*e.g.*, Bentley and Whittaker (1979); Whittaker (1979, 1982); Bentley *et al.* (1980); Dirzo and Harper (1982); Cottam (1985) and Brown *et al.* (1987a). These effects are mainly manifest by differences in plant growth and performance and are often mediated through competitive effects. Changes in natural plant communities, resulting from insect herbivory, have seldom been reported and some authors have indicated that there are indeed no measurable effects (Hairston *et al.* 1960). Recent work at Silwood Park, Berkshire, U.K. and Wytham, Oxfordshire, U.K. (*e.g.*, Brown 1982, 1984, 1985; Brown *et al.* in press; Gibson *et al.* 1987a) and elsewhere (*e.g.*, McBrien *et al.* 1983) has demonstrated beyond doubt that insect herbivory can influence both the texture and structure of the plant community (sensu Barkman 1979). These studies compared various characteristics of the plant community in the field when subjected to natural levels of insect herbivory and when herbivory was reduced by the application of insecticide.

Vegetation structure, in its broadest sense, refers to the spatial (horizontal, vertical and temporal) arrangement of different morphological elements (*i.e.*, its texture) (Barkman 1979). Part of this general concept of vegetation structure has been considered in detail in Southwood *et al.* (1979); Lawton (1983); Stinson (1983); Stinson and Brown (1983) and Brown and Southwood (1987). In particular, these works emphasized the importance of plant structure to insect communities and identified two measures of plant structural diversity. Spatial diversity was defined

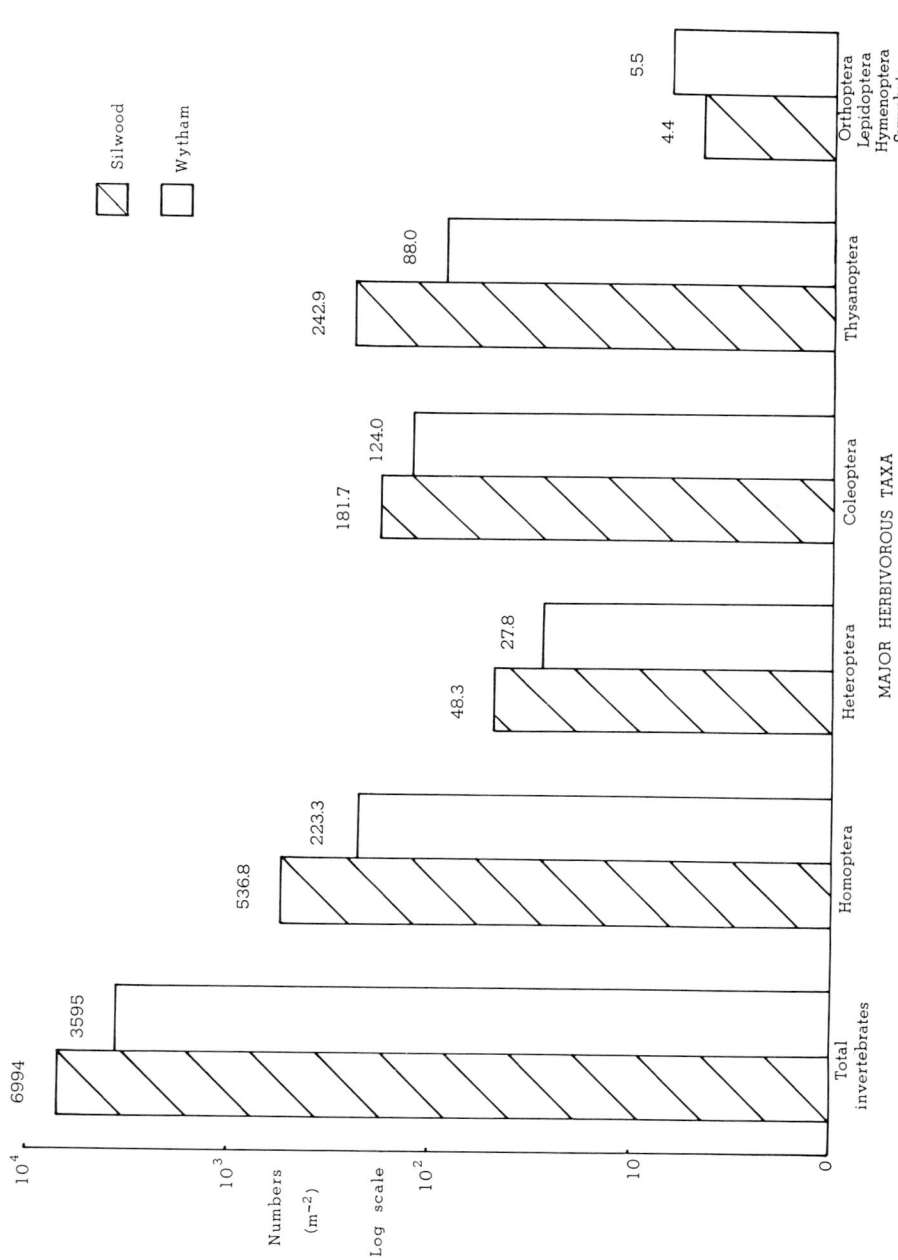

Fig. 1. Density of above-ground invertebrates and major herbivorous taxa, sampled by D Vac suction at two sites in southern Britain. Data are for August 1985.

as the distribution of plant structures in space above ground level and was quantified in terms of spatial (height) profiles of the vegetation or as an index of diversity (see Southwood et al. 1979). On the other hand, architectural diversity was defined as the distribution of different *types* of plant structure (e.g., leaves, stems, flowers, fruits etc.). This provides a measure of resource availability, since many insects are specifically associated (for feeding, oviposition or resting) with a particular type of plant structure (Brown and Southwood 1987). Brown (1982, 1985) and Brown et al. (1987a) have indicated how insect herbivory may modify plant architecture by reducing the number of reproductive structures as well as altering the phenology of the plant. Brown et al. (1987a) have also demonstrated changes in certain vegetative structures in response to herbivory. However, differences in spatial attributes resulting from insect herbivory have received only superficial mention (see Brown 1985).

Here, we describe a comparative study involving two sites with similar land history but differing in soil type and therefore nature of vegetation. The same experimental procedure and sampling techniques are used at both sites, thereby affording a direct comparison of the effects of insect herbivory on vegetation structure. We first look at the composition and abundance of the major types of insect herbivores and then at the structural changes in the vegetation which result from manipulating the levels of insect herbivory. Although vertical structure receives most attention, horizontal structure (assessed in terms of vegetation cover) is included. Furthermore, seasonal cycles in these two attributes afford a measure of temporal change in vegetation structure.

The experimental sites

The sites were formerly arable land on which regular cultivation ceased five years ago (1982). The site at Silwood Park is on Bagshot Sands with sandy, acid soil and that at Wytham has shallow soil overlying Jurassic corallian limestone. Details of the two sites and experimental design are given in Brown et al. (1987a, b) and Gibson et al. (1987b) respectively. Natural levels of insect herbivory were reduced by the regular application of a non-persistent insecticide (Malathion-60) applied at the standard agricultural rate. The limitations of the method and precautions, cited in Brown et al. (1987c), were heeded. Each experimental treatment plot was 3×3 m. At Silwood each of the two treatments (referred to as insecticide-treated and control) were applied to 10 3×3 m plots and at Wytham 12. Treatment began in May 1985 and data are presented up to the present (June 1987).

The insect herbivores

The above-ground invertebrates were sampled by suction (D Vac) at monthly intervals from May–October. Figure 1 provides an example of the data for a single sampling occasion (August 1985) and shows that the total density of invertebrates was higher at the Silwood site. Furthermore, several of the major herbivorous taxa followed this trend. In particular, the sap-feeding Homoptera and Thysanoptera were more abundant at Silwood than at Wytham. Only the Coleoptera and Heteroptera had densities at Wytham approaching those at Silwood, although the latter

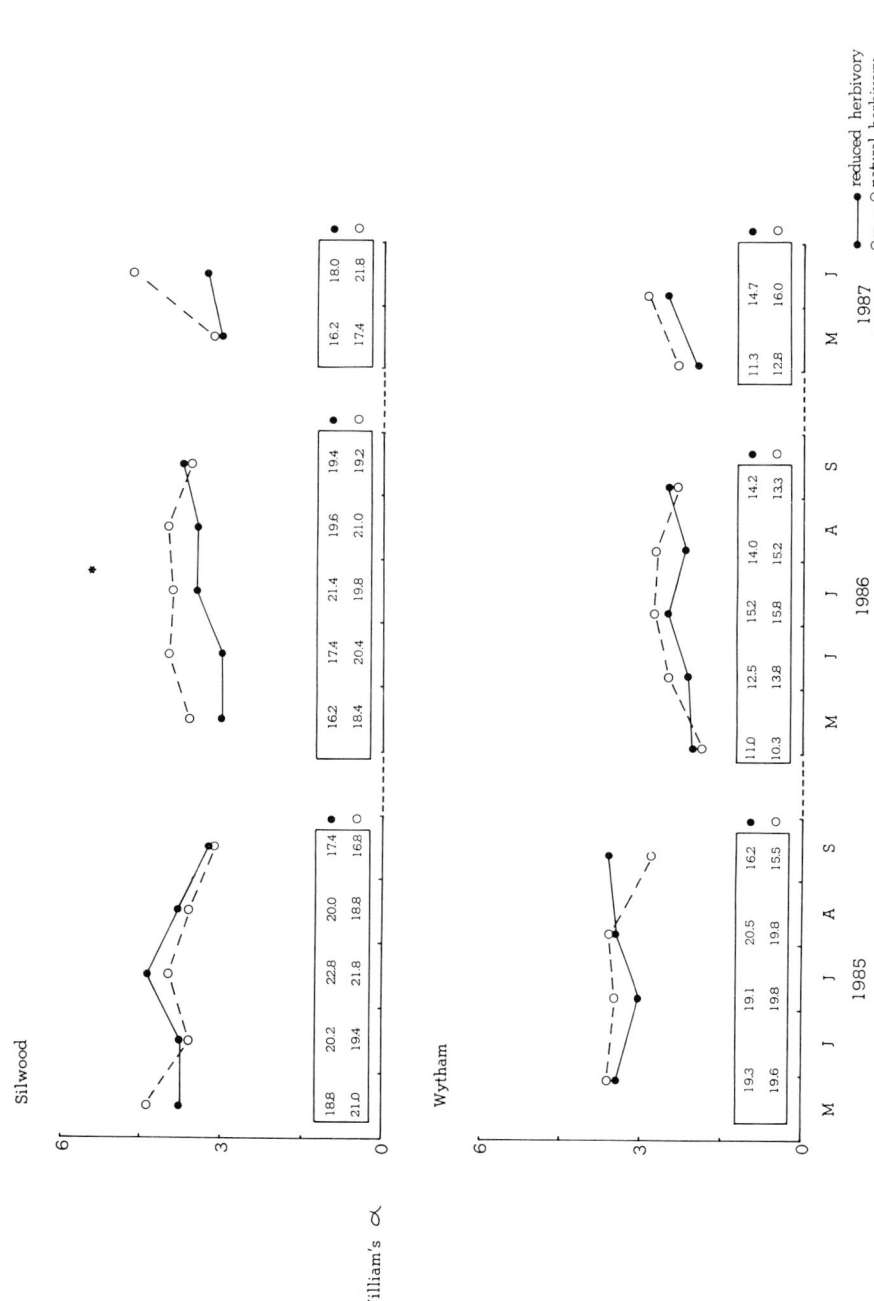

Fig. 2. Plant species diversity, expressed as Williams' α under natural and reduced levels of insect herbivory at two sites in southern Britain. Inset: mean species richness per 50 pins. Significance of difference between treatments (analyzed for 1985 and 1986 only) given as follows $^*p < 0.05$, $^{**}p < 0.01$, $^{***}p < 0.001$.

group was scarce at this stage of succession. The larger insect herbivores (Orthoptera, larval Lepidoptera, larval Hymenoptera: Symphyta) are not well sampled by suction, but direct observation of these taxa demonstrated that their densities were low in these sites. Interestingly, the other invertebrate guilds (predators, parasitoids and detrivores) were also more common on the acidic grassland system at Silwood.

The vegetation

This is described in terms of floristic composition (species richness and diversity) cover and structural attributes. Major plant groupings, based on life-history characteristics (an element of vegetation texture, sensu Barkman 1979) are described in detail in Gibson *et al.* (1987b), and are considered together with dominant plant species which make significant contributions to the differences in overall vegetation structure.

The vegetation was monitored by placing at random in each plot five linear 38 cm frames containing 10 equally-spaced vertical 3.0 mm diameter point quadrat pins. All touches of living plant material touches were recorded to species and by 2 cm (below 10 cm) or 5 cm (10 cm and above) height intervals. Data for five samples during the season (May–October) are discussed. These data provide information on:

i. Plant species richness and diversity, the latter assessed by Williams' α.

ii. The total number of touches of a plant species at all height intervals which provides a measure of cover abundance. This may be considered as a measure of horizontal structure, but is also dependent on the vertical complexity of the vegetation.

iii. A weighted mean height for a species derived from the number of touches at different height intervals as:

$$\sum_{i=1}^{N} (h_i \times n_i) \bigg/ \sum_{i=1}^{N} (n_i)$$

where h_i = mid point of height class i, n_i = number of touches at height class i, and N = number of height classes represented in the sample.

Trends in these three measures were analyzed by analysis of variance. Before analysis, data were transformed to square root (see Gibson *et al.* (1987b)). Complete analyses are cited only for 1985 and 1986.

iv. The structural profile of the sward and of individual plant species were based on the number of touches at each height interval. The sward profile shape was investigated by taking a random sample of touches of all species in a plot. Touches for different height classes in the sample were pooled in a manner appropriate for subsequent χ^2 analysis. Individual species profiles were also subjected to χ^2 analysis.

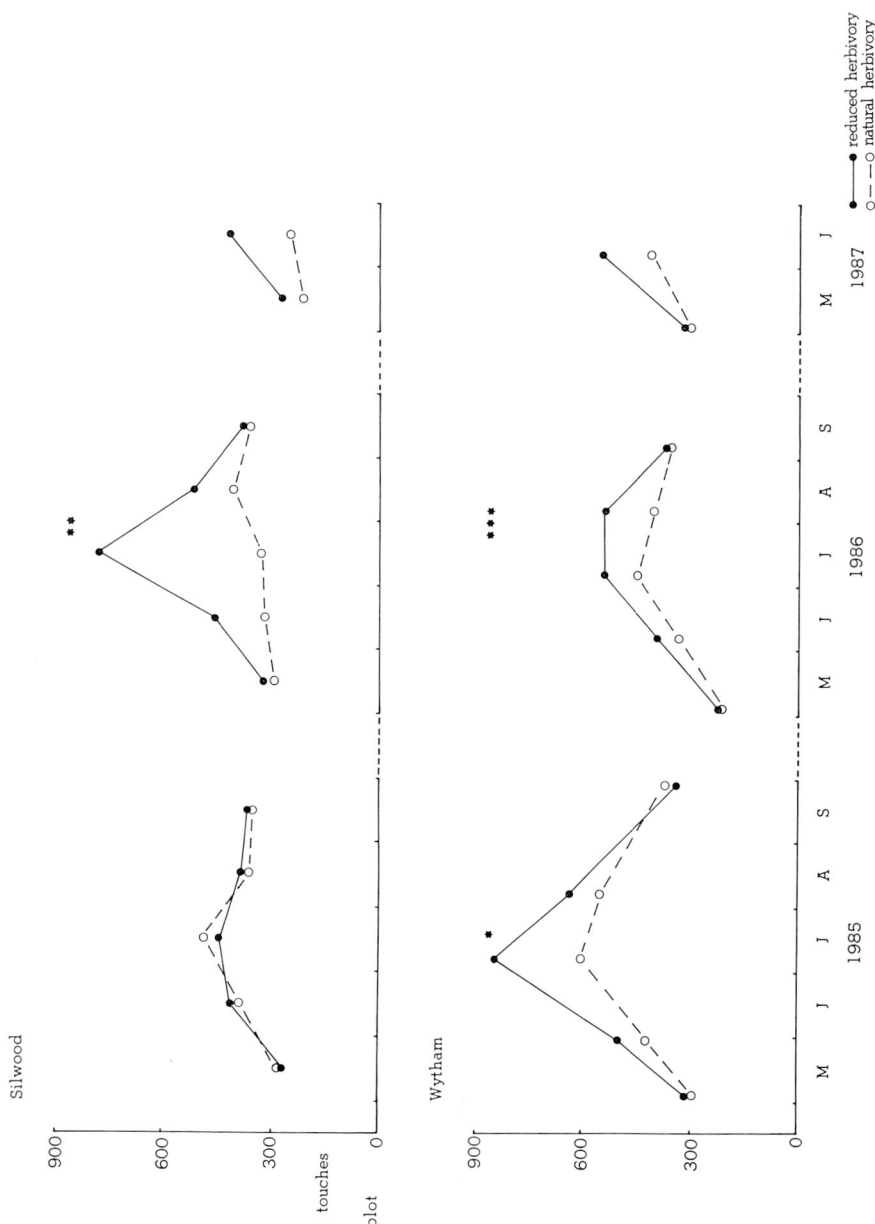

Fig. 3. Cover abundance of vegetation, expressed as mean number of vegetation touches on point quadrat pins, under natural and reduced levels of insect herbivory at two sites in southern Britain. Levels of significance as in Fig. 2.

Floristic composition

A total of 49 plant species have been found within the experimental plots at Silwood and 90 at Wytham since treatment began in 1985.

A reduction in insect herbivory had no significant effect on number of plant species (expressed as number of species per 50 pins in inset in Fig. 2), although there was a tendency for this to be higher in control plots subjected to natural levels of herbivory. This trend appears to be accelerating in the third season of treatment. A similar pattern emerged when species diversity (calculated as Williams' α) was compared (Fig. 2). At the Silwood site, species diversity was significantly increased by natural levels of herbivory in the second season ($p < 0.05$).

Vegetation structure

i. Community effects

The cover abundance of the sward was affected by insect herbivory, but in a different way in the two sites (Fig. 3). At the Wytham site vegetation cover was significantly reduced by insect herbivory from the first year of treatment, while at the Silwood site this effect only appeared in the second season. These differences can be explained by considering trends in the major plant groupings at this stage of succession (*e.g.*, Fig. 4). Perennial herbs and perennial grasses were not influenced significantly by natural levels of insect herbivory in the first season at either site. In the second season the herbs and particularly the grasses at Silwood were enhanced by a reduction in herbivory. At Wytham the perennial herbs showed no consistent significant effect (even though there is some indication that their cover may be enhanced by reduced levels of herbivory), while from the middle of the first season the cover of grasses was increased greatly by a reduction in herbivory. The plant grouping contributing most to the difference between treatments in the first season at Wytham was hardly represented at Silwood, namely the short-lived perennial herbs. The cover of this group (dominated by *Medicago lupulina* L.) was greatly reduced by insect herbivory (Fig. 4, inset $p < 0.001$).

The difference in mean sward height (expressed as an index – see before) resulting from a reduction in herbivory mimicked that for cover, namely, an increase in sward height from the first year of treatment at the Wytham site, but only from the second year of treatment at the Silwood site (Fig. 5). These differences appear to be accelerating in the third year of treatment. The sward profile of the vegetation follows these general patterns and the difference in shape of the profile, expressed by χ^2 values, throughout the 1986 growing season can be seen in Fig. 6. Differences in shape were marked at both sites, but particularly so at the Silwood site in mid season. However, at the end of the season, vegetation die back reduced the difference to a non-significant level.

The major plant-groupings (perennial herbs and grasses) also conformed to this trend (Fig. 7). At Wytham the mean height of perennial grasses and herbs was increased by reduced herbivory, even though the cover of the latter group was decreased. Again the difference was more marked in the second year of treatment and appeared even more so in the third year. At Silwood, a difference in mean height only occurred in the grasses from the second season of treatment. There

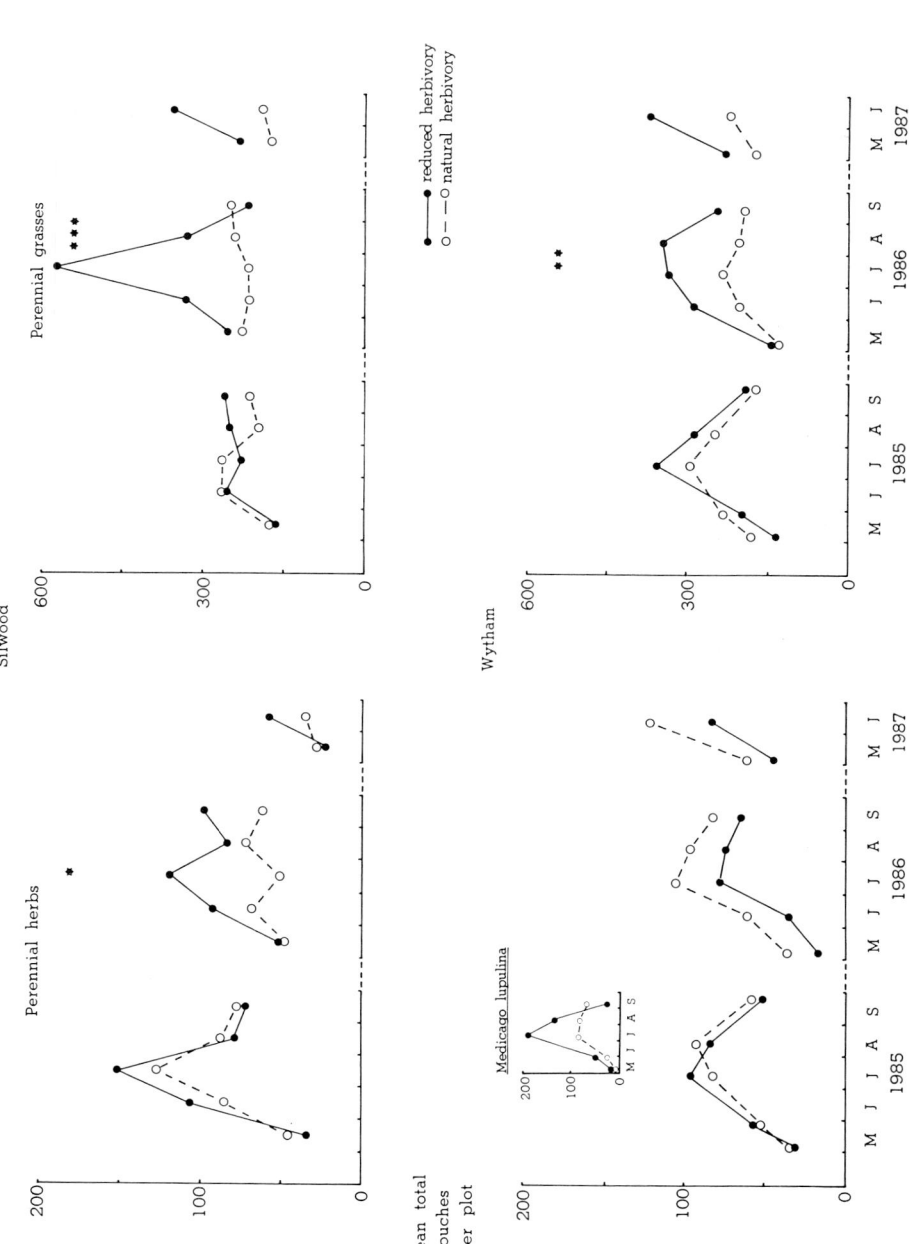

Fig. 4. Cover abundance of perennial herbs and grasses, expressed as in Fig. 3. Inset: data for short-lived perennial herb, *Medicago lupulina* at Wytham in 1985.

Insect herbivory and vegetational structure 271

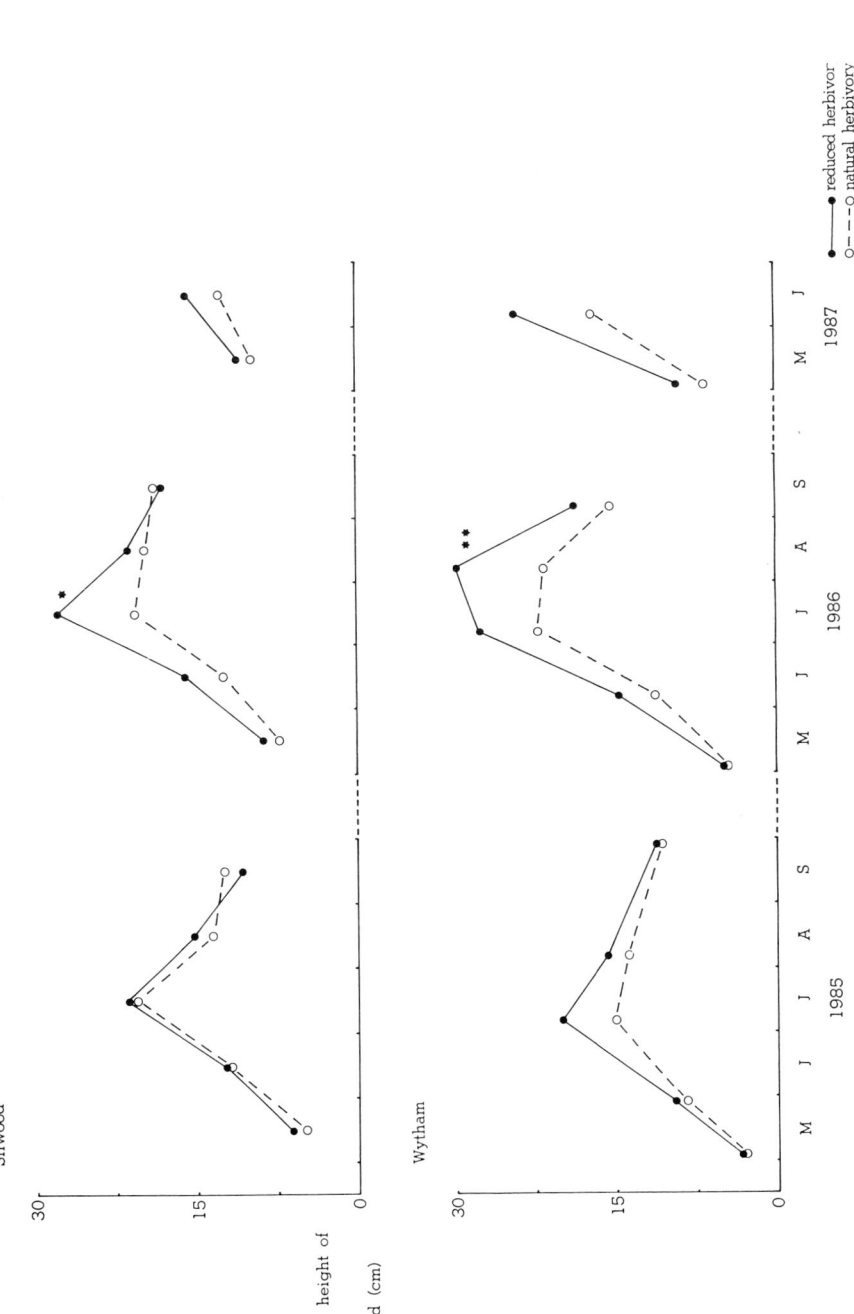

Fig. 5. Mean sward height, expressed as weighted mean height (see text) under natural and reduced herbivory at two sites in southern Britain. Levels of significance as in Fig. 2.

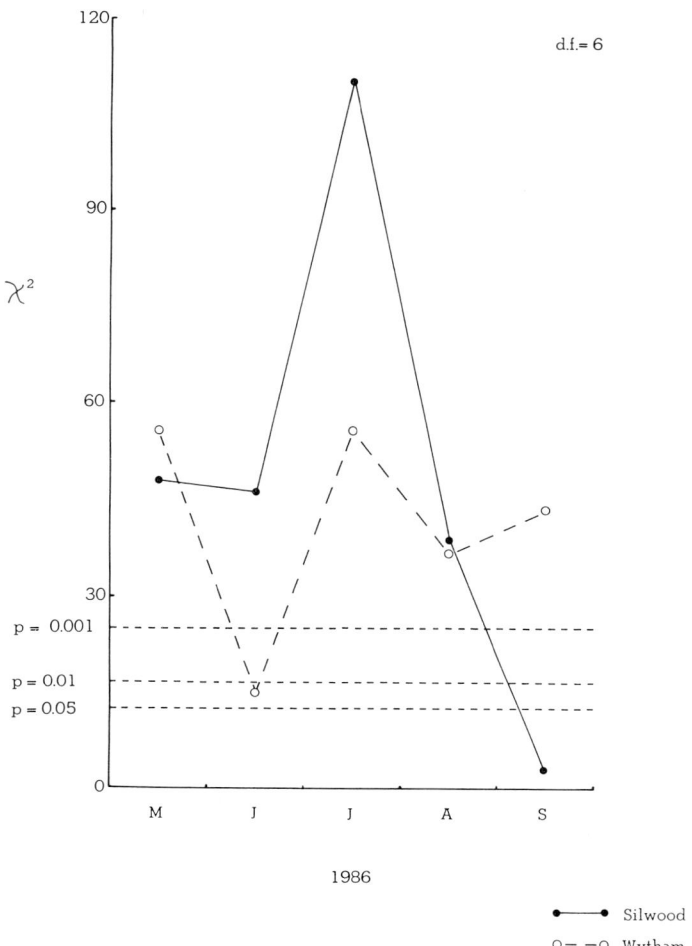

Fig. 6. χ^2 values of difference in structural profiles of vegetation resulting from reduced herbivory at two sites. Data for 1986.

was no increase in height comparable to the increase in cover seen in the perennial herbs.

ii. Trends in single species

Since grasses in the same species or genus were common to the two sites, we have selected these for more detailed consideration. *Holcus lanatus* L. occurred at both sites, but was more abundant on the acid, sandy soil at Silwood (Fig. 8b). However, it responded to reduced herbivory in the same way at the two sites: no significant difference in the first season of treatment, but a significant increase in cover from the second season. *Agrostis capillaris* L. and *A. stolonifera* L. also occurred at both sites, though the former was rare at Wytham and the latter rare at Silwood. The more abundant species at each site showed a similar response to reduced herbivory as *H. lanatus* (Fig. 8a). The seasonal changes in structural pro-

Insect herbivory and vegetational structure

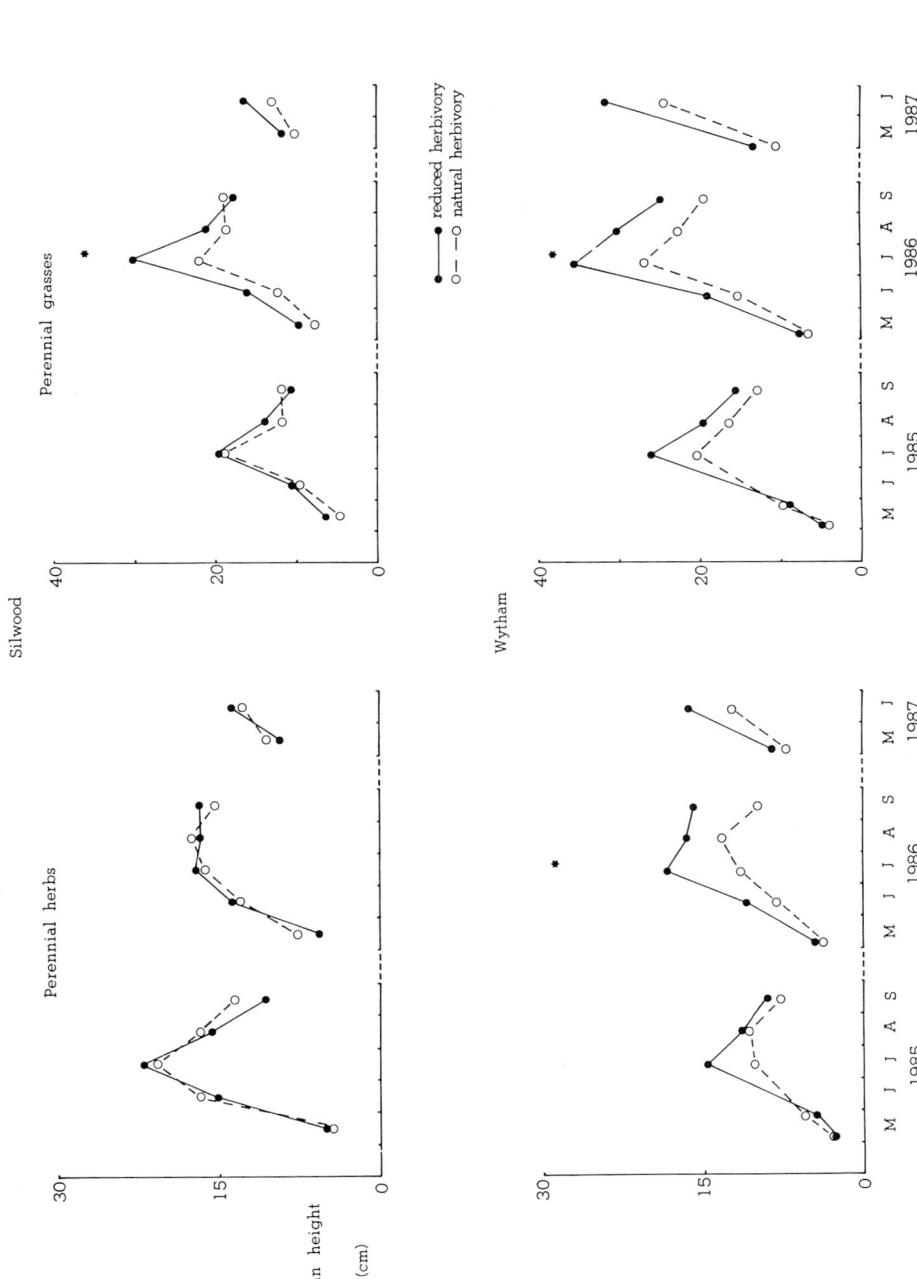

Fig. 7. Mean height of perennial herbs and grasses under natural and reduced herbivory at two sites in southern Britain. Levels of significance as in Fig. 2.

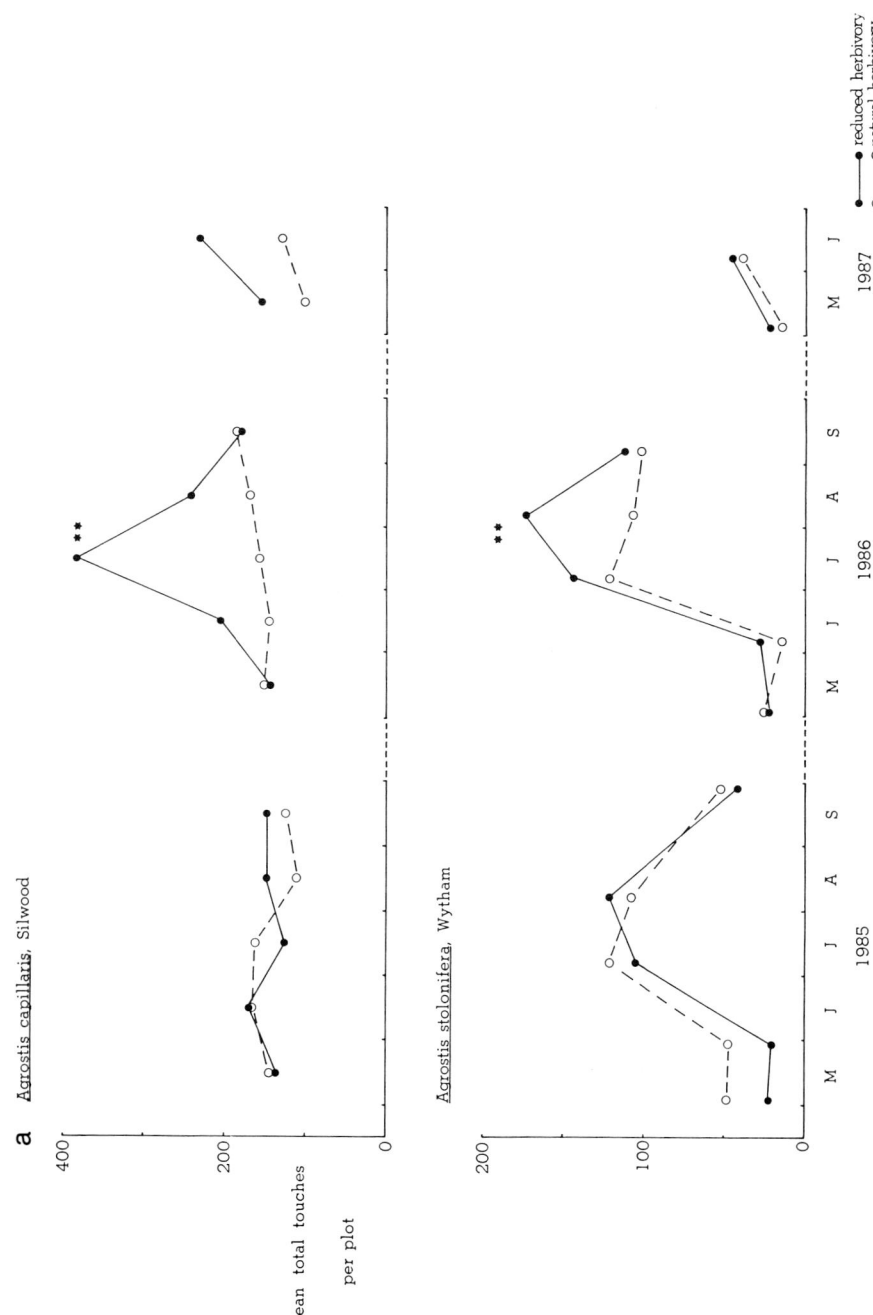

Fig. 8. Cover abundance for a) *Agrostis capillaris* (Silwood) and *Agrostis stolonifera* (Wytham) and b) *Holcus lanatus* at two sites under natural and reduced herbivory. Levels at significance as in Fig. 2.

Insect herbivory and vegetational structure

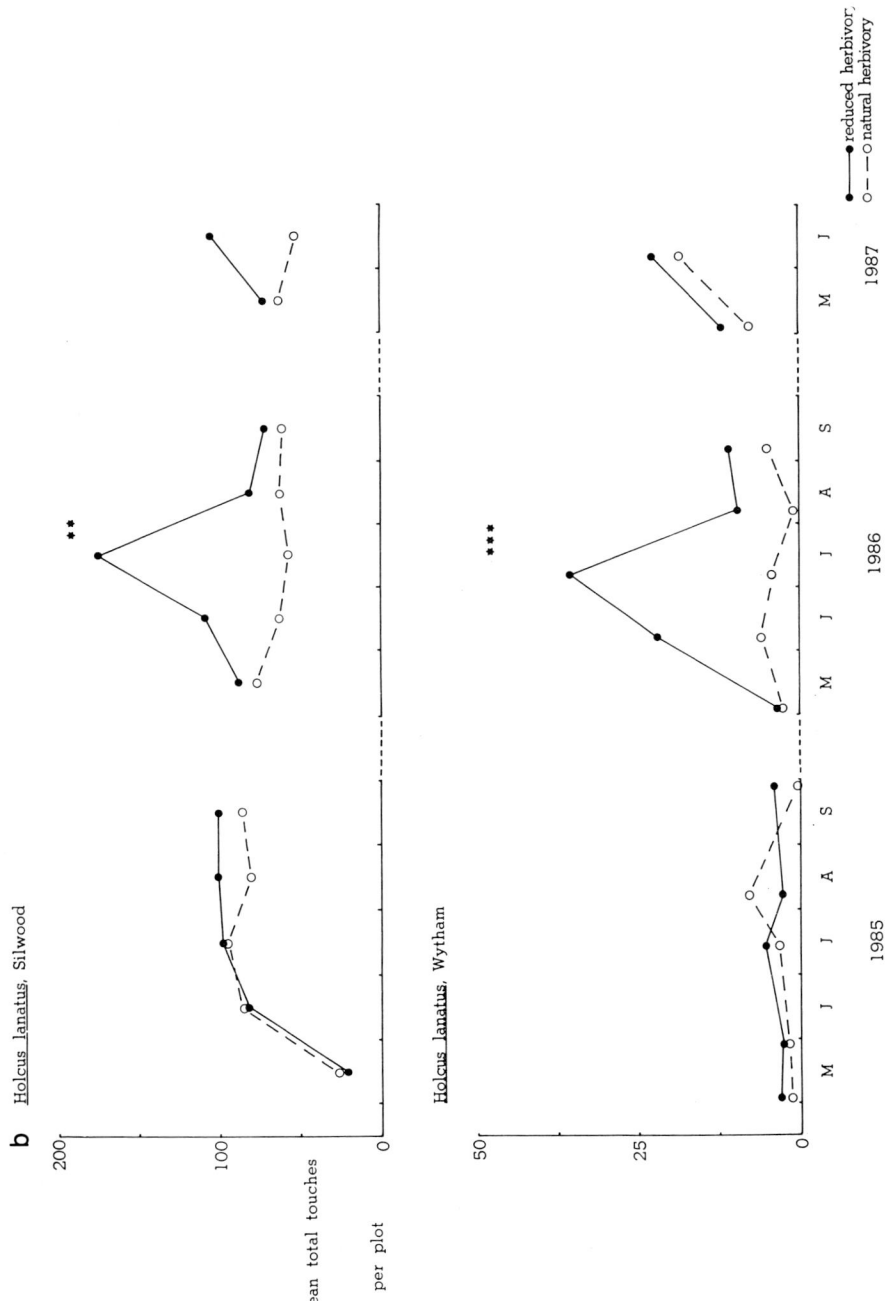

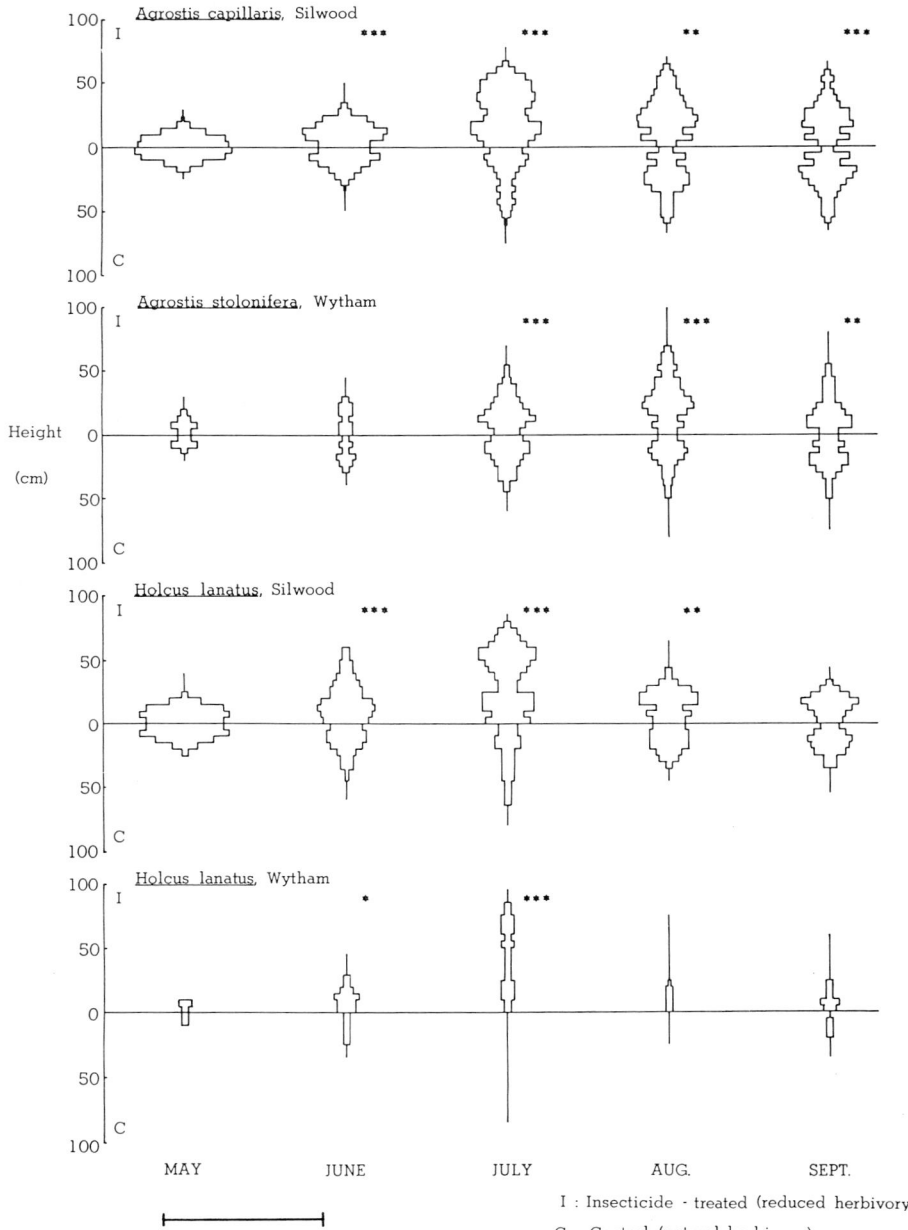

Fig. 9. Structural profiles for *Holcus lanatus* at two sites and *Agrostis capillaris* (Silwood) and *Agrostis stolonifera* (Wytham) under natural and reduced herbivory. Data for 1986. Differences assessed by χ^2 analysis. Levels of significance as in Fig. 2. Scale bar = 500 touches.

files of the species can be seen in Fig. 9. The difference in profile shape, tested by χ^2 techniques, was highly significant for most of the sampling dates. Only early in the season when the sward was short, or late in the season when *H. lanatus* was dying back, were the profile shapes not significantly different.

Discussion

The numbers of insects associated with different plant species has been attributed to a wide range of variables (Southwood 1961; Lawton and Schroder 1977; Strong and Levin 1979; Claridge and Wilson 1982). In addition, Lawton and McNeill (1979) predicted that the insect herbivore load on ruderal, annual plant species, characteristic of early succession, should be higher than on perennial species found later in succession. In earlier work, one of us (VKB) established that insect herbivory appeared to be a major driving force in the establishment and development of very early successional natural plant communities (Brown 1982) and suggested that the effects of insects on later (successional) stages must be the next question to pose. It is this question that we address here.

By selecting two sites in southern Britain with similar land history, but with different physical characteristics, we aimed to show that the effects of insect herbivory are not merely site specific. Within the limitations of manipulative field experiments (see Brown *et al.* 1987c), we have demonstrated that insect herbivory does have a considerable effect on both the texture and structure of a perennial-dominated plant community, prior to the establishment of shrubs and trees (Tormala 1982; Brown and Southwood 1987).

Among the more marked changes at the community level was the increase in mean sward height and the change in the shape of the structural profile resulting from reduced herbivory. Such structural changes may well enable subtle textural changes to occur. We suggest, even at this relatively early stage of an ongoing project, that the tendency towards a lower plant species richness and diversity (and cover of perennial herbs at one site) may be due to the enhanced performance of certain plant groupings. The differential response to herbivory of the seven plant groupings recognized by Gibson *et al.* (1987b) is relevant here. By the fifth year of succession annual herbs and grasses were too rare to show consistent patterns (Brown and Southwood 1987) and short-lived perennials, mosses, and shrubs and trees were also low in abundance. Thus, it was the balance between perennial herbs and grasses that was critical. From the current study, the latter group was consistently enhanced in cover and height by a reduction in herbivory. Such trends were mainly due to the substantial increase in flowering stems, which greatly increased the number of touches on point quadrat pins. Interestingly, it was particularly some of the fast-growing, invasive grasses which were enhanced, e.g., *Agrostis* species and *Agropyron repens* P.B. (at Wytham) (see Brown *et al.* in press). Such grass growth obscures microsites for the colonization of other species and in time leads to a reduction in species richness and diversity. The mechanism underlying this difference may well be the relative abundance of the insect herbivores. The predominantly grass-feeding Homoptera:Auchenorrhyncha were the numerically dominant herbivorous taxon and were especially abundant at the Silwood site, where the effect of herbivory on grasses was particularly marked.

By changing the balance between different plant groupings, not only the rate but also the direction of succession was altered. Looking at these results in a different, more strictly applied light, it can be seen that natural levels of insect herbivory are important in the maintenance of a herb species-rich sward and indeed to a plant community of diverse texture (sensu Barkman 1979). Thus, in management, conservation or habitat restoration programs the effects of insect herbivory should not be overlooked. Indeed, the indiscriminant use of pesticides might well have an adverse effect on a plant community and even counteract a positive strategy for increasing the quality of vegetational structure.

Acknowledgments

This work was supported by a grant to V.K. Brown from N.E.R.C. (Silwood site) and the Nuffield Foundation (Wytham site). We thank the Wytham Management Committee, Oxford University for permission to use the Wytham site. H.C. Dawkins advised on experimental design and analysis. Anne Storr and Martyn Jepsen assisted diligently in the field as did Charlotte Ford and Deborah Proctor for short periods financed by a B.E.S. Small Projects Grant. To them all we are extremely grateful.

References

Barkman, J.J. 1979. The investigation of vegetation texture and structure. In: M.J.A. Werger (ed.), The Study of Vegetation, pp. 125–160. Junk, The Hague.
Bazely, D.R. and Jefferies, R.L. 1986. Changes in the composition and standing crop of salt-marsh communities in response to the removal of a grazer. J. Ecol. 74: 693–706.
Belsky, A.J. 1986. Revegetation of artificial disturbances in grasslands of the Serengeti National Park, Tanzania. 1. Colonisation of grazed and ungrazed plots. J. Ecol. 74: 419–437.
Bentley, S. and Whittaker, J.B. 1979. Effects of grazing by a chrysomelid beetle *Gastrophysa viridula* on competition between *Rumex obtusifolius* and *Rumex crispus*. J. Ecol. 67: 79–90.
Bentley, S., Whittaker, J.B. and Malloch, A.J.C. 1980. Field experiments on the effects of grazing by a chrysomelid beetle (*Gastrophysa viridula*) on seed production and quality in *Rumex obtusifolius* and *Rumex crispus*. J. Ecol. 68: 671–674.
Brown, V.K. 1982. The phytophagous insect community and its impact on early successional habitats. Proceedings of the 5th International Symposium on Insect-Plant Relations, Wageningen 1982, pp. 205–213. Pudoc, Wageningen.
Brown, V.K. 1984. Secondary succession: insect-plant relationships. BioScience 34: 710–716.
Brown, V.K. 1985. Insect herbivores and plant succession. Oikos 44: 17–22.
Brown, V.K., Gange, A., Evans, I.M. and Storr, A.L. (1987a). The effect of insect herbivory on the growth and reproduction of two annual *Vicia* spp., at different stages of plant succession. J. Ecol. 75: 1173–1189.
Brown, V.K., Hendrix, S.D. and Dingle, H. 1987b. Plants and insects in early old-field succession: comparison of an English site and an American site. Biol. J. Linn. Soc. 31: 59–74.
Brown, V.K., Jepsen, M. and Gibson, C.W.D. Insect herbivory: effects on early old field succession. Oikos (in press).
Brown, V.K., Leijn, M. and Stinson, C.S.A. 1987c. The experimental manipulation of insect herbivore load by the use of an insecticide (Malathion-60): the effect of application on plant growth. Oecologia 72: 377–381.

Brown, V.K. and Southwood, T.R.E. 1987. Secondary succession: patterns and strategies. In: A.J. Gray, M.J. Crawley and P.J. Edwards (eds), Colonisation Succession and Stability, pp. 315–337. Blackwell Scientific Publications, Oxford.

Claridge, M.F. and Wilson, M.R. 1982. Insect herbivore guilds and species-area relationships: leaf-miners on British trees. Ecol. Entomol. 7: 19–30.

Cottam, D.A. 1985. Frequency-dependent grazing by slugs and grasshoppers. J. Ecol. 73: 925–934.

Dirzo, R. and Harper, J.L. 1982. Experimental studies on slug-plant interactions. iv. The performance of cyanogenic and acyanogenic morphs of *Trifolium repens* in the field. J. Ecol. 70: 119–138.

Gibson, C.W.D., Brown, V.K. and Jepsen, M. 1987a. Relationships between the effects of insect herbivory and sheep grazing on seasonal changes in an early successional plant community. Oecologia 71: 245–253.

Gibson, C.W.D., Dawkins, H.C., Brown, V.K. and Jepsen, M. 1987b. Spring grazing by sheep: effects on seasonal changes during early old field succession. Vegetatio 70: 33–43.

Gibson, C.W.D., Guilford, T.C., Hambler, C. and Sterling, P.H. 1983. Transition matrix models and succession after release from grazing on Aldabra stoll. Vegetatio 52: 151–159.

Gomez Sal, A., De Miguel, J.M., Casado, M.A. and Pineda, F.D. 1986. Successional changes in the morphology and ecological responses of a grazed pasture system in central Spain. Vegetatio 67: 33–44.

Hairston, N.G., Smith, F.E. and Slobodkin, L.B. 1960. Community structure, population control and competition. Am. Natur. 144: 421–425.

Lawton, J.H. 1983. Plant architecture and the diversity of phytophagous insects. Ann. Rev. Entomol. 28: 23–29.

Lawton, J.H. and McNeill, S. 1979. Between the devil and the deep blue sea: on the problem of being a herbivore. In: R.M. Anderson, B.D. Turner and L.R. Taylor (eds), Population Dynamics, pp. 223–244. 20th Symposium of the British Ecological Society. Blackwell Scientific Publicatons, Oxford.

Lawton, J.H. and Schroder, D. 1977. Effects of plant type, size of geographical range and taxonomic isolation on number of insect species associated with British plants. Nature 265: 137–140.

McBrien, H., Harmsen, R. and Crowder, A. 1983. A case of insect grazing affecting plant succession. Ecology 64: 1035–1039.

Merton, L.F.H., Bourn, D.M. and Hnatiuk, R.J. 1976. Giant tortoise and vegetation interactions on Aldabra atoll. Part 1: inland. Biol. Conserv. 9: 293–316.

Patton, D.L.H. and Frame, J. 1981. The effect of grazing in winter by wild geese on improved grassland in west Scotland. J. Appl. Ecol. 18: 311–325.

Southwood, T.R.E. 1961. The number of species of insect associated with various trees. J. Anim. Ecol. 30: 1–8.

Southwood, T.R.E., Brown, V.K. and Reader, P.M. 1979. The relationships of plant and insect diversities in succession. Biol. J. Linn. Soc. 12: 327–348.

Stinson, C.S.A. 1983. Effects of Insect Herbivores on Early Successional Habitats. Unpublished Ph.D. thesis, University of London.

Stinson, C.S.A. and Brown, V.K. 1983. Seasonal changes in the architecture of natural plant communities and its relevance to insect herbivores. Oecologia 56: 67–69.

Strong, D.R. Jr. and Levin, D.A. 1979. Species richness of plant parasites and growth form of their host. Am. Natur. 114: 1–22.

Tormala, T. 1982. Structure and dynamics of reserved field ecosystem in central Finland. Biol. Res. Rep. Univ. Jyvaskyla, Finland 8: 1–58.

Watt, A.S. 1981. A comparison of grazed and ungrazed grassland in East Anglian Breckland. J. Ecol. 69: 499–508.

Whittaker, J.B. 1979. Invertebrate grazing, competition and plant dynamics. In: R.M. Anderson, B.D. Turner and L.R. Taylor (eds), Population Dynamics, pp. 207–222. 20th Symposium of the British Ecological Society. Blackwell Scientific Publications, Oxford.

Whittaker, J.B. 1982. The effect of grazing by a chrysomelid beetle, *Gastrophysa viridula*, on growth and survival of *Rumex crispus* on a shingle bank. J. Ecol. 70: 291–296.

PATTERN AND PROCESS IN ARCTIC COASTAL VEGETATION IN RESPONSE TO FORAGING BY LESSER SNOW GEESE

R.L. JEFFERIES
Department of Botany, University of Toronto, Toronto, Canada, M5S 1A1

Abstract

Foraging activities by a herbivore (Lesser Snow Goose) modify the structure and composition of vegetation of coastal wetlands in the eastern Canadian Arctic. The patterns of vegetational development of these areas are interpreted in relation to the subtle interplay between the effects of foraging, of isostatic uplift of the coastal lowlands and of climatic change on the vegetation. Vegetational dynamics of both saltwater and freshwater marshes are strongly influenced by the relative intensities of spring grubbing of roots and rhizomes and summer grazing of aboveground vegetation by the geese.

Introduction

Equilibrium theory of vegetational structure predicts that a community will show little tendency to loose species with time when at or approaching equilibrium (Chesson and Case 1986). The community has the capacity to recover from short-term events that temporarily reduce the density of species to a low level. As the system moves towards equilibrium it 'forgets' previous abundances; the environmental perturbations have no lasting effects (Chesson and Case 1986). Theories based on a stable equilibrium of vegetational structure have been questioned, because environmental and biotic effects on population structure and on the density of individuals are to be expected (Sale 1977; Wiens 1977; Grubb 1977; Connell 1978). Not all species appear to have the attributes necessary for their populations to adjust rapidly, so that stable equilibria can develop (Davis 1986). The mean of environmental fluctuations and the frequency of biotic interactions (*e.g.*, herbivory) do not remain constant over any time scale that may be considered ecological time. Long-lived organisms, in particular, may respond to changes in environment with a considerable time lag (Davis 1986). Life history characteristics and dispersal abilities of species have a profound influence on the capacity of species to track environmental change (Davis 1986). A hierarchy of reproductive and dispersal efficiencies exist among plant species, such that rates of colonization of new habitats or patches in existing habitats are not in step for all species. Differential displacement of species within existing assemblages of species occurs. 'Suites of species present today may have an historical imprint that can only be uncovered by looking back in time' (Chesson and Case 1986). Since the assemblages of plant species are continually adjusting to new conditions brought about by physical or biotic agencies, they never complete the adjustment before conditions change again. Past abundances of species remain relevant in interpreting present abundances.

The thesis developed in this paper is that the structure and composition of arctic

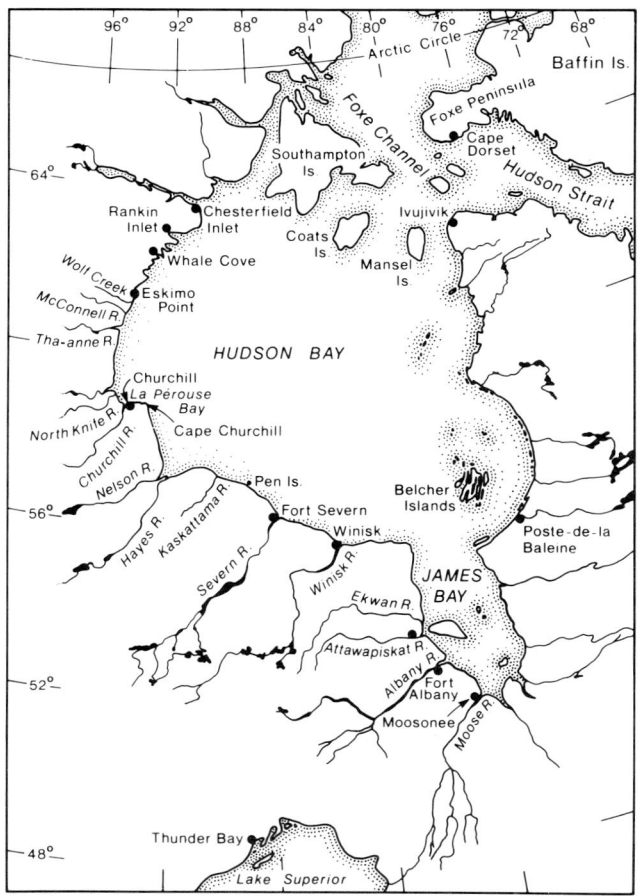

Fig. 1. Map of the Hudson Bay and surrounding coastal regions (modified from a frontpiece map in volume 109 of Le Naturaliste Canadien [1982] devoted to a symposium on James and Hudson Bays).

coastal wetland vegetation in much of the eastern North American Arctic, and the Hudson Bay region in particular, can only be interpreted in relation to non-equilibrium theories, which assume that species densities do not remain constant over time at a given location (Chesson and Case 1986). The combined effects on plant assemblages of climatic change during and since the Little Ice Age (ca. 1250 A.D. to 1880 A.D.), of isostatic uplift, and of foraging by large numbers of Lesser Snow Geese (*Chen caerulescens caerulescens* (L.) Gundl.) and by other animals reinforce each other in determining patterns of vegetational development of wetlands. The vegetation in these areas is dominated by perennials, particularly graminoids, many of which are capable of extensive clonal growth, but which undergo sexual reproduction infrequently. Annuals are extremely rare or absent. Changes in climate and geomorphology can be expected to result in changes in the species composition of plant communities. However, the role of an avifauna as a determinant of vegetational structure within a region has received little attention. The relationship between geese and vegetation can only be interpreted within the context of a hierarchical analysis of the structure and function of these wetlands, involving interactions between the herbivore and individuals, and between populations and communities of plants (Jefferies 1988).

Lesser Snow Geese: their numbers, movements and foraging behavior

The eastern arctic population of Lesser Snow Geese in North America is about 2 million birds (Kerbes 1975; Boyd et al. 1982). Birds nest in colonies on the southern and western coasts of Hudson Bay, the southwest coast of Baffin Island, Southampton Island and the smaller islands in the Foxe Basin and Hudson Bay (Fig. 1) (Gaston et al. 1987). The birds arrive on the breeding grounds in May or early June. Initiation of egg-laying commences soon afterwards and the young hatch following an incubation period of about 21 days. Adult geese and goslings graze intensively after hatch until mid-August when the birds commence their southward migration. The geese winter on the Gulf coasts of Texas and Louisiana and in refugia along the Mississippi flyway. Numbers of Lesser Snow Geese are unusually high at present (Kerbes 1975, 1982; Boyd et al. 1982); Gaston et al. 1987; MacInnes and Kerbes 1987). The reasons for the high numbers are not well understood, but they probably reflect a decline in mortality during autumn and spring migrations and on the wintering grounds (Reed 1976; Boyd et al. 1982). At least some of the populations have been increasing exponentially in the eastern Canadian Arctic in recent years (Cooke et al. 1982; Kerbes et al. 1983; MacInnes and Kerbes 1987).

The dominant vegetation of the intertidal salt marshes in the region of Hudson Bay and the Foxe Basin on which the geese feed consists of two perennials, *Puccinellia phryganodes*, a stoloniferous grass and *Carex subspathacea* a rhizomatous sedge (Kershaw 1976; Jefferies 1977; Jensen and Abraham 1979).* Landward of the salt marshes are frequently extensive freshwater meadows dominated by different species of sedges, which are also utilized by the geese as forage. In spring before the onset of aboveground plant growth, but after snow melt, geese grub for roots and rhizomes of graminoid plants in the salt marshes, creating patches of disturbed sediment (Figs 2a,b). In freshwater sedge meadows, geese eat the swollen basal portions of shoots of carices and discard the remainder of the shoots (Fig. 3) In summer, adults and goslings graze the salt-marsh vegetation intensively and they also graze the leaves of different species of sedges in the freshwater sedge meadows (Kotanen 1987; Kotanen and Jefferies 1987). Although other species of geese are present in the coastal wetlands, the densities of these birds are much lower and they do not form breeding colonies.

There is evidence that most of the lowlands presently supporting colonies of Lesser Snow Geese and other waterfowl are the result of isostatic uplift during the

---

* Nomenclature follows Porsild and Cody (1980)

Fig. 2a. Photograph of the effects of grubbing by Lesser Snow Geese in early spring on the vegetation of the intertidal flats at La Pérouse Bay, Manitoba, Canada. (Photograph taken by J. Montagnes).

last two millennia (Andrews 1966, 1970, 1973; Hansell *et al.* 1983; Scott *et al.* 1987; Gaston *et al.* 1987). The rate of uplift appears to be between 0.5 and 1.20 m per century. High densities of geese (all species) are present on the newly emerged land in the Foxe Basin and the northern regions of the Hudson Bay (Fig. 4). A similar relationship exists between land that has emerged during the last 2000 years on the west and south coasts of Hudson Bay and the presence of breeding colonies of geese.

During the Little Ice Age it is unlikely that these wetlands were as capable of supporting geese as now. Around 1450 A.D. the ice caps on Baffin Island advanced and the permanent snow line fell to approximately 450 m (Locke and Locke 1977). The Little Ice Age was marked by a period of much greater snow accumulation than that which occurs at present (Falconer 1966; Locke and Locke 1977). Between 1715 and 1760, and 1820 and 1920, white spruce trees at Churchill grew poorly suggesting that cold arctic conditions prevailed in summer (Scott *et al.* 1987). Historical evidence indicates that the northern tree limit was further south in the latter half of the eighteenth century compared with its position today (Ball 1986). Coastal lowlands throughout the region probably experienced a much shorter snow-free season during the Little Ice Age. This is likely to have made

Fig. 2b. Photograph of the effects of grubbing by Lesser Snow Geese on graminoid vegetation in early spring in a low-lying, non-tidal area close to the coast at La Pérouse Bay, Manitoba, Canada. Note the presence of dead willows (*Salix brachycarpa*). (Photograph taken by H. Sadul).

breeding difficult or impossible (Gaston *et al.* 1987). Even in this present decade there have been three years (1983, 1986, 1987) when adverse weather conditions in the Foxe Basin and on Southampton Island have prevented breeding of populations of Lesser Snow Geese. The general amelioration of climatic conditions during the present century is coincident with changes in the feeding habits of geese along the migration routes and on the wintering grounds in the United States and with the apparent exponential increase in numbers of Lesser Snow Geese in the eastern Canadian Arctic. The high numbers of geese have dramatically affected, and continue to affect, the plant assemblages of these coastal wetlands in the region of Hudson Bay.

Pattern of coastal vegetational development in the Hudson Bay region

There have been very few studies of the vegetation of the coastal lowland surrounding the Hudson Bay and the Foxe Basin. Descriptions of the major vegeta-

Fig. 3. Photograph taken in early spring of the remains of shoots of *Carex aquatilis* which have been pulled by Lesser Snow Geese at La Pérouse Bay, Manitoba, Canada. The geese eat the basal portion of the shoots and discard the remainder. (Photograph taken by J. Montagnes).

tional types in this region with one or two exceptions (*e.g.*, Thompson *et al.* 1980) are restricted to a few studies along the southern shore of Hudson Bay, a distance of approximately 800 km, hence the information on patterns of vegetational development in the region is of a very preliminary nature.

All of the coastal lowlands are dominated by considerable deposits of fluvo-glacial material. A characteristic feature of the Hudson Bay coastal zone is the presence of numerous beach ridges and sand spits, and the occurrence of broad tidal flats which extend seaward for approximately 1.0 km below the upper limit of tides. Much of the coastline constitutes one of the longest low-gradient emergent shorelines in the world (0.5 to 1 m/km) (Glooschenko 1979; Martini 1982). Salt marshes frequently develop landward of these sand spits. As occurs elsewhere in the Arctic, the initial coloniser of intertidal flats and silt-and-clay-bound gravel islands is *Puccinellia phryganodes*. This stoloniferous grass, which is a sterile triploid, at least in the Hudson Bay region, has never been known to set seed (Bowden 1961; Dore and McNeill 1980). The plant reproduces by extensive clonal

Pattern and process in arctic coastal vegetation 287

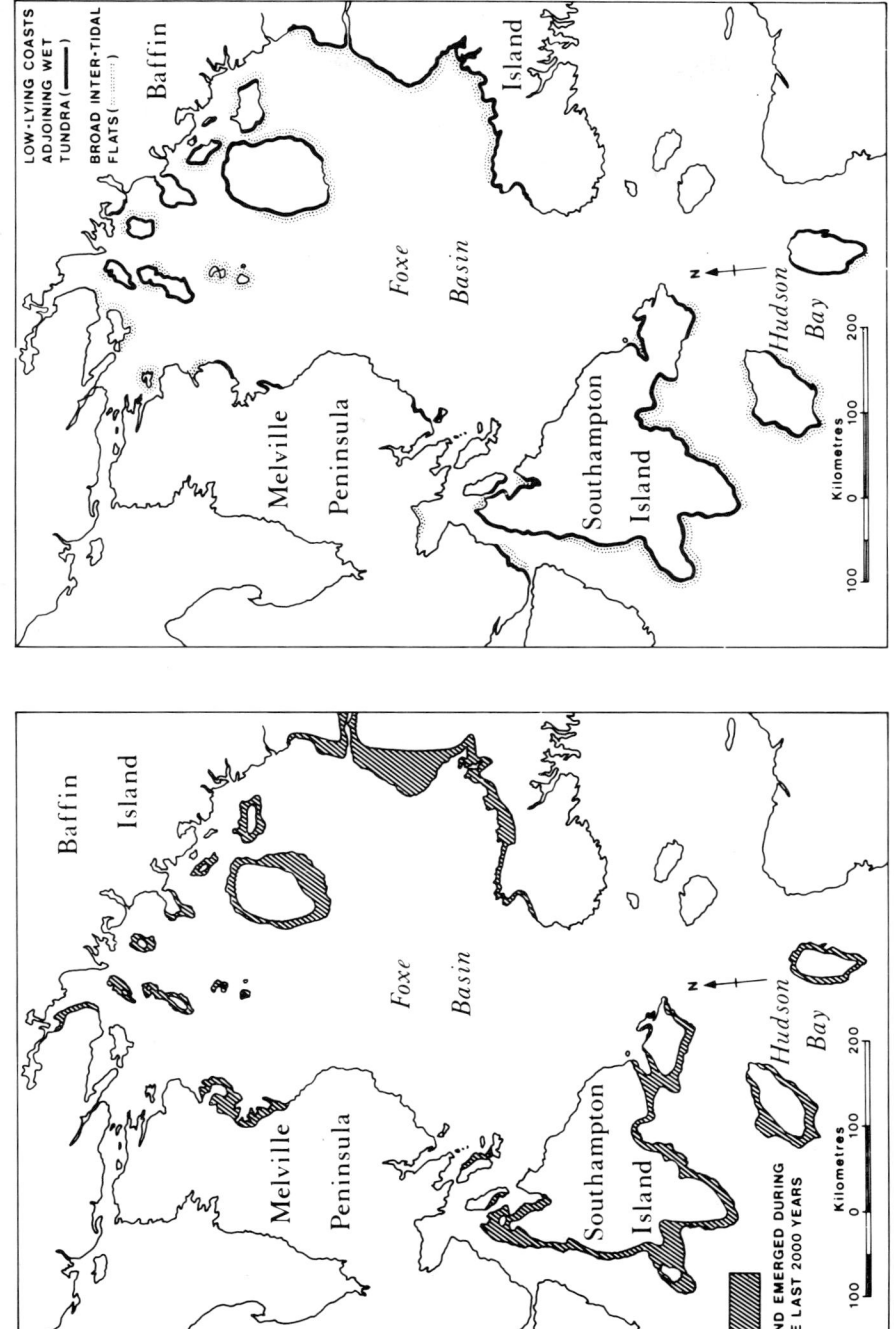

Fig. 4a. Distribution of land (hatched areas) that has emerged during the last 2000 years in the Foxe Basin and the northern regions of Hudson as a result of isostatic uplift (modified after Gaston, Decker, Cooch and Reed 1987).
Fig. 4b. Distribution of low-lying coasts (solid thick line) adjoining wet sedge meadow in the Foxe Basin and the northern regions of Hudson Bay. Broad intertidal flats are shown as dots (modified after Gaston, Decker, Cooch and Reed 1987).

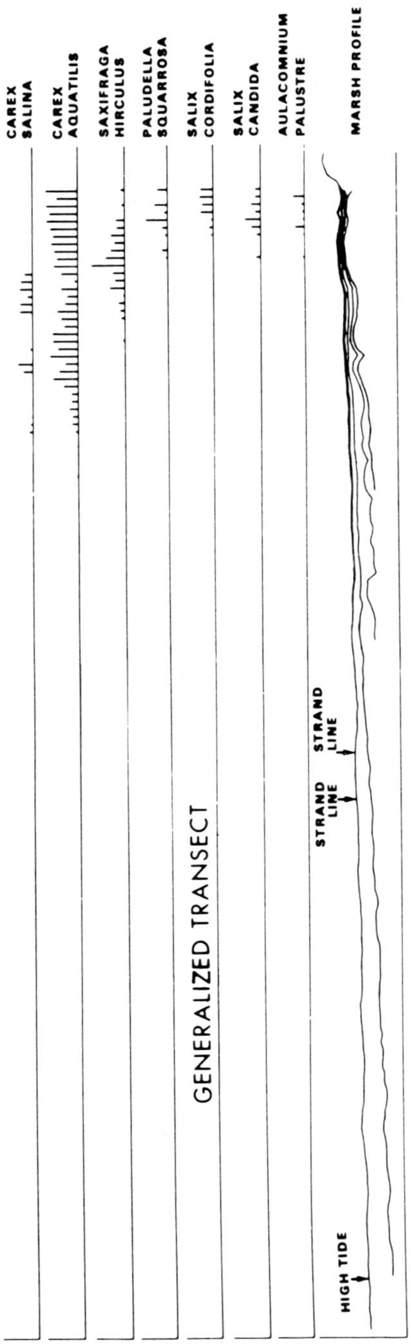

Fig. 5. A generalised transect of the vegetational zonation of the East Pen Island salt marshes, Hudson Bay. The percentage frequencies of species are displayed as histograms (from Kershaw 1976).

propagation. Within river deltas, especially along the southern shore of Hudson Bay where brackish conditions prevail, *Hippuris tetraphylla* is the initial coloniser of the unconsolidated clay sediments. Another species which is important in the early stages of colonization is the rhizomatous sedge, *Carex subspathacea*. As mentioned earlier, this species together with *Puccinellia phryganodes* are the two dominant graminoids of the coastal salt marshes (Polunin 1940, 1948; Kershaw 1976; Jefferies 1977; Glooschenko 1979; Glooschenko and Martini 1978; Jefferies et al. 1979). The transition to freshwater marsh is frequently abrupt. The graminoids of the salt marsh are replaced by *Carex aquatilis* var. *stans* or var. *aquatilis*, *Carex X flavicans* and *Eriophorum* species, depending on the region and the local conditions (Fig. 5). The sedge meadows often show the 'string' and 'flark' configuration. The sedges listed above are abundant around the edges of pools, or in the low lying areas (flarks) which may be flooded much of the summer. On the 'string' ridges species, such as *Salix arctica, S. arctophila, Betula glandulosa, Empetrum nigrum, Vaccinium vitis-idaea, Carex saxatilis, C. scirpoides*, and *C. rariflora*, predominate. Where the surface of the ground is elevated as a result of isostatic uplift and frost heave, *Calamagrostis deschampsiodes* and *Festuca rubra* replace the graminoids of the salt and freshwater marshes, at least in the southern and western coastal regions of Hudson Bay. At higher elevations the plant assemblages consist of low-lying or dwarf willow and birch shrubs, together with Ericaceous plants and sedges.

Locally, raised beach ridges separate salt-marsh communities from freshwater communities. The following species are common on these ridges: *Elymus arenarius, Festuca brachyphylla, Festuca rubra, Calamagrostis deschampsiodes, Saxifraga tricuspidata, Dryas integrifolia, Alectoria ochroleuca* and *Alectoria nigricans*.

Physiographic features of the landscape largely determine where the transitions from saltwater to freshwater communities and from wet mesic to dry xeric communities occur. The effects of isostatic uplift (both past and present), the presence of beach ridges, the development of permafrost and the associated impeded drainage of these lowlands, all contribute to the mosaic of plant communities in the coastal regions and largely account for the abrupt transition from one vegetation type to another. The spatial sequences of plant assemblages along transects inland from the coast show considerable variation from one locality to the next.

The effects of foraging by Lesser Snow Geese on the vegetation of the salt marshes of the intertidal flats

The foraging activities of the Lesser Snow Goose involving the grazing of leaves of graminoid plants in salt and freshwater marches and the grubbing of the underground organs of *Carex* and *Puccinellia* are separated in time. Grubbing and shoot pulling are largely confined to the early spring before the onset of aboveground growth, whereas grazing by adults and goslings of graminoid plants takes place during the post-hatch period in summer. Some grubbing also may occur in August, particularly in years when fall is early. Lesser Snow Geese may be regarded as a keystone species (Paine 1974, 1980) when describing the effects of their

Pattern and process in arctic coastal vegetation 291

Fig. 6. Intense grazing by Lesser Snow Geese produces a grazing lawn of *Carex subspathacea* and *Puccinellia phryganodes*. (Photograph taken by J. Montagnes).

foraging on species present in these coastal wetlands. The structure and composition of the vegetation is largely determined by the relative intensities of the two foraging activities, at least over an ecological time scale (Jefferies 1988).

Lesser Snow Geese nest in colonies at inland sites close to the salt marshes. The density of nests may exceed 2000 per km^2. Immediately after hatch between mid-June and early July, adults and goslings graze intensively on the vegetation of the salt marshes which is cropped to a height of about 2.5 cm. The adults regain their body weight lost during incubation (up to 40%) and the goslings increase in weight from about 80 g at hatch to 1500–1800 g seven weeks later before the fall migration in early August.

One effect in summer of the strong interaction between the grazer and the two graminoid forage species (*Carex subspathacea* and *Puccinellia phryganodes*) of the salt marsh is an acceleration of the nitrogen cycle. The geese increase the rate of breakdown of organic nitrogen, as a result of the passage of food through the gut, and soluble nitrogen is released in faeces (Bazely and Jefferies 1985). Since droppings accumulate at high densities throughout grazed swards (ca. 3 m^{-2} wk^{-1}), considerable impact on the vegetation may be expected. Net aboveground primary production (NAPP) in grazed swards is between 30% and 105% greater than that of ungrazed swards, depending on the year (Cargill and Jefferies 1984b; Bazely 1985; Bazely and Jefferies 1985; Jefferies 1988). Plant tissue is converted into goose biomass of faeces, only about 20 g m^{-2} accumulates as plant litter or standing dead tissue. The increased rate of turnover of nitrogen leads to an increase in the seasonal production of preferred forage species at a time when most required by the geese. Production of axillary tillers largely accounts for the increased NAPP in *Puccinellia phryganodes* (Bazely and Jefferies 198N), whereas in *Carex subspathacea*, the production of leaves on existing shoots is responsible for the increase in NAPP (Kotanen and Jefferies 1987). The results indicate the different growth strategies of the two species in response to grazing.

Within the intertidal flats intense grazing favours *Puccinellia* and *Carex* at the expense of dicotyledonous plants; this is unlike lightly grazed systems where numbers of species per unit area may be high (Fig. 6). The birds are generalists in their foraging behavior as both monocotyledons and dicotyledons are eaten. Three species of the latter group of plants present at low densities in the salt marshes are *Potentilla egedii, Ranunculus cymbalaria* and *Plantago maritima*. The removal by grazing of the apical region of stolons of *Potentilla* and *Ranunculus* results in no regrowth of existing stolons, or their replacement with new stolons during the growing season (Sadul 1987). As a result, the number of daughter individuals established from each parent is significantly lower in grazed plots. Few flowers of *Potentilla* or *Plantago* reach maturity in grazed swards and the populations of these species decreased significantly in numbers relative to those in ungrazed plots over a two-year period (Sadul 1987).

The species composition, biomass and height of these grazed plant assemblages of the salt marsh are therefore strongly linked to the colonial feeding behavior of the geese. When the grazer was excluded, dicotyledons such as *Potentilla egedii* and *Plantago maritima* increased in frequency. *Carex subspathacea* replaced *Puccinellia phryganodes* as the dominant graminoid. Over five years, eighteen species of angiosperms established in the exclosures, whereas permanent plots in grazed areas contained only six species (Bazely and Jefferies 1986). Rapid changes in species composition in the exclosures were associated with an increase in biomass and amount of plant litter, and adjustments in the competitive hierarchy of the species (Bazely and Jefferies 1986). When the fence of a two-year-old exclosure was removed in the spring of 1983 very little grazing of the modified sward occurred in 1983 or 1984. The changes in the vegetation in the absence of grazing represent a loss of preferred forage. In early 1987, the plot was still clearly visible, although much of the plant litter had been removed by tidal action and spring 'runoff' of melt water.

In these coastal arctic habitats disturbances appear necessary to achieve sustained biological production, as turnover times for movement of nutrients between and within abiotic and biotic compartments are thought to be long. The grazing animal regulates the release of soluble nutrients via faeces at the soil surface. Rates of net mineralization of nitrogen which occur in soils are strongly affected by the addition of faeces (Ladd and Paul 1973; Bosatta and Berendse 1984; Ruess and McNaughton 1987). The release of inorganic nitrogen into soil from faeces may be important in enabling the soil microbial flora to develop in soils where the amounts of organic nitrogen are low and the C/N ratio is high. The presence of faeces may result in sustained net mineralization of soil nitrogen which provides a nutrient source for plant growth, in addition to the soluble nitrogen present in faeces. In general, much of the nutrient input in arctic habitats is episodic and reflects lateral or vertical transfer to the soil surface by abiotic (*i.e.*, snow melt, spring runoff) and biotic agencies (*i.e.*, cyanobacteria, geese, caribou, muskoxen). The rearing of goslings and the summer moult of adults restricts the mobility of the geese. In such situations strong feedback processes are likely to have evolved between the herbivore and its forage plants, in order to sustain food production. The ability to seek alternative food resources is restricted as the birds are flightless. The overall effects of summer grazing by the geese result in enhanced aboveground primary production of forage, a rapid gain in weight of birds and the maintenance of the species composition of the intertidal grazing lawn.

As mentioned above, the geese grub in early spring, before the onset of the aboveground growth of vegetation (Jefferies 1988). The grubbing is patchy both in space and time (Jefferies 1988). The geese grub the slightly raised areas on the coastal salt marshes on the tidal flats which are above the level of melt water in spring. They create patches of 'bare' sediment in which, on average, 40% of the original shoots remain (Hik, pers. communication). A goose strips an area of about 1 m^2 in approximately one hour. The presence of 7000 pairs of breeding geese, together with unknown numbers of migrants, has resulted in the grubbing of large areas of tidal flats and shallow drainage channels landward of the high-water mark of spring tides each year at La Pérouse Bay and along the coastal flats adjacent to the McConnell River (Jefferies 1988; Jefferies and Kerbes 1985). An examination of changes in the vegetation and of the extent of grubbing along permanent transects of 400 metres in length at different sites on the intertidal flats at La Pérouse Bay indicates that most plant communities, at least on the raised areas, have been grubbed in the past (Jefferies, unpublished data). The scale of the patches on the intertidal flats varies from the size of individual beak marks to areas which are 300 m × 300 m in extent.

Small patches (< 10 cm × 10 cm) are recolonised by the production of axillary tillers of shoots of *Puccinellia* that remain after grubbing and by the inward growth of shoots from the perimeter of a cleared patch. This pattern and process (*cf.* Watt 1947) results in a complex vegetational mosaic in which patches of the sward of *Puccinellia* are of different ages. Preliminary studies indicate that swards of this grass that were grubbed in 1978, and which subsequently recovered have been regrubbed in 1986 and 1987. Whether an 8-10 year period is an average interval between grubbing events remains to be established.

In contrast, where grubbing occurs on a large scale and the remains of the organic veneer (ca. 5 cm deep) are subsequently removed by physical agencies, exposing the underlying marine silts and clays and glacial gravels, neither *Puccinellia* nor other species readily reestablish. The intertidal flats adjacent to the McConnell delta and those along the shore at La Pérouse Bay consist of large areas of bare sediment. At the seaward end of the flats at the latter location the grass is colonizing the soft sediments, but at the landward end the bare sediments mentioned above are evident. A consequence of grubbing is that large numbers of tillers with fine roots attached (total length of plant ca. 1 cm) are transported from the landward to the seaward end of the marsh during spring melt. Some of these tillers root in the soft sediment. The bare areas at the landward end of the flats dry out during the summer months, as the tides do not flood these sections of the flats until late summer. These physical conditions effectively prevent the reestablishment of propagules of *Puccinellia* in summer.

Three other processes also contribute to the formation of patches, at least at a local scale. Where geese roost on the raised areas in the intertidal flats, large piles of faeces accumulate. The grass underneath the dung dies and is replaced by moss.

When the bryophytes die bare patches are created in the sward which may be subsequently colonised by *Puccinellia* and *Carex* as a result of lateral vegetative growth from nearby clones. Accumulations of *Fucus* litter on the sward of *Puccinellia* also result in the death of the underlying grass and the formation of patches. Lastly, algal growth is rapid in shallow ponds which flood the surrounding *Puccinellia* sward on the intertidal flats at spring melt. Input of nutrients from goose faeces and the high water temperatures promote rapid algal growth in these ponds. As the water level recedes the mats of Chlorophyta (species of *Vaucheria* in particular) rest on the sward of *Puccinellia*. The grass dies as the algal mat decays creating a patch. Even this mode of patch formation appears dependent on the presence of the geese which provide an input of nutrients to the ponds via faeces.

Modification of freshwater communities by foraging by Lesser Snow Geese

The effects of foraging on the plant communities in freshwater meadows show both similarities and differences with respect to comparable effects on plant communities of saltwater marshes. As in the latter cases the effects are coupled with seasonal events in physical processes.

Foraging activities in spring are dominated by shoot pulling and grubbing, whereas in summer these activities are replaced by grazing (Kotanen 1987; Kotanen and Jefferies 1987). The pulling of shoots of *Carex aquatilis* and *C. X flavicans* is confined to those areas, which as a result of local topography and hydrology, either drain rapidly after melt, or else the areas are covered by shallow standing water (< 15 cm in depth). The birds seek out stands of these two species. A shoot is severed just above the sediment. The basal portion corresponding in length to the width of the bill is eaten, and the remainder of the shoot discarded. At this time of year the basal portion is high in nitrogen (ca. 4.5 to 5.0% of dry weight), whereas the upper sections of the lamina contain only 2 to 3% nitrogen. On average, 400 shoots per m^2 were taken each year over two successive years from permanent plots. In some plots the number of shoots removed was as high as 1500 per m^2 (Kotanen 1987). The extent of the removal of shoots is dependent on: (a) the number of breeding birds; (b) the number of migrants; (c) weather conditions further north; (d) local conditions of melt. This foraging activity, together with that of summer grazing, appears to weaken the clones of these species. The local populations of sedges around ponds disappear and are replaced by moss carpets. This only occurs where the summer water table is at or just above the surface, and where a layer of peat has formed above the underlying mineral layer. The moss communities may be 1–10 metres wide around the pools, depending on local topography. The moss carpets are widespread in the coastal regions of the west Hudson Bay where shallow ponds occur. Mosses are present at low frequency in stands of *Carex aquatilis* and *C. X flavicans* which are not utilized by the geese, but the bryophytes increase rapidly when stands of sedges are destroyed. The only angiosperms present in the moss carpets are *Potentilla palustris*, *Petasites sagittatus* and *Hippuris tetraphylla*, all of which are avoided by the geese. Circumstantial evidence from elsewhere indicates that the presence of moss carpets at other lowland sites, especially around lakes and ponds in the coastal areas of the Arctic, is a consequence of foraging by birds (Tikhomirov 1959; Bliss 1981; Giroux et al. 1984; Prop et al. 1984).

There is little evidence of the reestablishment of populations of *Carex aquatilis* or *C. flavicans* from seed in these moss carpets. We have not observed seedlings of those species at any site at La Pérouse Bay. Much of the extension of swards is the result of clonal growth of individual ramets which trap sediment. The species appear unable to establish directly in a peat layer, unlike species of *Eriophorum*.

The foraging activities of the geese, therefore, result in a change in the structure and species composition of the vegetation. The changes are effectively irreversible. Similar effects of foraging by geese have been reported for other sites in temperate regions (Glazener 1946; Lynch et al. 1947; Smith 1983; Giroux 1986). At present we are removing shoots and leaves of plants of *Carex aquatilis* in order to simulate grubbing and grazing of vegetation. Preliminary estimates, based on the results of this experiment which was started in 1985, indicate that given the present intensity of foraging, it takes about five years for the conversion of a sedge meadow to a moss carpet to occur.

Extensive grubbing of graminoid swards takes place at sites, which as a result of snow melt are exposed early in the season, but which drain rapidly. Frequently, these swards are in low-lying areas which form part of a drainage network. Melt water from the hinterland floods the sites in early spring. The drainage of the water may be blocked for several days by the presence of ice and snow seaward of the sites, resulting in the formation of transient lakes of melt water. When drained, the plant communities are exposed, although surrounding areas still may be covered with ice and snow. Some of these sites are relic salt-marsh plant assemblages interspersed with *Salix brachycarpa* and *Calamagrostis deschampsiodes* which grow on the highest ground. The communities are 2–3 km from the tidal flats at La Pérouse Bay. Other areas are sedge meadows dominated by *Kobresia hyperborea, Carex scirpoides, Calamagrostis deschampsiodes, Hierochloe pauciflora* and *Poa arctica*. This type of community is widespread between the McConnell River and Eskimo Point, N.W.T. On the 'strings' which cross these areas species of willow and birch and members of the Ericaceae make up the plant community of the hummocks. These inland sites are frequently intensive nesting areas for the Lesser Snow Geese, where the density of nests is greater than 1000 per km^2 and is frequently above 2000 nests per km^2.

The grubbing of the graminoid communities leads to the death of vegetation, including woody plants and the establishment of peat 'barrens'. The scale of destruction of the vegetation is measured in hundreds of meters and may extend intermittently over distances of 30 km or more, as occurs between McConnell River and Wolf Creek, N.W.T. The stripping of the active vegetation layer removes the insulating layer and exposes the underlying peat. This dries out in summer and is eroded at spring runoff, exposing glacio-marine deposits. The edaphic environment is unsuitable for the reestablishment of the original plant communities. Opportunistic species appear on these degraded sediments, including *Senecio congestus, Eleocharis acicularis* and *Salicornia borealis*; the last species grows in the more saline sites along the southern shores of Hudson Bay. Overall, the percentage plant cover is less than ten percent. Preliminary results of a study of the willows at La Pérouse Bay indicate that where marine deposits are close to the surface high salinities build up in the origincal layer, once the active graminoid layer is stripped. In addition, the roots of willow bushes are exposed upon stripping, as the plants are shallow-rooted. The combination of high salinity ($> 32°/oo$), exposure and desiccation of roots probably accounts for the death of the willows.

The destruction of existing plant communities and the degradation of the habitat occurs over a time scale of five to 15 years. These changes are brought about by the presence of high densities of birds. Rapid colonization by plants and the subsequent development of new communities fail to occur, at least over comparable time spans. These results are similar to those obtained from the intertidal flats where grubbing leads to degradation of the habitat.

General discussion and conclusions

Because of the presence of high densities of birds a population of Lesser Snow Geese at a site may be expected to exert a considerable effect on the vegetation, in addition to effects associated with isostatic uplift and climatic change. Although there is a strong correlation between large populations of birds and areas of extensive coastal lowlands (Gaston *et al.* 1987), small areas of marsh associated with estuaries, creeks or minor river systems also support a few small groups of birds. Hanson and Jones (1976) reported finding Lesser Snow Geese present at many of the estuaries of rivers between the Magus River on the west coast of Hudson Bay and Fort Severn on the Ontario coast. At all of these locations the densities of birds are high. It is also evident from detailed studies of Lesser Snow Geese populations on the west coast of Hudson Bay and at La Pérouse Bay (Cooke *et al.* 1982; Kerbes *et al.* 1983; MacInnes and Kerbes 1987) that the populations move from one river system to another over a number of years. Although the causes of this movement are not well understood, it is invariably associated with the destruction of plant communities and the deterioration of habitats. High densities of geese and a finite food supply results in a mismatch between food supply and food demand (Drent and Prins 1987). The birds are forced to move elsewhere. The findings support the limit cycle hypothesis (Krebs 1988) which states that there is no attainable natural equilibrium between the herbivore and the supply of forage. Hence the populations of birds are very mobile and may be expected to utilize most wetland areas for forage with the ensuing effects on the vegetation of these areas over a period of decades. This 'pattern and process' appears to have played a major role in shaping the structure and composition of the plant assemblages of these coastal wetlands.

As in all discussions of this type, both spatial and temporal scales are important in analyzing patterns of vegetational development. The spatial scales of disturbance may be a metre or less to scales which exceed 1 km; corresponding temporal scales are a decade to tens of decades for the reestablishment of the vegetation. In part, this reflects the relative inefficiency of a flora of perennial plants to exploit large areas. The plants are well adapted to exploit small patches in the immediate vicinity of a plant (via clonal propagation), but are unable to colonise large areas quickly as the seed output from populations of these species is frequently low. Physical processes are continually changing the nature of these large areas of disturbance, such that rates of physical change far exceed the capacity of the organisms to exploit the degraded areas. There may well be a sharp spatial threshold as patch size increases, which sets limits on the ability of these different graminoid species to colonise sites. The interesting exception to this are the propagules (rooted tillers and leaves) of *Puccinellia phryganodes*, which may be regarded as equivalent to seeds of an annual. They are produced in large numbers as a consequence of grubbing, they are widely dispersed by water, they root readily in sediment, they have a limited life (ca. one season) and are replaced by axillary tillers from the parent plant (Bazely and Jefferies 198N). Even in the case of this species, where the intertidal habitat has been degraded, as has occurred in the upper levels of the intertidal flats, the propagules are unable to establish quickly, because of lack of soft sediment. The only other species which shows similar characteristics is *Hippuris tetraphylla*. The species is abundant in brackish sites where silt collects. Small pieces of rhizome break from the parent plant and are carried to new areas where they grow and trap silt.

The foraging activities of the geese lead to a hierarchy in the abilities of individual species to tolerate foraging. The persistence of a certain species is not only related to their growth habit and location of meristems, but also to the large amount of biomass which lies below ground. This may exceed 90% of the total biomass of a plant (*cf. Puccinellia phryganodes*), and it provides obvious resources for regeneration of aboveground growth. Other species, particularly dicotyledonous plants, are less tolerant of the effects of foraging and their continued presence in grazed marshes probably is dependent on a periodic renewal of the seed bank from elsewhere. At the time of spring runoff, for example, seeds of plants are common in melt water.

The episodic events of grubbing and grazing result in a complex mosaic of communities which are at different stages of development and are of different ages. Isolated patches (ca. 1 m^2) of a formerly extensive plant community are left as 'relic' communities within areas which are undergoing physical alteration and which have yet to develop a closed sward of vegetation. The pattern of vegetational development in these areas is not always predictable, as edaphic conditions have changed and continue to change as a result of isostatic uplift. The birds may delay or modify vegetational development, they cannot arrest the inevitable changes in patterns of vegetation associated with the transition from saltwater to freshwater conditions. As some species fail to set seed (as in *P. phryganodes*) and populations of other species undergo sexual reproduction rarely, there is unlikely to be close coupling between the rate of environmental change (both abiotic and biotic influences) and the ability of populations of all species to generate propagules in sufficient numbers, so that areas are colonised rapidly. Non-equilibrium theories of community structure offer the only conceptual basis to interpreting 'pattern and process' in these arctic coastal plant assemblages. The high frequency of disturbance events relative to the time required for the establishment of a closed sward of vegetation, the opportunity for colonization of open, disturbed, habitats, and the relative unpredictability of the species composition of plant communities which may develop at these sites, are all attributes of non-equilibrium conditions.

Acknowledgments

I thank students and staff at the La Pérouse Bay Field Station and R.H. Kerbes of the Canadian Wildlife Service for valuable discussions about topics discussed in this paper.

The Natural Sciences and Engineering Research Council of Canada, the Canadian Wildlife Service and the Department of Indian and Northern Affairs provided financial support for the study. Ms. L. May kindly typed the manuscript.

References

Andrews, J.T. 1966. Patterns of coastal uplift and deglaciation. West Baffin Island, N.W.T. Geogr. Bull. 8: 174–193.

Andrews, J.T. 1970. Differential crustal recovery and glacial chronology (6700 B.P.), West Baffin Island, N.W.T., Canada. Arct. Alp. Res. 2: 115–134.

Andrews, J.T. 1973. The Wisconsin Laurentide ice-sheet: dispersal centres, problems of rates of retreat, and climatic interpretations. Arct. Alp. Res. 5: 185–189.

Ball, T.F. 1986. Historical evidence and climatic implications of a shift in the boreal forest – tundra transition in central Canada. Climatic Change 8: 121–134.

Bazely, D.R. 1984. Responses of salt-marsh vegetation to grazing by Lesser Snow Geese (*Anser caerulescens caerulescens*). M.Sc. Thesis, University of Toronto, Canada.

Bazely, D.R. and Jefferies, R.L. 1985. Goose faeces: a source of nitrogen for plant growth in a grazed salt marsh. J. Appl. Ecol. 22: 693–703.

Bazely, D.R. and Jefferies, R.L. 1986. Changes in the composition and standing crop of salt marsh communities in response to the removal of a grazer. J. Ecol. 74: 693–706.

Bazely, D.R. and Jefferies, R.L. 198N. Leaf and shoot demography of an arctic stoloniferous grass, *Puccinellia phryganodes*, in response to grazing. J. Ecol. (submitted).

Bliss, L.C. 1981. North American and Scandinavian tundras and polar deserts. In: L.C. Bliss, O.W. Heal and J.J. Moore (eds), Tundra Ecosystems: A Comparative Analysis, pp. 8–24. Cambridge University Press, London.

Bosatta, E. and Berendse, F. 1984. Energy or nutrient regulation of decomposition: implications for the mineralization – immobilization response to perturbations. Soil Biol. Biochem. 16: 63–67.

Bowden, K.M. 1961. Chromosome numbers and taxonomic notes on northern grasses IV. Tribe Festuceae: *Poa* and *Puccinellia*. Can. J. Bot. 39: 123–128.

Boyd, H., Smith, G.E.J. and Cooch, F.G. 1982. The Lesser Snow Geese of the eastern Canadian Arctic. Canadian Wildlife Service Occasional Paper. Number 46, Ottawa, Canada. 24 pp.

Cargill, S.M. and Jefferies, R.L. 1984. The effects of grazing by Lesser Snow Geese on the vegetation of a sub-arctic salt marsh. J. Appl. Ecol. 21: 669–686.

Chesson, P.L. and Case, J.J. 1986. Overview: non-equilibrium community theories: chance, variability, history, and co-existence. In: J. Diamond and T.J. Case (eds), Community Ecology, pp. 229–239. Harper and Row, New York.

Connell, J.H. 1978. Diversity in tropical rain forests and coral reefs. Science 199: 1302–1310.

Cooke, F., Abraham, K.F., Davies, J.C., Findlay, C.S., Healey, R.F., Sadura, A. and Segin, R.J. 1982. The La Pérouse Bay Snow Goose Project – a 13-year report. 194 pp. Department of Biology, Queen's University, Kingston, Ontario.

Davis, M.B. 1986. Climatic instability, time lags, and community disequilibrium. In: J. Diamond and T.J. Case (eds), Community Ecology, pp. 269–284. Harper and Row, New York.

Dore, W.G. and McNeill, J. 1980. Grasses of Ontario. Monograph 26, Research Branch, Agriculture Canada, Ottawa. 566 pp.

Falconer, G. 1966. Preservation of vegetation and patterned ground under a thin ice body in Northern Baffin Island, N.W.T. Geogr. Bull. 8: 194–200.

Drent, R.H. and Prins, H.H.T. 1987. The herbivore as prisoner of its food supply. In: J. van Andel, H.H.T. Prins and R.W. Snaydon (eds), Disturbance in Grasslands, pp. 131–147. Junk, Dordrecht.

Gaston, A.J., Decker, R., Cooch, F.G. and Reed, A. 1987. The distribution of larger species of birds

breeding on the coasts of Foxe Basin and northern Hudson Bay, Canada. Arctic 39: 285–296.
Giroux, J.F. 1986. Utilisation des marais a scirpe de l'estuaire du Saint-Laurent par la grande oie blanche. Ph.D. Thesis, L'Université Laval, Quebec, Canada. 205 pp.
Giroux, J.F., Bedard, Y. and Bedard, J. 1984. Habitat use by greater snow geese during the brood period. Arctic 27: 155–160.
Glazener, W.C. 1946. Food habits of wild geese on the Gulf coast of Texas. J. Wildl. Manage. 10: 332–329.
Glooschenko, W.A. and Martini, I.P. 1978. Hudson Bay lowlands baseline study. Coastal zone 1978. Proceedings of the Symposium on technical, environmental, socio-economic and regulatory aspects of coastal zone management, pp. 663–679. American Society of Chemical Engineers, San Francisco, CA.
Glooschenko, W.A. 1979. Coastal ecosystems of the Hudson Bay / James Bay area of Canada. Geoscience and Man. Louisiana State University Publication in Geography and Anthropology. 12 pp.
Grubb, P.J. 1977. The maintenance of species – richness in plant communities: the importance of regeneration niche. Biol. Rev. 52: 107–145.
Hansell, R.I.C., Scott, P.A., Staniforth, R. and Svoboda, J. 1983. Permafrost development in the intertidal zone at Churchill, Manitoba. A possible mechanism for accelerated beach uplift. Arctic 36: 198–203.
Hanson, H.C. and Jones, R.L. 1976. The biochemistry of Blue, Snow and Ross' Geese. Illinois Natural History Survey, Special Publication, Number 1. 281 pp.
Jefferies, R.L. 1977. The vegetation of salt marshes at some coastal sites in arctic North America. J. Ecol. 65: 661–672.
Jefferies, R.L. 1988. Vegetation mosaics, plant – animal interactions and resources for plant growth. In: L.D. Gottlieb and S.K. Jain (eds), Plant Evolutionary Biology, pp. 341–369. Chapman & Hall, London.
Jefferies, R.L., Jensen, A. and Abraham, K.F. 1979. Vegetational development and the effect of geese on vegetation at La Pérouse Bay, Manitoba. Can. J. Bot. 57: 1439–1450.
Jefferies, R.L. and Kerbes, R.H. 1985. The effects of foraging by geese on the vegetation in the coastal vicinity of Eskimo Point, N.W.T. Report to the Canadian Wildlife Service. 8 pp.
Kerbes, R.H. 1975. The nesting populations of Lesser Snow Geese in the eastern Canadian Arctic. A photographic inventory of June 1973. Report series nunber 35, Canadian Wildlife Service, Environment Canada, Ottawa. 47 pp.
Kerbes, R.H. 1982. Lesser Snow Geese and their habitat on west Hudson Bay. Naturaliste Can. (Rev. Ecol. Syst.) 109: 905–911.
Kerbes, R.H., McLandress, M.R., Smith, G.E.J., Beyersbergen, G.W. and Godwin, B. 1983. Ross' Goose and Lesser Snow Goose colonies in the central Canadian arctic. Can. J. Zoo. 61: 168–173.
Kershaw, K.A. 1976. The vegetation zonation of the East Pen Island salt marshes, Hudson Bay, Can. J. Bot. 54: 5–13.
Kotanen, P.M. 1987. A comparison of the growth responses of three sedges by foraging by Lesser Snow Geese. M. Sc. Thesis, University of Toronto, Canada. 221 pp.
Kotanen, P.M. and Jefferies, R.L. 1988. The leaf and shoot demography of grazed and ungrazed plants of *Carex subspathacae*. J. Ecol. 75: 961–975.
Krebs, C.J. 1988. The message of ecology. Harper and Row, New York. 195 pp.
Ladd, J.N. and Paul E.A. 1973. Changes in enzymic activity and distribution of acid-soluble amino acid-nitrogen in soil during immobilization and mineralization. Soil Biol. Biochem. 5: 825–840.
Locke, C.W. and Locke, W.W. III. 1977. Little ice age snow cover extent and palaeoglaciation thresholds: North-central Baffin Island, N.W.T., Canada. Arct. Alp. Res. 9: 291–300.
Lynch, J.J., O'Neill, T.O. and Lay, D.W. 1947. Management significance of damage by geese and muskrats to Gulf coast marshes. J. Wildl. Manage. 11: 50–76.
MacInnes, C.D. and Kerbes, R.H. 1987. Growth of the Snow Goose, *Chen caerulescens*, colony at McConnell River, Northwest Territories: 1940–1980. Can. Fld. Nat. 101: 33–39.
Martini, I.P. 1982. Geomorphological features on the Ontario coast of Hudson Bay. Naturaliste Can. (Rev. Ecol. Syst.) 109: 415–429.
Paine, R.T. 1974. Intertidal community structure. Oecologia 15: 93–120.
Paine, R.T. 1980. Food webs, linkage, interaction strength and community infrastructure. J. Anim. Ecol. 49: 667–685.
Polunin, N. 1940. Botany of the Canadian eastern Arctic. Part I. Pteridophyta and Spermatophyta. Natl. Mus. Can. Bull. 92: 1–408.

Polunin, N. 1948. Botany of the Canadian eastern Arctic. Part III. Vegetation and Ecology. Natl. Mus. Can. Bull. 104: 1–304.

Porsild, A.E. and Cody, W.J. 1980. Vascular plants of continental northwest territories, Canada. National Museums of Canada. 667 pp.

Prop, J., Van Eerden, M.R. and Drent, R.H. 1984. Reproductive success of the Barnacle goose *Branta leucopsis* in relation to food exploitation on the breeding grounds, western Spitzbergen. Norsk polarinstitut skritter 181: 87–117.

Reed, A. 1976. Geese, nutrition and farmland. Wildfowl 27: 153–156.

Ruess, R.W. and McNaughton, S.J. 1987. Grazing and the dynamics of nutrient and energy regulated microbial processes in the Serengeti grasslands. Oikos 49: 101–110.

Sadul, H. 1987. The effects of Lesser Snow Geese grazing on sub-arctic plant populations and communities. M. Sc. Thesis, University of Toronto, Canada. 172 pp.

Sale, P.F. 1977. Maintenance of high diversity in coral reef fish communities. Am. Nat. 111: 337–359.

Scott, P.A., Hansell, R.I.C. and Fayle, D.C.F. 1987. Establishment of white spruce populations and responses to climatic change at the treeline, Churchill, Manitoba, Canada. Arct. Alp. Res. 19: 45–51.

Smith III, T.J. 1983. Alteration of salt marsh plant community composition by grazing snow geese. Hol. Ecol. 6: 204–210.

Tikhomirov, B.A. 1959. Relationship of the Animal World and the Plant Cover. Publication of the Botanical Institute, Academy of Science of the USSR, Moscow and Leningrad. Translated by E. Issakoff and T.W. Barny. Edited by W.A. Fuller, the Boreal Institute, Calgary, Canada.

Thompson, D.C., Klassen, G.H. and Cihlar, J. 1980. Caribou mapping in the southern district of Keewatin, N.W.T.: An application of digital landsat data. J. Appl. Tech. 17: 125–138.

Watt, A.S. 1947. Pattern and process in the plant community. J. Ecol. 34: 1–22.

Weins, J.A. 1977. On competition and variable environments. Am. Sci. 65: 590–597.

MODIFICATION OF VEGETATION STRUCTURE AND ECOSYSTEM PROCESSES BY NORTH AMERICAN GRASSLAND MAMMALS

APRIL D. WHICKER[1] and JAMES K. DETLING[1,2]
[1]Natural Resource Ecology Laboratory, [2]Department of Range Science, Colorado State University, Fort Collins, CO 80523, USA

Abstract

Historically, prairie dogs (*Cynomys* spp.), small colonial mammals, and bison (*Bison bison*), large migratory ungulates, were among the most infuential grazers in North American grasslands. Our studies in a northern mixed-grass prairie in Wind Cave National Park, South Dakota, indicate that colonization and intensive grazing by prairie dogs greatly alters the ecosystem patch structure and increases the attractiveness of colonized areas to other herbivores such as bison. The objectives of this research have been to (1) investigate how these native grazers influence structural and functional properties of grasslands over time, and (2) determine how these and other herbivores respond to induced changes in the ecosystem.

Soon after colonizing grassland areas, prairie dog populations usually increase and occupy available habitats where they dig burrows and graze surrounding vegetation to a height of several cm. Thereafter, prairie dogs and other herbivores keep the plants clipped off such that aboveground biomass is maintained at one-third to two-thirds that of adjacent, uncolonized areas. Over time, there is usually a shift in community dominance from grasses to forbs and dwarf shrubs, and belowground plant biomass decreases markedly. As these changes occur, utilization of the colonies by other herbivores increases. Aboveground, bison feed preferentially on graminoid-dominated areas of the colonies, while pronghorn (*Antilocapra americana*), smaller ungulates, feed principally in areas in which forbs and dwarf shrubs have become dominant. Belowground, populations of root-feeding nematodes often increase on heavily grazed prairie dog colonies, and they consume a greater proportion of the annual net root production than on lightly grazed uncolonized areas. The combined activities of these other aboveground and belowground herbivores likely accelerate structural and functional changes in the vegetation initiated by prairie dogs.

Introduction

It is commonly recognized that some herbivores affect plant community structure and composition (Crawley 1983). For example, native grazers as diverse as ungulates (McNaughton 1984), geese (Cargill and Jefferies 1984), and tortoises (Merton *et al.* 1976) can modify the vertical structure by reducing plant canopy height and creating 'grazing lawns'. Many herbivore-related processes, including both selective and non-selective grazing, can increase, or decrease, plant species diversity (Harper 1969). Such impacts may result in patches or vegetational mosaics which are obviously different from the surrounding community. It is perhaps less obvious that plant morphology, chemical composition, and physiology may also have been affected. Concomitant changes in animal densities and species not only may follow, but also may be vectors of further modification of the patch structure. Finally, such major changes in vegetation structure and species composition of animal communities may also lead to functional differences within the system

Plant form and vegetation structure, pp. 301–316
edited by M.J.A. Werger, P.J.M. van der Aart, H.J. During and J.T.A. Verhoeven
©1988 SPB Academic Publishing, The Hague, The Netherlands

(Floate 1981). Several papers in this volume present evidence for such difference in functioning.

Very few comprehensive studies exist which address the complex interactions between herbivores and the plant community. Data must be collected from a variety of hierarchical levels of organization (*e.g.*, population, ecosystem), from many taxa, and usually for many years, to gain an insightful interpretation of the observations. During the last decade, we have been studying structural and functional properties of a grassland ecosystem within which exist patches that are created, and whose processes are perhaps controlled, by native herbivores. Our research focuses on black-tailed prairie dogs (*Cynomys ludovicianus* Ord) in the northern mixed grass prairie at Wind Cave National Park, South Dakota, USA.

Prairie dogs are relatively large (~ 1 kg adult mass), burrowing rodents. They are highly gregarious and form colonies covering several to thousands of hectares with prairie dog densities averaging 10 to 55 animals per ha (O'Meilia *et al.* 1982; Knowles 1986; Archer *et al.* 1987; Brizuela 1987). Historically, the range of black-tailed prairie dogs coincided with the extensive grasslands of North America, especially the Great Plains. However, over the last century, they have been viewed as forage competitors with cattle, and have been systematically exterminated. Present populations are probably less than 2% of those of less than a century ago (Summers and Linder 1978). In 1978 at Wind Cave National Park (WCNP), 11 active prairie dog colonies covered approximately 6% of the park area and ranged in size from 4 to 250 ha (Dahlsted *et al.* 1981). Large native ungulates (bison, *Bison bison* L.; elk, *Cervus elaphus* L. and pronghorn antelope, *Antilocapra americana* Ord) are free ranging within the confines of the 11,400 ha park and, as will be discussed later, are often found associated with prairie dog colonies.

The objectives of this research have been to (1) investigate how prairie dogs influence structural and functional properties of grasslands over time, and (2) determine how these and other herbivores respond to induced changes in the ecosystem. Information presented in this paper includes both published data and hitherto unpublished work. In the studies presented in this review, prairie dog colonies were divided into areas of various ages, or states of colonization, each of which was aged relative to other states within a particular colony (Fig. 1): (a) an older area (b) a relatively young area and (c) an edge. The terms 'on colony' or 'central colony' refer to either (a) or (b), or to an area where no distinction was made between them. An edge may have been very recently colonized, or may have been a very lightly used border adjacent to uncolonized areas. The (d) uncolonized prairie ('off colony') adjacent to a colony was used as a baseline, control site.

Plant community – patches within the ecosystem

The visual impact of prairie dog activity is not subtle. Large areas of the landscape are disrupted by mounds of bare soil (diameter, 1–3 m; Fig. 2) excavated during burrowing. Densities of these mounds within colonies typically range from 100 to 300 per ha (King 1955; O'Meilia *et al.* 1982; Archer *et al.* 1987; Brizuela 1987). Taller vegetation is cropped, probably both for food and facilitation of predator

Vegetation structure and ecosystem processes 303

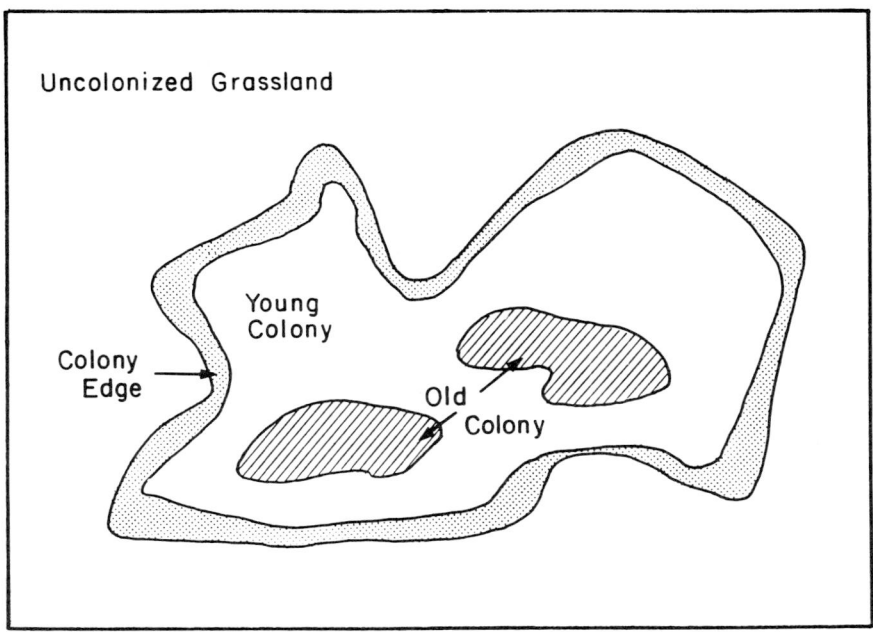

Fig. 1. Diagrammatic view of a prairie dog colony showing states of colonization referred to in text.

Fig. 2. Prairie dog (*Cynomys ludovicianus*) on a soil mound around burrow, Wind Cave National Park, South Dakota, USA.

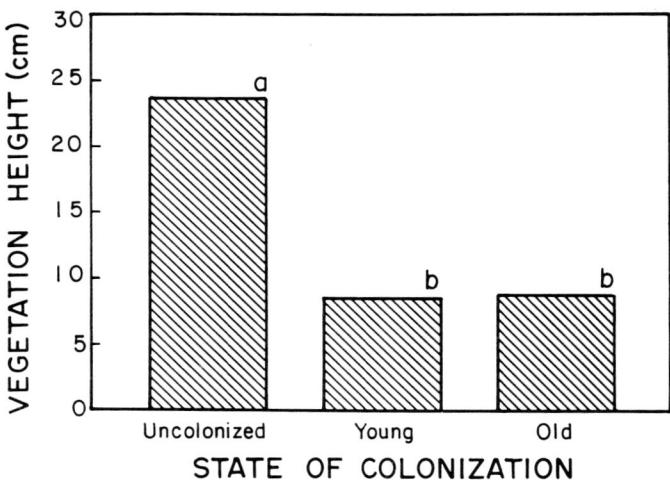

Fig. 3. Mean vegetation height as a function of state of colonization. Data are from two prairie dog colonies sampled in June, July, and August of 1987. (Detling and Whicker, unpubl.). Bars with different letters are significantly different at $p \leq 0.05$.

detection (King 1955). Frequently, the colony area appears greener than the surrounding vegetation. As a result, prairie dog colonies appear as distinct patches within the general matrix of mixed grass prairie. Even within a colony, a mosaic of vegetation types exists, somewhat corresponding to degree of impact over time (Fig. 1). The effects of prairie dog colonization are concentrated, continual, and long-term.

Vertical structure and composition

Following occupation by prairie dogs, overall canopy height is reduced by continual defoliation and is maintained in that state as long as prairie dogs are present. In two recently studied sites, the seasonal mean height of vegetation in uncolonized areas was 24 cm; on colonies occupied for several years to decades, mean height was 9 cm (Fig. 3). In another colony, the mean canopy height decreased 62% during the first two years of colonization, but changed little thereafter (Archer *et al.* 1987).

Prairie dogs directly cause the change in canopy structure by repeatedly clipping and grazing plants, never allowing the shoots to reach full growth. Grazing also acts indirectly to change the competitive balance of plants within the colony and to modify the microhabitat such that some plants are better adapted than others. Both of these effects may result in a reduced canopy because (1) many of the taller species in the plant community are replaced by shorter species and (2) within some species, there exist morphological differences whereby grazing tolerant ecotypes are shorter and more prostrate than less tolerant ones.

Reduced vegetation height and change in plant species composition have been observed on many prairie dog colonies (Osborn and Allen 1949; King 1955; Koford 1958; Bonham and Lerwick 1976; Agnew *et al.* 1986), but have not been

Fig. 4. Morphological differences expressed in representative plants of *Agropyron smithii* collected from a heavily grazed prairie dog colony and a grazing exclosure, and grown for three years in a common environment.

documented in detail. Coppock *et al.* (1983a) and Archer *et al.* (1987) studied the rate of plant species change, replacement, and diversity on separate colonies in WCNP. The trends were similar between colonies, although the rate of change, controlled in part by grazing pressure of prairie dogs and other herbivores, initial community composition, soil type, and weather, did vary. Usually, more than two years of colonization were necessary before species composition appreciably changed relative to uncolonized prairie. After three years, shifts in plant dominance and composition had begun (Coppock *et al.* 1983a) or had rapidly progressed (Archer *et al.* 1987). Mid-height grasses, dominant in the uncolonized prairie, were being replaced by shortgrass species and annual forbs. Several more years of continued heavy grazing resulted in final dominance by a few species of forbs or dwarf shrubs. Plant species diversities were similar, but low, off the colonies and in the older portions of each colony. Diversity was greatest in those areas occupied for intermediate lengths of time.

We have also shown that ecotypic differentiation occurs in several species of grasses when heavily grazed by prairie dogs (Detling and Painter 1983; Detling *et al.* 1986; Jaramillo and Detling 1988; Painter 1987). For example, sod blocks containing western wheatgrass (*Agropyron smithii* Rydb.) were collected from an intensively grazed prairie dog colony and from within a large, permanent grazing exclosure erected more than 40 years earlier. Nine months after being transplanted to a common greenhouse environment, significant morphological differences persisted in plants from the two populations, suggesting that these populations

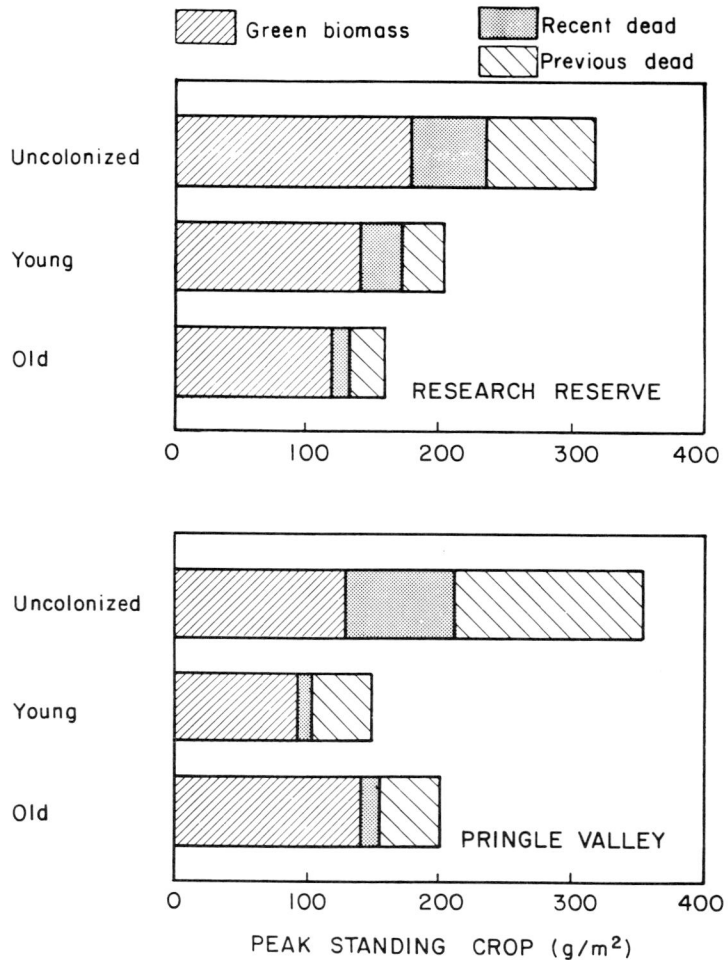

Fig. 5. Peak seasonal standing crop of vegetation on two sites of uncolonized grassland and nearby prairie dog colonies (Research Reserve and Pringle Valley in Wind Cave National Park), in 1987. Recent dead material died during the current growing season, previous dead material died prior to the current growing season (Detling and Whicker, unpubl.).

were genetically distinct. Plants from the prairie dog colonies were shorter, had more tillers per plant, fewer leaves per tiller, shorter and narrower leaves, higher blade/sheath ratios, and were more prostrate than plants from ungrazed populations (Fig. 4). These differences also remained more than two years later (Cid 1985). Both the 'tall' and 'short' morphs exist in the uncolonized grassland, but the tall morphs are dominant (Detling *et al.* 1986). Grazing has apparently modified the selection pressures and competitive balance that existed in the ungrazed populations, thereby causing a shift in dominance to an ecotype that may be more grazing resistant or, because of its shorter stature, be less intensively grazed. Many examples of grazing induced morphological and ecotypic changes, especially stature, can be found in the literature (see review in Painter 1987).

Plant biomass and cover

Concomitant with reduction in the vertical structure of the canopy and plant species replacements following prairie dog occupation are changes in total plant biomass, proportion of biomass in various plant functional groups, and plant cover.

In 1987, peak standing crop (live plus recent dead) was greatest in the uncolonized areas bordering two colonies (> 200 g.m^{-2}) and 25–50% lower within the colonies (Fig. 5). Grasses comprised 89–93% of this biomass off the colonies, but only 15–20% of the standing crop in the oldest, central parts of the colonies. Total peak standing crop (including remaining standing dead from previous years) was over 300 g.m^{-2} in both sites of uncolonized prairie and 150–200 g.m^{-2} within the colonies (Fig. 5). These trends were very similar to what Krueger (1986) found earlier for seasonal mean aboveground biomass on two other colonies. Similarly, in 1979, Coppock et al. (1983a) observed peak live biomass in a young colony to be about half (95 g.m^{-2}) that of uncolonized prairie (190 g.m^{-2}) and the old part of the colony (170 g.m^{-2}).

The mean seasonal proportion of total green biomass in various functional groups changed significantly among states of colonization, but not between colonies in 1987 (Fig. 6). For example, proportion of C_3 (cool season) graminoids declined from uncolonized grassland to young to old portions of the colonies. Proportion of C_4 (warm season) grasses was greatest in the young areas, but did not differ between uncolonized and old colony sites. Dwarf shrubs were a significantly greater proportion of total green biomass in the old colony than in other areas, as generally were forbs. Forbs, however, showed an interactive response between colonies and states of colonization.

The colonized areas typically have a greater proportion of live biomass relative to standing dead biomass than the uncolonized prairie (Fig. 5 and Coppock et al. 1983a). This accounts for the 'greener' appearance of prairie dog colonies mentioned above. Whether or not prairie dogs use it for food, they continually clip the vegetation; therefore, relatively little of it matures and dies. Thus, standing dead plant material does not accumulate in large quantities. This can affect the cover of the soil surface in at least two ways. First, the amount of material that eventually falls to the ground as litter is reduced (Coppock et al. 1983a), and bare ground increases. For example, Archer et al. (1987) found that rapid increases in bare ground occurred in the first two years following colonization, but by the third year, bare ground had stabilized at 35% (compared to ~ 10% initially) and litter cover had decreased to less than 10% (~ 20% initially). An alternative effect of repeated defoliation may be that the litter input remains similar to what would have eventually fallen to the ground without clipping, but bare ground increases due to reduced plant standing crop. This was observed in the two colonies most recently sampled. Litter cover was not significantly different among states of colonization (~ 12–18%), but bare ground increased from less than 5% on uncolonized sites and young areas of colonies to 18% on older areas (Fig. 6).

Plant nutritional dynamics and patch structure

Grazing or defoliation may change the nutrient levels and dynamics within plants,

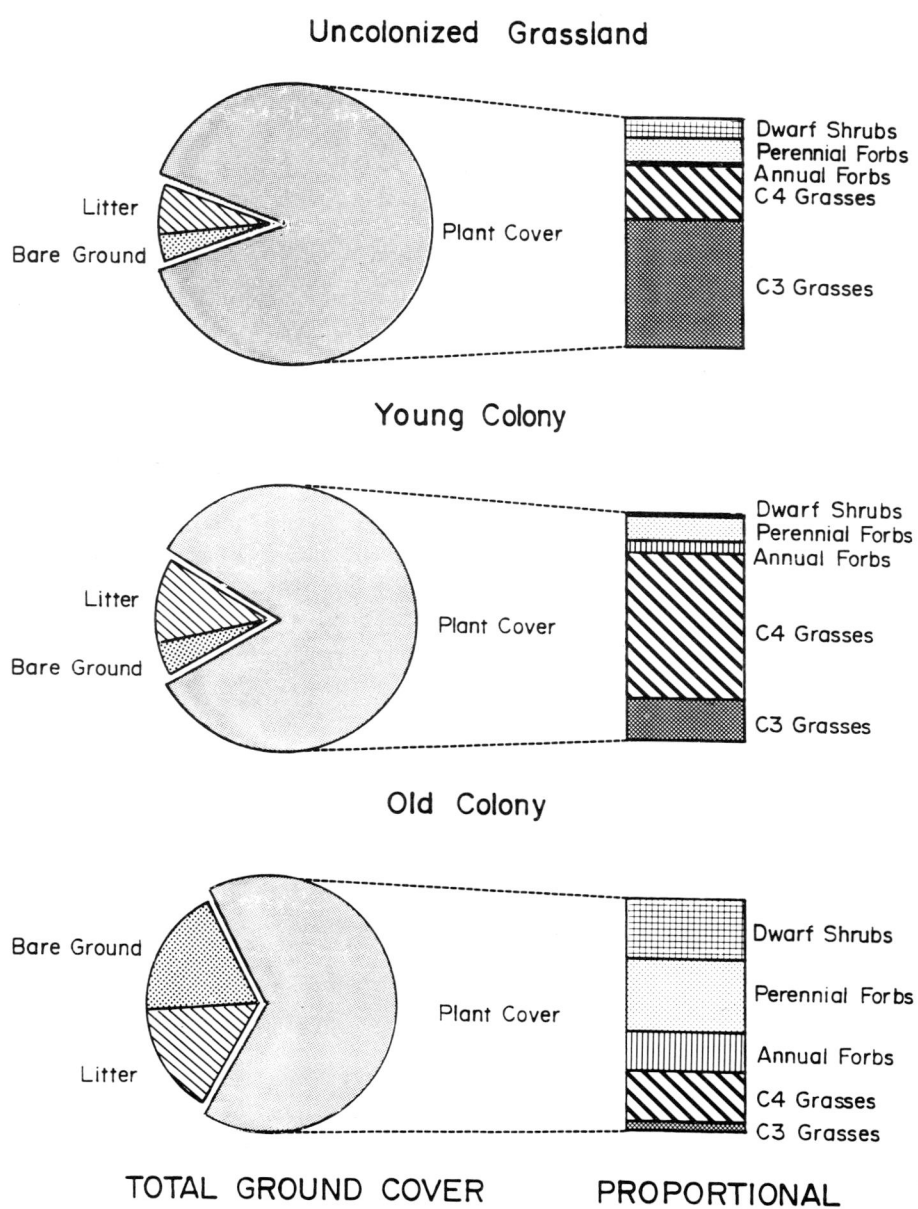

Fig. 6. Pie diagram indicating proportion of ground covered by standing plant material and litter. Of the total green biomass, bar graphs indicate the relative contribution by each of several plant functional groups. Data are seasonal means for June, July, and August, 1987, on two sites of uncolonized grassland and nearby prairie dog colonies at different states of colonization. (Detling and Whicker, unpubl.).

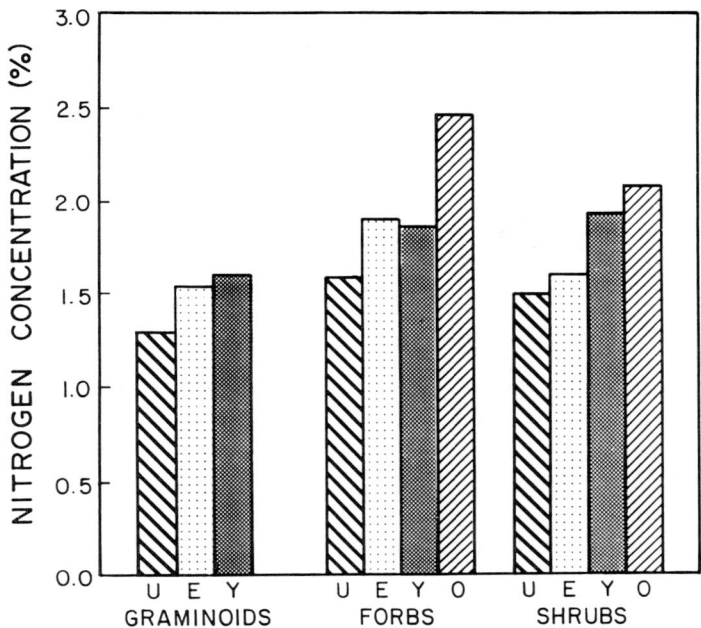

Fig. 7. Mean seasonal nitrogen concentration in shoots of graminoids, forbs, and dwarf shrubs as a function of state of colonization. (U = uncolonized grassland, E = edge of colony, Y = young part of colony, O = old part of colony). Data from Coppock *et al.* (1983a).

populations, and communities. In general, as plants mature their nutritive value declines (Van Soest 1982). Grazing removes aging leaves and, especially in plants such as grasses with basal meristematic growth, new growth may be stimulated. The new tissues usually have higher nitrogen concentrations and are more digestible than those from ungrazed plants (McNaughton 1984). Field studies at WCNP indicate that prairie dogs have a directional effect on plant nutrition and positively influence forage quality by their grazing.

During the 1979 growing season, Coppock *et al.* (1983a) collected live material of six grass species (three C_3 species and three C_4 species), a composite of forb species, and a dwarf shrub, *Artemisia frigida* Willd., from sites representing several states of colonization. Shoot nitrogen concentrations were generally lowest in plants from the uncolonized grassland, and increased with the length of time an area had been occupied (Fig. 7), although there was a progressive decline in plant N during the season. Digestibility of grasses followed a pattern similar to that of nitrogen concentration: digestibility declined as the season progressed; grasses from the uncolonized area had lower digestibilities than those from the edge or young colony; cool season grasses were more digestible than warm season ones.

In recent laboratory studies on two dominant grass species at WCNP (*A. smithii* and *Bouteloua gracilis* (H.B.K.) Griffiths, blue grama), plant responses and adaptations to defoliation have been further investigated (Polley 1987; Jaramillo and Detling 1988) using populations that had apparently genetically differentiated under heavy grazing pressure on prairie dog colonies. Initially, aboveground shoot nitrogen concentrations did not differ between on and off colony populations of either species, suggesting that population differences in N concentrations observed in the field resulted from factors other than genetic differences. The responses to defoliation, however, differed between populations and species. Following defoliation, plants of *A. smithii* from a prairie dog colony had high rates of resource capture, a strategy of grazing tolerance, compared to off colony plants (Polley 1987). In particular, defoliated plants from the colony exhibited greater total N accumulation and higher relative N accumulation rates compared to off colony plants. In contrast, although on and off colony plants of *B. gracilis* differed in many aspects of resource allocation, plants of the two populations responded similarly to defoliation (Jaramillo and Detling 1988). For example, whole plant (shoot and root) nitrogen concentration was greater in off colony plants but defoliation increased this parameter as well as root specific nitrogen uptake in both on and off colony populations. These two studies and others (Chapin 1980; Floate 1981; Ruess *et al.* 1983) suggest that defoliation and, by implication, grazing enhance nutrient uptake, increase nutrient concentration in repeatedly grazed grasses, and accelerate nutrient cycling by the actions of herbivores.

Another response to grazing is increased silica concentration in leaves of grasses. It has been suggested (McNaughton *et al.* 1985), based on studies in the African savanna, that this may be a defense against herbivores, because silica decreases digestibility and palatability and promotes tooth wear (Van Soest 1982). In field studies, Brizuela *et al.* (1986) found consistently higher silicon concentrations in tillers of *A. smithii* and *Schizachyrium scoparium* (Michx.) Nash (little bluestem) from heavily grazed prairie dog colonies than from lightly grazed areas. However, repeated defoliation actually decreased silicon concentration in controlled field (Brizuela *et al.* 1986) and laboratory (Cid 1985) experiments. Thus higher whole tiller silicon concentrations from prairie dog colony plants may be explained, at least partly, by higher silicon concentrations in leaf blades compared to sheaths (Cid 1985) and the higher blade/sheath ratios in on – versus off – colony plants (Detling and Painter 1983; Cid 1985).

Prairie dogs and the belowground ecosystem

Grazing-induced structural changes in the plant community can, of course, occur belowground as well as aboveground. For example, cattle grazing usually reduces root biomass (Schuster 1964). In grasslands, much of the primary producer's dynamics occurs belowground (Coleman *et al.* 1976) where soil invertebrates, especially nematodes, are major consumers of plant material. Thus aboveground grazing by prairie dogs may directly influence the belowground responses of plants, and this may indirectly influence other animals.

Numerous earlier laboratory studies on North American grassland species have shown that defoliation of aboveground shoots reduces root biomass (see review in Ellison 1960). In all prairie dog colonies sampled at WCNP, we typically have observed a marked decline in total root biomass from uncolonized areas to older parts of the colonies (Fig. 8). Ingham and Detling (1984) found seasonal mean root

Vegetation structure and ecosystem processes 311

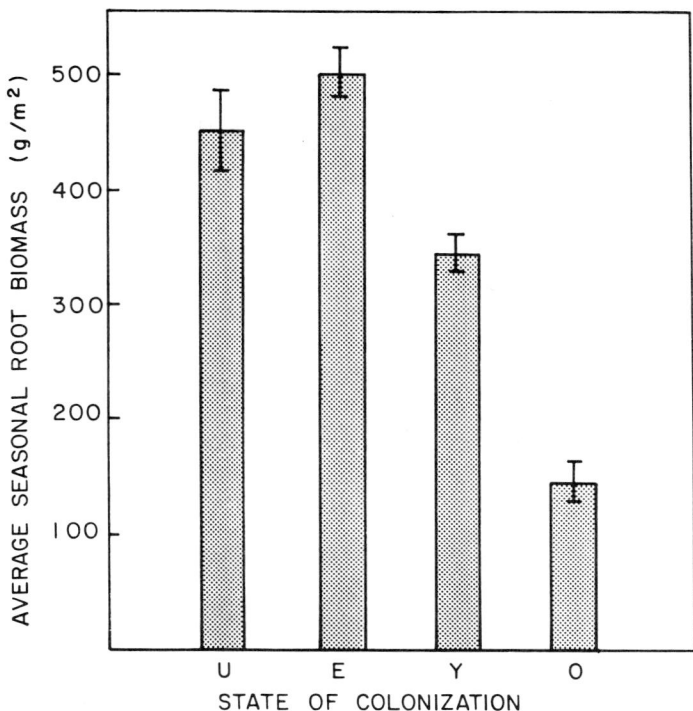

Fig. 8. Mean seasonal root biomass (ash-free) in the upper 10 cm of soil during 1984 for Research Reserve prairie dog colony as a function of state of colonization. Symbols (U, E, Y, and O) are defined as in Fig. 7. (Detling and Whicker, unpubl.).

biomasses of *A. smithii* and *S. scoparium* from a heavily impacted section of a prairie dog colony were 20-30% lower than off the colony, but total nematode densities under those grasses were 30-60% higher on the colony than off (Fig. 9a). Although annual net root production was about 40% lower on the colony than in uncolonized grassland, nematode consumption, as a proportion of net root production, more than doubled (Fig. 9b).

The population structure and species composition of nematodes also changed with prairie dog colonization (Fig. 9a). Both the plant parasitic Dorylaimida, and Rhabditida which are not plant parasites, were significantly more abundant on the colonies than off. Densities of the non-parasitic Dorylaimida and parasitic Tylenchida did not change between sites. Root biomass alone is insufficient to explain these trends, because the direction of population changes of herbivorous versus non-herbivorous nematodes on and off colonies was not consistent. It is most likely that complex indirect effects of grazing, such as changes in soil microclimate (Archer and Detling 1986) or soil or plant chemistry, may be responsible for the structural changes in nematode populations (Ingham and Detling 1984).

Regardless of the mechanisms involved, it appears that the activity of prairie dogs not only has a substantial impact on belowground vegetation structure but also indirectly enhances rates of energy and material flow through belowground consumers.

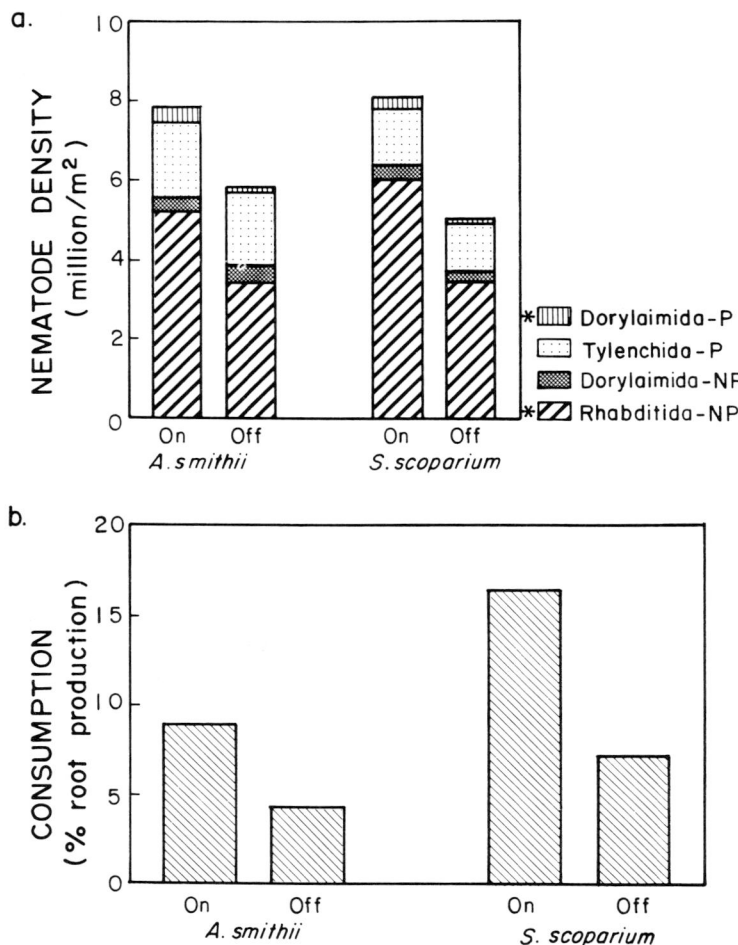

Fig. 9. (a) Densities of four groups of plant parasitic (P) and non-plant parasitic (NP) nematodes from on and off a prairie dog colony. *'s indicate significant density differences between populations on and off the colony, $p \leq 0.05$. (b) Relative consumption of root production of two grass species by nematodes on and off a prairie dog colony. Data from Ingham and Detling (1984).

Prairie dog colonies and other animals

We have discussed how populations and community composition of very non-mobile animals such as nematodes can be changed by prairie dog activities. These changes probably result from differential reproduction and survival of the various nematode species. However, the distribution of more mobile animals may also be influenced by the presence of colonies. Differential survival may certainly be a factor, but population and community changes are probably strongly influenced by immigration and emigration changing the relative proportion of breeding residents. For example, small, relatively mobile, animals such as birds (Agnew *et al.* 1986) could quickly respond and move into or out of a suitable habitat. Very large

Vegetation structure and ecosystem processes

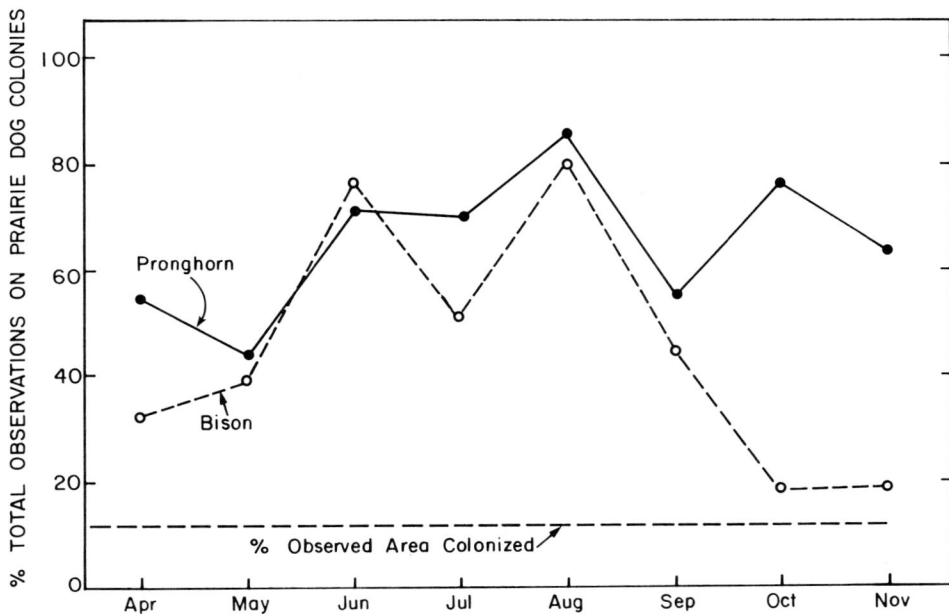

Fig. 10. Utilization of prairie dog colonies for feeding by bison and pronghorn. Points occurring above horizontal dashed line (random use) indicate selection for colony site. Data from Krueger (1986).

mobile animals, especially bison, elk, and pronghorn, probably do not become permanent residents or change population levels or reproductive output because of prairie dog colonies; however, they do have the ability to modify activity patterns and site selection over very large areas. The questions, then, are: (a) do large ungulates respond to the vegetational patches created by prairie dogs and, (b) if so, why?

Several active prairie dog colonies are dispersed throughout WCNP. All colonies are readily accessible to ungulates; bison and pronghorn move freely throughout the park and have long been known to be associated with prairie dog colonies (King 1955; Koford 1958). If animals randomly use whatever habitat they encounter, the frequency of observations of those animals in a habitat should approximate the proportion of that habitat in the park. Selection for, or preferential use of, an area is indicated if observations there are greater than would be expected by chance alone. Recent research has verified that there is significant selection for prairie dog colonies by both bison and pronghorn, especially during the growing season (Fig. 10, Coppock et al. 1983b; Krueger 1986; Wydeven and Dahlgren 1985).

In addition to parkwide selection for colonies, there is also preferential use of areas within each colony (Coppock et al. 1983b). For example, over one growing season, bison used the (a) younger, grass-dominated portion of a colony for both grazing and resting (3.0 and 2.7 times expected, respectively), the (b) edge primarily for grazing (2.5 times expected), and the (c) forb/dwarf shrub-dominated older areas for resting (2.5 times expected). By contrast, they used the adjacent uncolonized prairie only 20% of the expected time, indicating that this area was avoided in preference for the colony.

Krueger (1986) found that the pattern of use within the colonies differed between bison and pronghorn. While her results for bison were similar to those of Coppock *et al.* (1983b), she found that 57-97% of the pronghorn feeding on colonies were on the forb-shrub dominated centers. Diets were very similar between bison and prairie dogs in the grass dominated areas and between pronghorn and prairie dogs in the centers. However, interspecific competition for forage was not detectable.

Why might ungulates choose to forage on prairie dog colonies? As discussed above, prairie dogs create patches within the grasslands where plant material has a greater live to dead ratio (albeit lower standing crop), a higher crude protein (nitrogen) level, and a greater digestibility than plant material from the uncolonized prairie (Coppock *et al.* 1983a). This implies better available nutrition per unit of forage consumed on the colonies compared to off colonies (Coppock *et al.* 1983b; Krueger 1986). Vanderhye (1985) simulated potential nutritional benefits accrued to bison by feeding on colonies using a model (Swift 1983) driven by dietary information. The output suggested that if mature cows randomly use prairie dog colonies for feeding, they will gain an additional 2 kg (7% of total seasonal weight gain) of body weight over what they would gain by not feeding on colonies at all. Typical usage (39% of total growing season feeding) of colonies confers an additional 18% weight gain. For yearling bison, typical use of colonies could add a 46% weight gain over not grazing on colonies. Thus, prairie dogs apparently have modified the environment, creating a favorable feeding and resting habitat for other animals.

Patch changes influenced by other animals

These changing patterns of site use by other animals in response to patches created by prairie dogs may in themselves be vectors of further change. It has been estimated that, over a growing season, bison foraging on prairie dog colonies would consume about as much plant biomass as prairie dogs (Coppock *et al.* 1983b). This consumption, by altering the vegetation structure, might exacerbate vegetation change initiated by prairie dogs and actually accelerate the expansion of colonies into uncolonized areas (Osborn and Allan 1949). In addition, because of localized use patterns, nutrients, deposited as feces and urine, may also be concentrated on colonies and become more available to plants. On a small scale, increases in small mammal and avian densities on colonies (O'Meilia *et al.* 1982; Agnew *et al.* 1986) may differentially affect seed availability and germination, thereby influencing the rate or direction of vegetation change.

Prairie dogs obviously cause physical disruption of the soil during burrowing, mound building, and digging for roots, but bison contribute to it as well. Bison create many bare areas, wallows, several meters in diameter on colonies. Heavily used trails and other trampled areas can cause soil compaction. All of these impacts change vegetational and soil properties that influence soil temperature and moisture and surface microclimate (Ellison 1960; Archer and Detling 1986; Day 1988).

As part of natural grassland ecosystems, prairie dogs enhance certain features of the vegetation and create favorable habitat patches for other animals. Therefore, for situations such as in Wind Cave National Park, a limited number of prairie dog colonies scattered throughout the grasslands may actually improve the health and increase the diversity of other wildlife species. However, continued extensive utilization of prairie dog colonies by large herds of ungulates may ultimately lead to vegetation changes of sufficient magnitude to reduce suitability of these sites for bison, prairie dogs, and other animals as well.

Acknowledgments

Preparation of this manuscript was supported by grant BSR-84-06660 from the U.S.A. National Science Foundation and contract NAS-5-28766 from the National Aeronautics and Space Administration. We are indebted to Mr. Rich Klukas, Park Biologist, and Mr. Ernest Ortega, Park Superintendent, for logistical support and encouragement to conduct research at Wind Cave National Park.

References

Agnew, W., Uresk, D.W. and Hansen, R.M. 1986. Flora and fauna associated with prairie dog colonies and adjacent ungrazed mixed-grass prairie in western South Dakota. J. Range Manage. 39: 135–139.
Archer, S. and Detling, J.K. 1986. Evaluation of potential herbivore mediation of plant water status in a North American mixed-grass prairie. Oikos 47: 287–291.
Archer, S., Garrett, M.G. and Detling, J.K. 1987. Rates of vegetation change associated with prairie dog (*Cynomys ludovicianus*) grazing in North American mixed-grass prairie. Vegetatio 72: 159–166.
Bonham, C.D. and Lerwick, A. 1976. Vegetation changes induced by prairie dogs on shortgrass range. J. Range Manage. 29: 221–225.
Brizuela, M.A. 1987. Prairie dog feeding behavior: Response to colonization history and fire. Ph.D. Dissertation. Colorado State University, Ft. Collins, Colorado.
Brizuela, M.A., Detling, J.K. and Cid, M.S. 1986. Silicon concentration of grasses growing in sites with different grazing histories. Ecology 67: 1098–1101.
Cargill, S.M. and Jefferies, R.L. 1984. The effects of grazing by lesser snow geese on the vegetation of a sub-arctic salt marsh. J. Appl. Ecol. 21: 669–686.
Chapin, F.S. 1980. Nutrient allocation and responses to defoliation in tundra plants. Arct. Alp. Res. 12: 553–563.
Cid, M.S. 1985. Effects of grazing exposure history and defoliation on silicon concentration of *Agropyron smithii*. MS Thesis. Colorado State University, Ft. Collins, Colorado.
Coleman, D.C., Andrews, R., Ellis, J.E. and Singh, J.S. 1976. Energy flow and partitioning in selected man-managed and natural ecosystems. Agro-Ecosystems 3: 45–57.
Coppock, D.L., Detling, J.K., Ellis, J.E. and Dyer, M.I. 1983a. Plant-herbivore interactions in a North American mixed-grass prairie I. Effects of black-tailed prairie dogs on intraseasonal aboveground plant biomass and nutrient dynamics and plant species diversity. Oecologia 56: 1–9.
Coppock, D.L., Ellis, J.E., Detling, J.K. and Dyer, M.I. 1983b. Plant-herbivore interactions in a North American mixed-grass prairie II. Responses of bison to modification of vegetation by prairie dogs. Oecologia 56: 10–15.
Crawley, M.J. 1983. Herbivory: The Dynamics of Animal-Plant Interactions. University of California Press, Berkeley. 437 pp.
Dahlsted, K.J., Sather-Blair, S., Worcester, B.K. and Klukas, R. 1981. Application of remote sensing to prairie dog management. J. Range Manage. 34: 218–223.

Day, T.A. 1988. Modification of individual plant and community water and nitrogen relations by grassland herbivores. Ph.D. Dissertation. Colorado State University, Ft. Collins, Colorado.

Detling, J.K. and Painter, E.L. 1983. Defoliation responses of western wheatgrass populations with diverse histories of prairie dog grazing. Oecologia 57: 65–71.

Detling, J.K., Painter, E.L.and Coppock, D.L. 1986. Ecotypic differentiation resulting from grazing pressure: evidence for a likely phenomenon. In: P.J. Joss, P.W. Lynch and O.B. Williams (eds), Rangelands: A Resource under Siege, pp. 431–433. Australian Academy of Science Canberra.

Ellison, L. 1960. Influence of grazing on plant succession of rangelands. Bot. Rev. 26: 1–78.

Floate, M.J.S. 1981. Effects of grazing by large herbivores on nitrogen cycling in agricultural ecosystems. In: F.E. Clark and T. Rosswall (eds). Terrestrial Nitrogen Cycles. Ecological Bulletin (Stockholm) 33: 585–601.

Harper, J.L. 1969. The role of predation in vegetational diversity. In: Diversity and Stability in Ecological Systems. Brookhaven Symposium in Biology 22: 48–62.

Ingham, R.E. and Detling, J.K. 1984. Plant-herbivore interactions in a North American mixed-grass prairie III. Soil nematode populations and root biomass on *Cynomys ludovicianus* colonies and adjacent uncolonized areas. Oecologia 63: 307–313.

Jaramillo, V.J. and Detling, J.K. 1988. Grazing history, defoliation, and competition: effects on shortgrass production and nitrogen accumulation. Ecology (in press).

King, J.A. 1955. Social behavior, social organization, and population dynamics in a black-tailed prairie dog town in the Black Hills of South Dakota. Contributions from the Laboratory of Vertebrate Biology No. 67. University of Michigan. Ann Arbor. 123 pp.

Knowles, C.J. 1986. Some relationships of black-tailed prairie dogs to livestock grazing. Great Basin Nat. 46: 198–203.

Koford, C.B. 1958. Prairie dogs, whitefaces, and blue grama. Wildl. Monogr. No. 3. 78 pp.

Krueger, K. 1986. Feeding relationships among bison, pronghorn, and prairie dogs: an experimental analysis. Ecology 67: 760–770.

McNaughton, S.J. 1984. Grazing lawns: animals in herds, plant form, and coevolution. Am. Nat. 124: 863–886.

McNaughton, S.J., Tarrants, J.L., McNaughton, M.M. and Davis, R.H. 1985. Silica as a defense against herbivory and a growth promoter in African grasses. Ecology 66: 528–535.

Merton, L.F.H., Bourn, D.M. and Hnatiuk, R.J. 1976. Giant tortoise and vegetation interaction on Aldabra Atoll. I. Inland. Biol. Conserv. 9: 293–304.

O'Meilia, M.E., Knopf, F.L. and Lewis, J.C. 1982. Some consequences of competition between prairie dogs and beef cattle. J. Range Manage. 35: 580–585.

Osborn, B. and Allan, P.F. 1949. Vegetation of an abandoned prairie-dog town in tall grass prairie. Ecology 30: 322–332.

Painter, E.L. 1987. Grazing and intraspecific variation in four North American grass species. Ph.D. Dissertation. Colorado State University, Ft. Collins, Colorado.

Polley, H.W. 1987. Grazing history, competition, and soil nitrogen effects on plant responses to defoliation. Ph.D. Dissertation. Colorado State University, Ft. Collins, Colorado.

Ruess, R.W., McNaughton, S.J. and Coughenour, M.B. 1983. The effects of clipping, nitrogen source and nitrogen concentration on the growth responses and nitrogen uptake of an east african sedge. Oecologia 59: 253–261.

Schuster, J.L. 1964. Root development of native plants under three grazing intensities. Ecology 45: 63–70.

Summers, C.A. and Linder, R.L. 1978. Food habits of the black-tailed prairie dog in western South Dakota. J. Range Manage. 31: 134–136.

Swift, D.M. 1983. A simulation model of energy and nitrogen balance for free-ranging ungulates. J. Wildl. Manage. 47: 620–645.

Van Soest, P.J. 1982. Nutritional Ecology of the Ruminant. O&B Books, Corvallis, Oregon.

Vanderhye, A.V.R. 1985. Interspecific nutritional facilitation: Do bison benefit from feeding on prairie dog towns? MS Thesis, Colorado State University, Ft. Collins, Colorado.

Wydeven, A.P. and Dahlgren, R.B. 1985. Ungulate habitat relationships in Wind Cave National Park. J. Wildl. Manage. 49: 805–813.

THE EFFECT OF HERBIVORY ON VEGETATION STRUCTURE

O.E. SALA
Department of Ecology, Faculty of Agronomy, University of Buenos Aires, Av. San Martin 4453, Buenos Aires 1417, Argentina

Abstract

Grazing by large herbivores is one of the major determinants of the structure of grasslands. The objectives of this paper are: (1) to assess the effect of grazing by large domestic herbivores upon the structure of grasslands using as an example the subhumid grassland of the flooding pampa in Argentina; (2) to evaluate the mechanisms involved in the changes of structure determined by grazers; (3) to compare the effect of grazing upon vegetation structure along different temporal and spatial scales.

One of the largest responses to grazing in the flooding pampa was observed in the vertical and horizontal distribution of leaves in the canopy. Total values of basal cover or leaf area changed very little with grazing. Changes in structure were accounted for by the changes in vigor as well as by changes in species composition. Grazing resulted in the introduction of a large number of exotic species.

In subhumid grasslands the response to grazing is related to changes in the light quality underneath the canopy which modifies the rate of tillering, tiller elongation and seed germination.

Responses to herbivory at different time scales are not independent but form a hierarchy. Slow responses, in evolutionary time constrain fast responses in ecological and ontogenetical time.

Based upon evolutionary grazing history and moisture conditions a model is presented of the effect of grazing upon vegetation structure. Contrasting examples in which grazing increased or decreased spatial heterogeneity are presented along with hypotheses explaining these differences.

Introduction

Herbivores are one of the major determinants of the structure of vegetation. Their effect varies across vegetation types and it is maximum in grasslands. Unlike forests, where loss of biomass to herbivores is usually less than 10%, grasslands are frequently characterized by herbivory rates of 50% or more of the aboveground net primary production (McNaughton 1976, 1985) and perhaps 25% of the belowground productivity (Coleman *et al.* 1976; Ingham and Detling 1984; Scott *et al.* 1979). Herbivores affect the structure of vegetation in at least two ways: directly through consumption and indirectly through a variety of mechanisms such as regeneration of limiting nutrients or changes in the light environment (Deregibus *et al.* 1985; Sterner 1986).

Most of the grassland management techniques modify vegetation structure (Harper 1971; Stoddart *et al.* 1975). Burning, chaining or spraying selective herbicides are often aimed at decreasing the shrub component, which supposedly will increase grass production. Fertilization, depending upon the concentration of major elements, usually affects the life-form composition besides modifying total productivity.

There is a tight link between the structure of grasslands and their function. Vegetation structure is the result of adaptations to the environment by the biotic components and their influences on the abiotic conditions. Vegetation structure

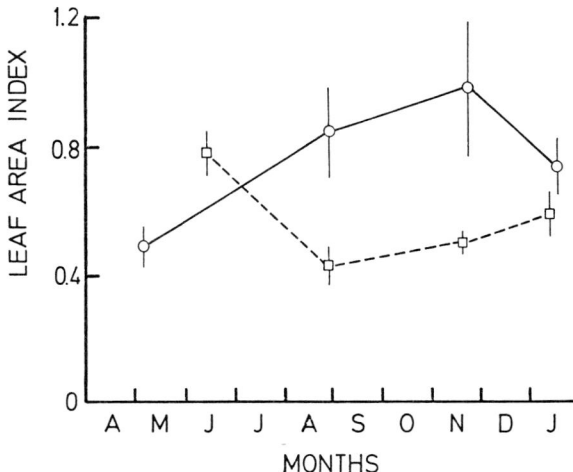

Fig. 1. Leaf area index (± 1 standard error) through the year for the grazed (□ ‒ ‒ ‒) and ungrazed areas (○——).

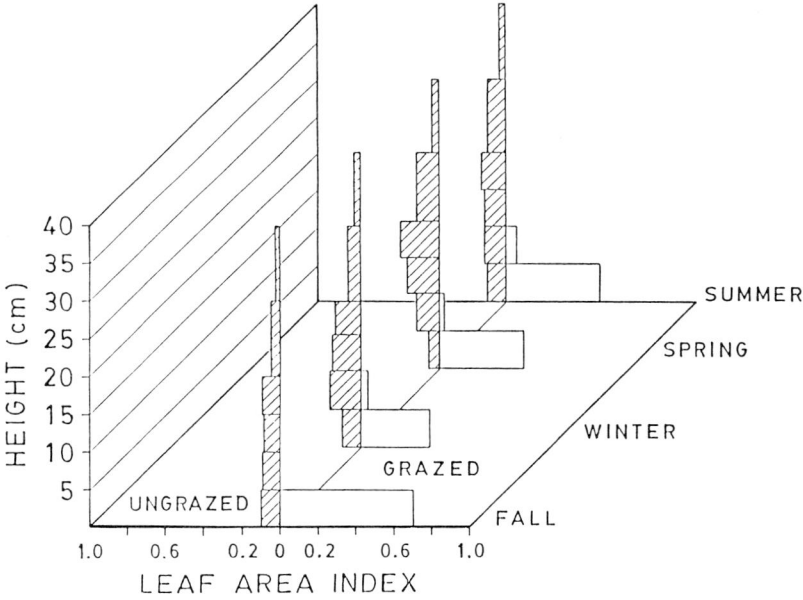

Fig. 2. Leaf area index for different layers from soil level up to 40 cm for the ungrazed areas (shaded bars) and grazed areas (empty bars) along the four seasons.

is one of the influential determinants of ecosystem processes which in turn shape the structure of vegetation.

The objectives of this paper are: (1) to describe the effects of grazing by large domestic herbivores upon vegetation structure; (2) to evaluate the mechanisms involved in the changes in structure determined by grazers and (3) to compare the effect of grazing upon vegetation structure along different temporal and spatial

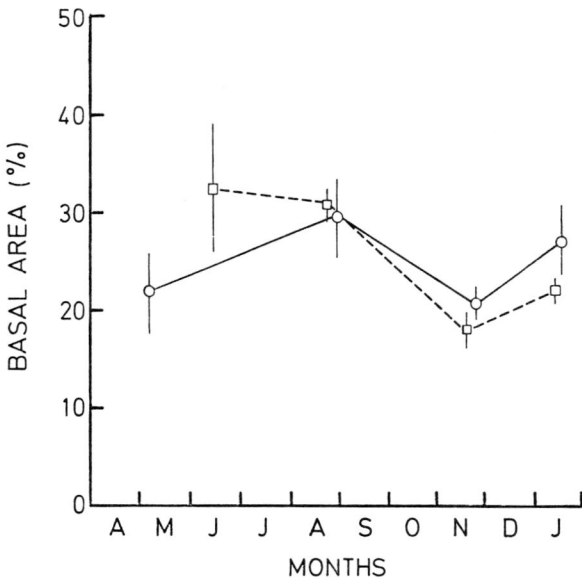

Fig. 3. Basal area (± 1 standard error) through the year for the grazed (□---) and ungrazed areas (o——).

scales. To accomplish these objectives the paper will focus on a particular grassland located in the flooding pampa in Argentina. First, I will describe the effect of large herbivores upon leaf area index, basal area and species composition as well as their vertical and horizontal distribution. Second, I will look at possible mechanisms which may account for a fraction of the observed changes in structure. Thirdly, I will compare the response of grasslands to grazing at three time scales and two spatial scales emphasizing the relationships among scales.

A description of the effects of herbivores upon vegetation structure: the case of the flooding pampa

The flooding pampa is a region in the eastern portion of Argentina of approximately 5 million ha which is covered in 80% of its area by native grasslands (León *et al.* 1984). It is a temperate region with a mean annual precipitation of 920 mm and mean monthly temperatures ranging from 7°C in July–August to 22°C in January. Mesic conditions and the small range between winter and summer temperatures are mainly the result of its proximity to the ocean. Flat topography, along with low hydraulic conductivity in the B horizon, determine the frequent occurrence of both flooding events and droughts.

Vegetation is composed of an intricate mosaic of plant communities associated to small differences in topographical position (15–20 cm). Two communities mainly form this mosaic; one characterized by *Piptochaetium montevidense, Ambrosia tenuifolia, Eclipta bellidiodes* and *Mentha pulegium* which is located on

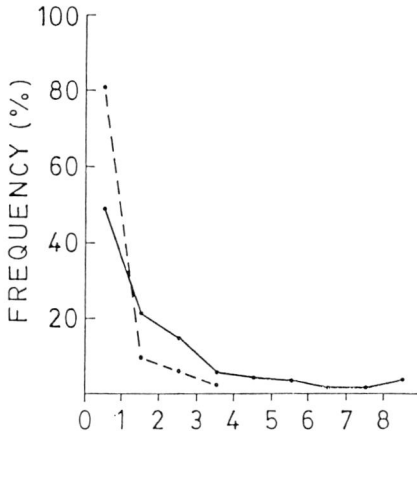

Fig. 4. Frequency distribution of size classes of the interception of individual plants with the line used to estimate basal area. Frequency distribution of size classes of interception is related with the frequency distribution of classes of diameter of individuals at the base level. For the grazed (□ ---) and ungrazed areas (o——).

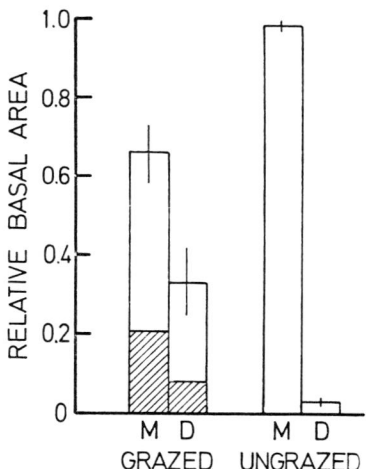

Fig. 5. Relative basal area of monocots (M) and dicots (D), annuals (shaded bars) and perennials (empty bars) for the grazed and ungrazed areas.

the uplands and the other by *Mentha pulegium*, *Leontodon nudicaulis* and *Paspalidium paludivagum* located on the lowlands (León 1975; nomenclature follows this publication and Cabrera 1968).

Cattle usually graze these grasslands year long. This has been a pervasive practice throughout the region, leaving no relict grasslands. Experiments aimed at as-

Table 1. Basal area by species (± 1 standard error) for grazed and ungrazed areas of the upland community. Boxes show the species which decrease in the ungrazed condition. Species with no values entered were not recorded in the spring sampling but in another season.

Species	Grazed Spring	Ungrazed Spring
Natives		
Cool-season species		
Stipa neesiana		0.6 ± 0.6
Danthonia montevidensis	1.8 ± 0.7	3.8 ± 1.2
Heleocharis sp.	0.4 ± 0.1	0.3 ± 0.3
Briza subaristata	0.9 ± 0.1	3.5 ± 0.7
Carex phalaroides	1.0 ± 0.1	4.3 ± 3.3
Stipa papposa	0.2 ± 0.2	
Sisyrinchium platense	0.2 ± 0.1	
Gamochaeta spicata	0.6 ± 0.6	
Trifurcia lahue	0.1 ± 0.1	
Cypella herbertii		
Eryngium ebracteatum	0.1 ± 0.1	
Lilaea scilloides	0.1 ± 0.1	
Warm-season species		
Sporobolus indicus	0.6 ± 0.1	1.1 ± 0.8
Bothriochloa laguroides	2.2 ± 0.6	0.1 ± 0.1
Leersia hexandra		0.1 ± 0.1
Paspalum vaginatum		
Panicum gouinii	0.1 ± 0.1	
Ambrosia tenuifolia	0.5 ± 0.1	0.3 ± 0.1
Paspalidium paludivagum		
Distichlis scoparia		4.8 ± 2.2
Panicum milioides		0.2 ± 0.1
Stenotaphrum secundatum	0.1 ± 0.1	
Dichondra repens	0.2 ± 0.1	
Eclipta bellidiodes	0.2 ± 0.1	
Aster squamatus	0.1 ± 0.1	
Paspalum dilatatum		
Chaetotropis elongata		
Setaria geniculata	0.1 ± 0.1	
Gerardia communis		
Spilanthes stolonifera	0.1 ± 0.1	
Phyla canescens	0.1 ± 0.1	
Apium leptophyllum		
Exotics		
Cool-season species		
Plantago lanceolata	1.5 ± 0.3	
Leontodon taraxacoides	0.8 ± 0.6	
Mentha pulegium	0.8 ± 0.8	
Lolium multiflorum	1.7 ± 0.5	
Oxalis sp.		
Vulpia dertonensis	1.5 ± 0.2	
Briza minor	0.1 ± 0.1	
Linum usitatissimum	0.2 ± 0.1	
Medicago polymorpha		
Lythrum hyssopifolia	0.1 ± 0.1	
Warm-season species		
Bupleurum tenuissimum		

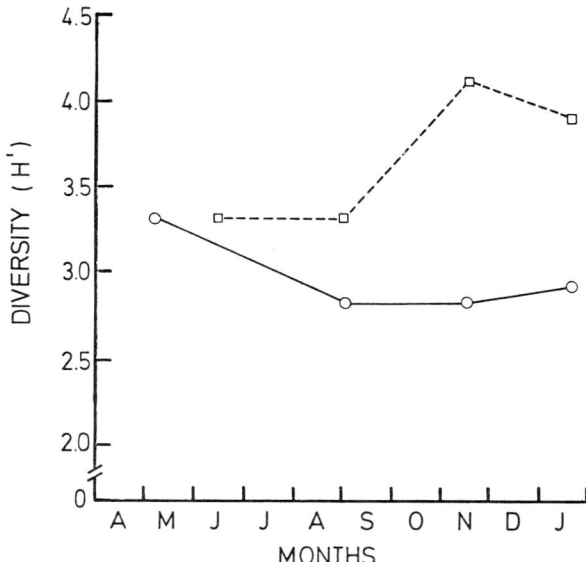

Fig. 6. Species diversity (H') through the year for the grazed (□ - - -) and ungrazed areas (o——).

sessing the effect of grazing upon vegetation structure were based on the comparison between grazed areas and exclosures.

One of the effects of excluding large domestic herbivores has been a small increase in total leaf area index (LAI) (Fig. 1) (Sala et al. 1986). Annual average LAI increased 30% after four years of exclosure. The distribution of leaves in the canopy presented a larger effect of grazing than total LAI (Fig. 2). In the grazed area, most of the green material was concentrated in the 0-5 cm layer whereas in the ungrazed area the largest portion of the leaf area was in the 10-30 cm layer. The small difference in total LAI and biomass between the grazed and ungrazed areas along with the large difference in vertical distribution resulted in a sharp decrease of biomass concentration (2.3 to 0.5 mg cm^{-3}, respectively).

Total basal area presented no differences between grazed and ungrazed areas (Fig. 3). The major effect of grazing upon basal area was observed, as in the case of LAI, not on the total value but on its distribution. The average size of tussocks was higher on ungrazed than on grazed areas (Fig. 4). Grazing exclusion resulted in the replacement of a large number of small tussocks by few large ones.

Changes in LAI and basal area due to grazing were accounted for by changes in vigor and growth form as well as by changes in species composition. As a result of the exclusion of large herbivores, monocotyledoneae increased from 65% to 95% of total cover (Fig. 5). This compares well with results of Willems (1983) from chalk grasslands where monocotyledoneae contributed 40% to total aboveground phytomass in sheep-grazed sites as against 85% in exclosures. In the exclosure in the flooding pampa annuals, both dicots and monocots, were replaced by perennials. The major effect of the exclosure upon species composition was observed in a drastic reduction of all exotic species along with native planophiles (Table 1). In contrast, there was an increase of major native perennial grasses. The

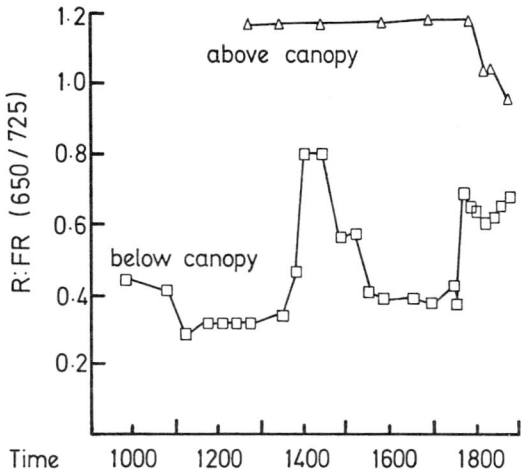

Fig. 7. R:FR quantum flux ratio above and below the canopy of a temperate grassland on a clear day in the fall. Redrawn from Deregibus et al. 1985.

same species which occurred on the ungrazed area, occurred on the grazed area. The difference between these two areas was given by a set of species presumably adapted to grazing which appeared on the grazed conditions. The exotic species which responded to grazing were cool-season species originating in the Mediterranean region. Diversity was higher on grazed areas than on ungrazed areas (Fig. 6). There are several examples in which grazing determines higher species diversity (McNaughton 1979; Picket 1980; During and Willems 1984).

Mechanisms of the changes in structure caused by grazing

The major changes in structure caused by grazing in the flooding pampa were observed in the distribution of leaf area and basal area and they were the result of changes in growth form and species composition. Changes in growth form are mainly the result of indirect effects of herbivory. Grazing determines better water supplies, better nutrient supplies and, at the same time, increases light intensity and modifies its quality for the remaining tissues. These indirect effects lead to compensatory growth (McNaughton 1983a). The importance of these three factors, water, nutrients and light, varies among systems and among processes within a system (McNaughton et al. 1983).

The effect of herbivory upon growth form in the flooding pampa occurs mainly as a result of alterations of light quality. There is evidence for crops and grasslands that light intensity and quality change throughout the canopy (Holmes and Smith 1977; Deregibus et al. 1985). The light beneath the canopy is poorer in red than above the canopy. For an ungrazed grassland in the flooding pampa, which intercepted 80% of the incoming radiation, the red (R): far red (FR) ratio had a daily average of 1.2 above the canopy and 0.4 beneath the canopy (Fig. 7). The same

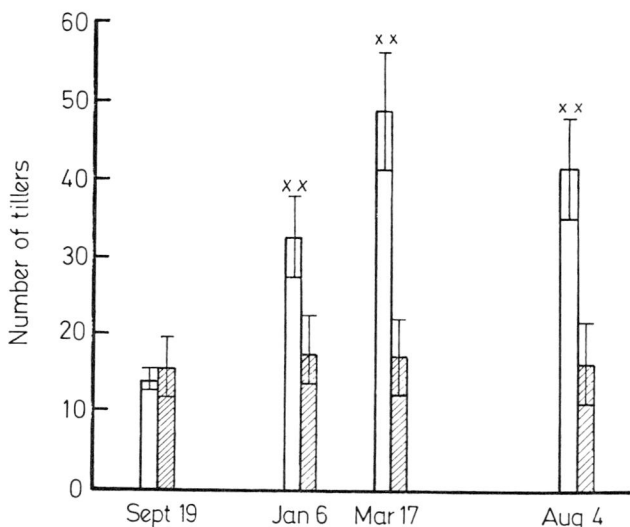

Fig. 8. Tiller populations per plant (± 1 standard error) for *Paspalum dilatatum* as affected by increased red light conditions beneath the canopy. Significant treatment differences ($p < 0.01$) are indicated by xx. Additional red light empty bars and control shaded bars. Redrawn from Deregibus et al. 1985.

values were measured in an ungrazed chalk grassland in the Netherlands (Willems 1983). This sharp decrease of the ratio throughout the canopy is caused by the preferential absorption by green leaves of R compared to FR wavelengths. In contrast, the R:FR ratio at the soil level of the grazed area averaged 1.0.

There is evidence indicating that the light quality received at the base of plants influences the rate of tillering in grasses (Deregibus et al. 1983). An increase in the R:FR ratio, without significantly modifying the photosynthetically active radiation, resulted in an increase in tiller production in *Lolium* plants cultivated in a growth chamber. The enrichment with red light beneath the canopy of an ungrazed grassland also resulted in an increase in the tillering rate of two native grasses, *Paspalum dilatatum* and *Sporobolus indicus* (Fig. 8) (Deregibus et al. 1985).

These changes in the spectral composition of light which occur throughout the canopy profile also modify shoot extension in grasses (Casal et al. 1987). A low R:FR of light reaching the base of grasses resulted in longer shoots and leaves.

An increase in grazing intensity increases the R:FR ratio of the light reaching the base of plants as a consequence of the direct effects of removal and trampling. This change in light quality results in shorter leaves and more numerous tillers which will further increase the R:FR ratio. There is a shift in the allocation of resources. As a result of grazing the investment in new tillers increases and the investment in elongation of the tillers already formed decreases.

The effect of herbivory upon species composition is the result of changes in mortality rate and in the establishment of new individuals. There are morphological and physiological traits which confer capacity to tolerate or evade grazing (Coughenour 1985). Differential mortality under grazing conditions is the result of presence or absence of these traits.

Grazing affects establishment of new individuals by controlling seed production and availability of safe-sites (Harper 1971). Changes in light intensity and quality due to grazing may regulate germination rate from seed banks (Silvertown 1980; Fenner 1980; Verkaar 1986). Results from the flooding pampa indicated that grazing exclusion decreased seedling establishment for a rosette plant *Leontodon taraxacoides* and for one of the dominant grasses *Danthonia montevidensis* (Oesterheld and Sala unpublished). These experiments indicated the relative importance of availability of seeds and safe-sites under different grazing conditions. For example, the seed bank of the rosette species was depleted after six years of exclosure as a consequence of the negative balance of seed production, germination and disappearance.

The effect of herbivory at different scales

Time scales

Up to this point, I have discussed changes in vegetation structure which occur in months or years. However, the effects of herbivory occur in a broader range of time scales, from days to centuries. McNaughton (1983a) described responses to herbivores in ontogenetical, ecological and evolutionary time. In the first class are those responses which occur within an individual and during its life cycle. The most important responses at this time scale include compensatory growth and defensive chemical mobilization. The ecological time includes changes in population demographics and encompasses successional studies related to grazing. Changes in evolutionary time occur at a coarser scale. The effect of herbivory is reflected in traits which evolve throughout long periods of time and are engraved in genetic codes.

These responses to herbivory at different time scales are not independent of each other but form a hierarchy. Responses are ordered in this hierarchy from fast to slow. After establishing a hierarchy, deductions can be drawn from hierarchy theory and applied to this case (Allen and Starr 1982; O'Neill *et al.* 1986). Slow responses at the higher level in the hierarchy constrain fast responses which occur at lower levels in the hierarchy. Responses in evolutionary time constrain responses in ecological and ontogenetical time.

Evolution set the limits of what can occur in plant demographics. One of the aspects of succession is a sequential replacement of species. These species possess traits which have already evolved. Similarly, responses in ontogenetical time depend upon species composition which is the result of succession.

The occurrence of dissimilar responses to herbivory in different grasslands led Milchunas *et al.* (1988) to develop a conceptual model of the effect of herbivores on grassland structure. The model is focused on the effects of herbivory in ecological time. It uses two explanatory variables: evolutionary grazing history and environmental moisture.

Sites with different evolutionary grazing history will have different responses to herbivory in ecological and ontogenetical time. As an example, I will compare the response in ecological time of two grasslands with presumably different evolutionary grazing history, the flooding pampa and the Serengeti grasslands. Grasslands of the flooding pampa evolved under light grazing conditions (Webb 1978) and so remained until the arrival of Europeans in the 16th century who introduced domestic herbivores and the frequent use of fire. In contrast, grasslands of the Serengeti have a long evolutionary grazing history. In both grasslands, grazing exclusion resulted in large changes in canopy structure. In the Serengeti, these changes were accounted for primarily by changes in the growth form of native species since exotic species were not observed neither in grazed nor in ungrazed areas (McNaughton 1979, 1983b). In the grasslands of the flooding pampa, structural changes were accounted for by changes in the growth form of native species as well as by the presence or absence of exotic species. It suggests that native grasslands of the flooding pampa did not have the ecotypes or species adapted to the heavy grazing conditions imposed by Europeans.

Adaptations that enhance survival in arid environments promote tolerance or avoidance of grazing. Coughenour (1985) examined how basal meristems, small stature, high shoot density, deciduous shoots, and belowground nutrient reserves allow grasses to evade or tolerate both arid conditions and grazing. Water stress and grazing stress are similar in that both periodically result in partial or total loss of organs. Tolerance to one necessarily includes tolerance to the other. With respect to avoidance, the same characteristic can conserve moisture as well as evade grazing. For example, basal meristems protected by basal sheaths may better withstand drought as well as avoid grazing (Stebbins 1972; Barlow *et al.* 1980).

Milchunas *et al.* (1988) based their model on the idea that adaptations to grazing are divergent from adaptations to humid environments. In contrast, selection pressures in arid grasslands that have coevolved with large grazers are convergent. A relaxation of the grazing pressure in a humid or subhumid grassland will result in a complete shift in the selection pressure. In contrast, excluding grazing in an arid or semiarid grassland will not change the direction of selection pressure. Large changes in the structure of grasslands as a result of grazing are expected in humid and subhumid regions and relatively small changes in arid to semiarid regions.

This model focuses on the effect of grazing on diversity. Arid and semiarid grasslands with a long grazing history are dominated by relatively short grasses that have developed in response to the convergent selection pressures of herbivory and aridity. The semiarid shortgrass steppe of the North American Great Plains is an example of this type of grassland. Grazing has relatively small effects on species composition and diversity, decreasing slightly with increased grazing intensity, as rare and less-grazing tolerant species are eliminated.

Humid and subhumid grasslands with a long grazing history are composed of short, mid- and tallgrasses as a result of divergent selection for grazing tolerance and canopy dominance. In the absence of grazing, a few tall species dominate the community. Moderate grazing creates a mosaic pattern, with shortgrasses predominating in heavily grazed patches, mixture of grasses in moderately grazed patches, and tallgrasses where ungrazed growth deter grazing. Grasslands of the Serengeti are one example of grasslands of this type.

Arid and semiarid grasslands with a short grazing history are dominated by short- and midgrasses that have been selected for their drought tolerance which

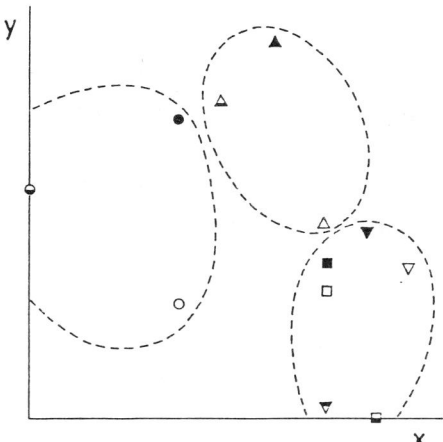

Fig. 9. Ordination of vegetation samples collected in fall (open symbols), in spring (half shaded), and summer (dark symbols) for ◻ = upland community grazed; ○ = upland community ungrazed; ▽ = lowland community grazed, and △ = lowland community ungrazed.

also confer some degree of grazing resistance. Structural changes due to grazing are moderate, less than what occurs in the humid-short evolutionary grazing history type, but greater than what occurs in arid grasslands with a long history of grazing. Diversity exhibits a small increase at low grazing intensities and slow decrease throughout most of the range of grazing conditions. Examples of grasslands of this type are the semiarid steppe of northwestern United States and the Pantagonian steppe in southern Argentina (Mack and Thompson 1982; Markgraf 1985).

Humid or subhumid grasslands with a short evolutionary grazing history have the greatest potential for being altered by grazing. Grasslands of the flooding pampa are an example of this type of grasslands (Sala *et al.* 1986). Species were selected primarily for canopy dominance. Communities are made up of tall-grasses that are not tolerant of grazing. The increase of exotic species produce large increases in diversity at relatively low grazing intensities. Diversity declines rapidly thereafter as grazing intensity increases.

Spatial scales

Throughout this paper the discussion and observations of the effect of herbivory upon vegetation structure have been concentrated at the scale of plant or stand. In this section I will explore the effect of herbivory upon vegetation at a coarser scale. I will analyze the effect of herbivory at the scale of landscape.

The grasslands of the flooding pampa are made up of a mosaic of mainly two plant communities which are located in different topographical positions. Although topographical differences are only 15–20 cm, they are enough to determine differences in the duration of floods, in soil properties and in vegetation.

Under conditions of light grazing the two communities were floristically very different (Fig. 9). For example, the upland community was dominated by cool-season species whereas the lowland community was dominated by warm-season species. Grazing erased the floristic differences between these two communities or phases of the mosaic (Fig. 9). The differences due to environmental factors such as topography and frequency or duration of floods were overshadowed by the effect of grazing.

The Serengeti offers a contrasting example; spatial pattern apparently is caused and maintained by grazing animals (McNaughton 1983b; Belsky 1986). This is also the case of a two-phase grassland mosaic. The *Andropogon* phase is strongly dominated by the grass *Andropogon greenwayi* Napper, which forms large, dense mats while the *Chloris pycnothrix* Trin. phase is sparsely vegetated by a species-rich mixture. Overgrazing and disturbances by large herbivores produce the mosaic by breaking up continuous *Andropogon greenwayi* mats (McNaughton 1983b). Sown grasslands in The Netherlands provide another example of grazing dependent pattern. In this case, sheep grazing in a very small pasture created small scale spatial pattern (Bakker *et al.* 1984).

McNaughton (1983b) suggested that landscape pattern may facilitate foraging efficiency and promote coexistence of diverse grazing animals. If a preferred food source is concentrated in a stand instead of being randomly distributed in several stands, animals can walk from stand to stand to feed. In this manner animals save time and foraging efficiency is facilitated. Different grazing animals with different preferred foods can coexist in the same grassland with minimal competition, by exploiting different vegetation patches (McNaughton 1983b).

These apparently contrasting examples, from the flooding pampa and the Serengeti, suggest that there is a hierarchy of factors which create spatial heterogeneity. One of these factors is grazing. The hierarchy encompasses fine to coarse grained heterogeneities. The pattern of grazing can be finer or coarser than the pattern generated by environmental factors. If the pattern of grazing is coarser, grazing erases the environmental pattern. In contrast, grazing generates pattern if it is finer than the environmental pattern.

Large herbivores generate spatial patterns at different scales as a result of different activities (Senft *et al.* 1987). Fecal pats and the action of hooves create microscale pattern (Coffin and Lauenroth 1988). Preferential grazing in selected stands and seasonal movements of animals create large-scale pattern. Herbivory may be a factor generating pattern at one scale and erasing pattern at another scale.

Evidence that animals generate or erase pattern in grasslands with long or short evolutionary grazing history respectively, suggests that foraging facilitation resulting from the ability of animals to create and maintain pattern may have been a trait selected in evolutionary time. This may be a case similar to the development and maintenance of grazing lawns (McNaughton 1984).

Acknowledgments

This paper is based upon work supported by Consejo Nacional de Investigaciones Científicas y Técnicas of Argentina (PID 391 2102/85). I thank Martín Oesterheld for his invaluable assistance and Alberto Soriano, Rolando León and Daniel Milchunas for critical remarks on this paper.

References

Allen, T.F.H. and Starr, T.B. 1982. Hierarchy. Perspectives for Ecological Complexity. The University of Chicago Press, Chicago and London. 310 pp.
Bakker, J.P., de Leeuw, J. and van Wieren, S.E. 1984. Micro-patterns in grassland vegetation created and sustained by sheep-grazing. Vegetatio 55: 153–161.
Barlow, E.W.R., Munn, R.E. and Brady, C.J. 1980. Drought responses of apical meristems. Chapter 13. In: N.C. Turner and P.J. Kramer (eds), Adaptations of Plants to Water and High Temperature Stress. Wiley and Sons, New York.
Belsky, A.J. 1986. Population and community processes in a mosaic grassland in the Serengeti, Tanzania. J. Ecol. 74: 841–856.
Cabrera, A.L. 1968. Flora de la Provincià de Buenos Aires. INTA, Buenos Aires.
Casal, J.J., Sanchez, R.A. and Deregibus, V.A. 1987. The effect of light quality on shoot extension growth in three species of grasses. Ann. Bot. 59: 1–7.
Coffin, D.P. and Lauenroth, W.K. 1988. The role of gap phase dynamics in a shortgrass plant community. Ecology (in press).
Coughenour, M.B. 1985. Graminoid responses to grazing by large herbivores: Adaptations, exaptations, and interacting processes. Ann. Miss. Bot. Garden 72: 852–863.
Coleman, D.C., Andrews, R., Ellis, J.E. and Singh, J.S. 1976. Energy flow and partitioning in selected man-managed and natural ecosystems. Agroecosystems 3: 45–54.
Deregibus, V.A., Sanchez, R.A. and Casal, J.J. 1983. Effects of light quality on tiller production in *Lolium* spp. Plant Physiol. 72: 900–902.
Deregibus, V.A., Sanchez, R.A., Casal, J.J. and Trlica, M.J. 1985. Tillering responses to enrichment of red light beneath the canopy in a humid natural grassland. J. Appl. Ecol. 72: 199–206.
During, H.J. and Willems, J.H. 1984. Diversity models applied to a chalk grassland. Vegetatio 57: 103–114.
Fenner, M. 1980. The inhibition of germination of *Bidens pilosa* seeds by leaf canopy shade in some natural vegetation types. New Phytol. 84: 95–102.
Harper, J.L. 1971. Grazing, fertilizers and pesticides in the management of grasslands. In: E. Duffey and A.S. Watt (eds), The Scientific Management of Animal and Plant Communities for Conservation, pp. 15–33. Blackwell Scientific Publications, Oxford, London, Edinburgh.
Holmes, M.G. and Smith, H. 1977. The function of phytochrome in natural environment I. Characterization of daylight for studies in photomorphogenesis and photoperiodism. Photochem. and Photobiol. 25: 539–545.
Ingham, R.E. and Detling, J.K. 1984. Plant-herbivore interactions in a North American mixed-grass prairie. III. Soil nematode populations and root biomass on *Cynomis indovicianus* colonies and adjacent uncolonized areas. Oecologia 63: 307–313.
León R.J.C. 1975. Las comunidades herbaceas de la región Castelli-Pila. Monografias Comisiòn de Investigaciones Cientificas de la Pcia. de Bs. As. La Plata 5: 75–107.
Leòn, R.J.C., Rusch, G.M. and Oesterheld, M. 1984. Pastizales pampeanos: Impacto agropecuario. Phytocoenologia 12: 201–218.
Mack, R.N. and Thompson, J.N. 1982. Evolution in steppe with few large, hooved mammals. Am. Nat. 119: 757–773.
Markgraf, V. 1985. Late pleistocene faunal extinctions in southern Patagonia. Science 228: 1110–1112.
McNaughton, S.J. 1976. Serengeti migratory wildebeest: Facilitation of energy flow by grazing. Science 191: 92–94.
McNaughton, S.J. 1979. Grassland-herbivore dynamics. In: A.R.E. Sinclair and M. Norton Griffiths (eds), Serengeti: Dynamics of an Ecosystem. University of Chicago Press, Chicago, pp. 46–81.
McNaughton, S.J. 1983a. Physiological and ecological implications of herbivory. In: O.L. Lange, P.S. Nobel, C.B. Osmond and H. Ziegler (eds), Encyclopaedia of Plant Physiology, Vol. 12C, Physiological Plant Ecology III. Responses to the Chemical and Biological Environment, pp. 657–677. Springer, New York.
McNaughton, S.J. 1983b. Serengeti grassland ecology: The role of composite environmental factors and contingency in community organization. Ecol. Monogr. 53: 291–320.
McNaughton, S.J. 1984. Grazing lawns: Animals in herds, plant form, and coevolution. Am. Nat. 124: 863–885.
McNaughton, S.J. 1985. Ecology of a grazing system: The Serengeti. Ecol. Monogr. 55: 259–294.

McNaughton, S.J., Wallace, L.L. and Coughenour, M.B. 1983. Plant adaptation in an ecosystem context: effects of defoliation, nitrogen and water on growth of an African C4 sedge. Ecology 64: 307–318.

Milchunas, D.G., Sala, O.E. and Lauenroth, W.K. 1988. A generalized model of the effects of grazing by large herbivores on grassland community structure. Am. Nat. 132: 87–106.

O'Neill, R.V., De Angelis, D.L., Waide, J.B. and Allen, T.F.H. 1986. A hierarchical concept of ecosystems. Princeton University Press, New Yersey. 253 pp.

Picket, S.T.A. 1980. Non-equilibrium coexistence of plants. Torr. Bot. Club 107: 238–248.

Sala, O.E., Oesterheld, M., León, R.J.C. and Soriano, A. 1986. Grazing effects upon plant community structure in subhumid grasslands of Argentina. Vegetatio 67: 27–32.

Scott, J.A., French, N.R. and Leetham, J.W. 1979. Patterns of consumption in grasslands. In: N.R. French (ed.), Perspectives in Grassland Ecology, pp. 89–105. Springer Verlag, New York.

Senft, R.L., Coughenour, M.B., Bailey, D.W., Rittenhouse, L.R., Sala, O.E. and Swift, D.M. 1987. Large herbivore foraging and ecological hierarchies. BioScience 37: 789–799.

Silvertown, J. 1980. Leaf canopy-induced seed dormancy in a grassland flora. New Phytol. 85: 109–118.

Stebbins, G.L. 1972. The evolution of the grass family. In: V.B. Youngner and C.M. McKell (eds), The Biology and Utilization of Grasses, pp. 5–23. Academic Press, New York.

Sterner, R.W. 1986. Herbivores direct and indirect effects on algal populations. Science 231: 605–607.

Stoddart, L.A., Smith, A.D. and Box, T.W. 1975. Range Management. McGraw Hill Book Company, New York, 532 pp.

Verkaar, H.J. 1986. When does grazing benefit plants? TEE 1: 168–169.

Webb, S.D. 1978. A history of savanna vertebrates in the new world. II. South America and the great interchange. Ann. Rev. Ecol. Syst. 9: 393–426.

Willems, J.H. 1983. Species composition and above ground phytomass in chalk grassland with different management. Vegetatio 52: 171–180.

GRAZING, CANOPY STRUCTURE AND PRODUCTION OF FLOODPLAIN GRASSLANDS AT KAFUE FLATS, ZAMBIA

G.A. ELLENBROEK and M.J.A. WERGER
Department of Plant Ecology, University of Utrecht, Lange Nieuwstraat 106, 3512 PN Utrecht, The Netherlands

Abstract

At Kafue Flats, Zambia, the grasslands are annually flooded to depths varying from a few cm to several meters. The deepest flooded parts are also flooded longest. Kafue lechwe, a semi-aquatic antelope, is the main herbivore grazing shallowly inundated and emerged grasslands. There are four major types of seasonally flooded grasslands which also differ in grazing regime: A. *Vossia cuspidata – Echinochloa scabra* grassland on the deepest inundated parts; B. *Paspalidium obtusifolium* water meadows on rather deep flooded parts; C. *Panicum repens – Leersia denudata* water meadows along shallowly inundated floodplain edges; D. *Acroceras macrum* water meadows in flooded depressions above the high flood line.

In exclosures the above-ground primary production of the grassland communities varies between 4250 and 450 g.m^{-2} and correlates highly with leaf area index. Even when excluded from grazing the canopy structures of the communities are closely related to the prevailing grazing regimes of the Kafue lechwe under which they evolved, demonstrating the strong selective effect of grazing on plant form. As communities are less subject to grazing their canopies are lusher and productivity is higher. The temporarily grazed communities (A and B) develop emerged canopies. These communities contain mainly C_4-plants. The permanently grazed communities (C and D) develop submerged canopies which restricts productivity. In Community C about half of the species are C_3-plants and half C_4-plants. In Community D all main species are C_3-plants. It is suggested that this may partly explain the lower production values for Community C as compared to Community D.

Introduction

The Kafue Flats is a wide, seasonally inundated floodplain which borders some 400 km of the lower reaches of the Kafue River in Zambia. The area is covered mainly by various types of grasslands with different canopy structures and species composition. The total area of grassland measures some 7000 km² of which 5000 km² is annually flooded (Ellenbroek 1987). Rainfall averages 880 mm per year and is highly seasonal. Maximum flood levels occur at the end of the rainy and the beginning of the dry seasons (Fig. 1).

The Kafue Flats show a clear zonation of vegetation types. In the flooded area the zonation is strongly determined by the length of the inundation and the depth of flooding which can be up to several meters (Douthwaite and Van Lavieren 1977; Rees 1978; Werger and Ellenbroek 1980; Ellenbroek 1987). The seasonally changing pattern of flooding attracts a great number of birds and as for birdlife the two national parks on the Kafue Flats are listed among the 10 best-stocked sanctuaries in the world.

The seasonal pattern of flooding also determines the pattern of grazing of the Kafue lechwe, *Kobus leche kafuensis* Haltenorth, a semi-aquatic antelope endemic to the Kafue Flats. They occur in huge numbers in the two national parks in the

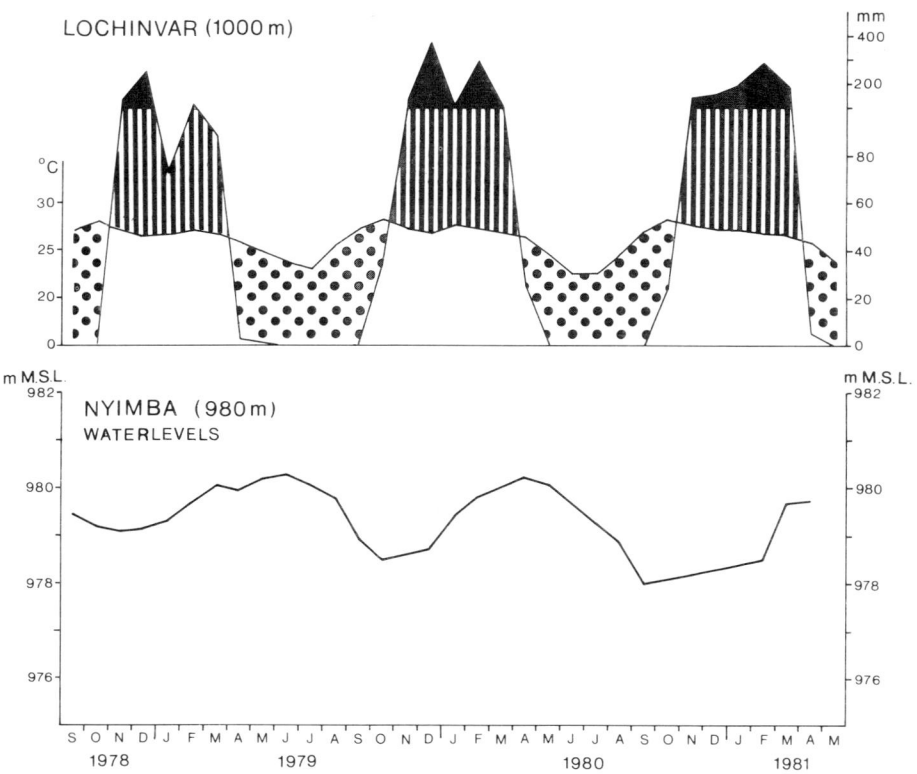

Fig. 1. Climatic diagram of Lochinvar National Park for the period September 1978 to March 1981 and water levels of the Kafue River at Nyimba, demonstrating the time lag between peak of the rainy season and maximum water levels.

area where they graze the floodplain grasslands. During the period of inundation they graze the emergent vegetation in areas flooded up to 50 cm deep (Ellenbroek 1987). Maintenance of these large herds is only possible thanks to a high annual production of the grasslands. The annual primary production of the various grassland communities is partly determined by their canopy structures. The grazing activities of the lechwe have a strong influence on the structure of the grassland canopies. Both total leaf area and spatial leaf arrangement are affected. This in turn affects the light climate and other variables inside the canopy that influence photosynthetic production. There is substantial evidence that large herbivores also affect other plant features, such as plant establishment and reproductive success (see *e.g.*, McNaughton 1984, 1985). The exploitation of the wet grasslands at Kafue by the lechwe has occurred over a very long time span. It is possible, therefore, that in evolutionary time, this grazing pressure has selected plant genotypes which allow developmental responses that are adaptive under the local grazing regime. Another possibility is that grazing pressure in ecological time has given sufficient advantage to certain species that already possessed a morphology that befitted the local grazing regime as a result of which these species acquired dominance in the community. While the first possibility implies a long and stable grazing regime under which the components of the system co-evolved, the second possibility does not necessarily require such co-evolution. If either of these possibilities applies, protection of the stands from grazing can be expected to have relatively little influence on the canopy structure of the stands on the short-term. However, if these possibilities are not relevant, protection from grazing may be expected to lead to rapid and drastic changes in the canopy structure of the stands as compared to the grazed stands.

In this study we investigate the effect of removal of grazing on the canopy structure and annual primary production of four floodplain grassland communities at Lochinvar National Park, an area in which the natural zonation of the floodplain communities has been particularly well-preserved. We measured the canopy structure and primary production inside and outside exclosures established before the growing season started. With this study we want to establish whether grazing has resulted in canopy structures of the stands determined by inherently fixed plant morphologies of the species, or whether the canopy structures are the immediate products of repeated defoliation or of protection during the growing season. In case the canopy structures result from inherently fixed plant morphologies our study does not allow to discrimate between either of the two possibilities mentioned above, *i.e.*, co-evolution in evolutionary time, or selection for dominance in ecological time.

Grassland communities and sampling methods

The following communities were studied:

Community A: floodplain grassland with *Vossia cuspidata* Griff. and *Echinochloa scabra* (Lam.) Roem. and Schult. This community occupies the extensive, long and deeply flooded areas of the floodplain (see also Thompson 1985). Parts of this community are flooded by several meters of water but our plots were flooded up to 190 cm deep. They are unavailable to the grazing lechwe for a long period of the year.

Community B: water meadow with *Paspalidium obtusifolium* (Delile) N.D. Simpson which occupies the depressions along incoming streams on the edge of the floodplain. These water meadows are flooded up to 100 cm deep but the period of deep flooding is shorter than in Community A and the community is available to grazing lechwe for a large part of the year.

Community C: water meadow with *Panicum repens* L. and *Leersia denudata* Launert covering the shallowly inundated edges of the floodplain. Flooding is up to 40 cm deep and occurs for shorter periods than in Communities A and B. Flooding occurs during the main growth season. The community is grazed by lechwe and other large herbivores throughout the year.

Community D: water meadow with *Acroceras macrum* Stapf occupying shallow depressions outside the direct floodplain. During the growth season it is flooded up to 40 cm deep, mainly by stagnant rain water and it is permanently grazed by lechwe and other large herbivores.

All four communities grow on heavy, fertile montmorillonitic clays which do not differ much in their soil properties between these communities (Ellenbroek 1987).

To determine canopy structure and primary production the grasslands were sampled, whenever, possible, at monthly intervals during the growing season and once or twice during the dry period. The communities were sampled in and outside exclosures to study the effects of grazing. Sampling consisted of harvesting the above-ground vegetation in quadrates of 1 m² by stratified clipping in layers of 20 cm high. Samples were taken in triplicate. Leaf area (two sides) and dry weights of leaves and stems were determined. Light climates above and inside the canopies, above and below the water surface were measured during peak standing crop by LI-185 quantum sensors sensitive to total photosynthetically active radiation (PAR) only. Grazing intensity of the lechwe was observed throughout the year. Further details on methods are described in Ellenbroek (1987). In this paper only the canopy structure of the communities during peak standing crop in the exclosures are depicted. More data on growing stages and canopy structures under grazing are given in Ellenbroek (1987).

Results and discussion

All communities are relatively poor in species (Table 1) and only one, two or seldom three dominant species determine the canopy structure to a very large extent. The stratification of leaf area and the penetration of PAR in the canopies of the four communities together with a schematic side view of the canopies, in exclosures, is shown in Fig. 2. Table 1 gives some characteristics of the structure of the four communities and their maximal flooding depths.

The deeply flooded Community A is largely protected from grazing. Its species produce a lush canopy, highly emerged above the water surface, and with a large LAI. About 90% of PAR is intercepted above the water surface and few leaves persist submerged. The tall, lush canopy produces a high net annual above-ground production (NAP), even though the relative investment in leaves (LAR, leaf area ratio, based on the dry weights of above-ground plant parts only) is not very high. A rather large part of the above-ground phytomass is made up of long stems which, supported by the water, keep the foliage up in the air.

Community B is grazed for a much longer part of the growing season. Species developing most of their leaves at or slightly above the water surface predominate. Though about 40% of full PAR penetrates into the canopy down to the water surface, this is largely intercepted by leaves at the water surface or in the uppermost cm of water; deeper down few leaves persist. LAI is consequently moderate to low and so is NAP. This corresponds to a LAR-value which is less than half of that in Community A.

Table 1. Characteristics of the floodplain grassland communities.

Community	A	B	C	D
Maximum water depth when flooded (cm)	190	100	40	40
Net annual above-ground production (g.m^{-2})	4250	1650	450	1050
Leaf Area Index (LAI) (m^2.m^{-2}) (two sided)	8.9	5.0 ± 0.9	0.7 ± 0.1	3.7 ± 1.0
Above-ground Leaf Area Ratio (LAR) (dm^2.g^{-1})	0.9	0.4 ± 0.07	0.5 ± 0.03	1.2 ± 0.10
Number of species (per m^2)	5	7	3	12

Grazing, canopy structure and production of floodplain grasslands

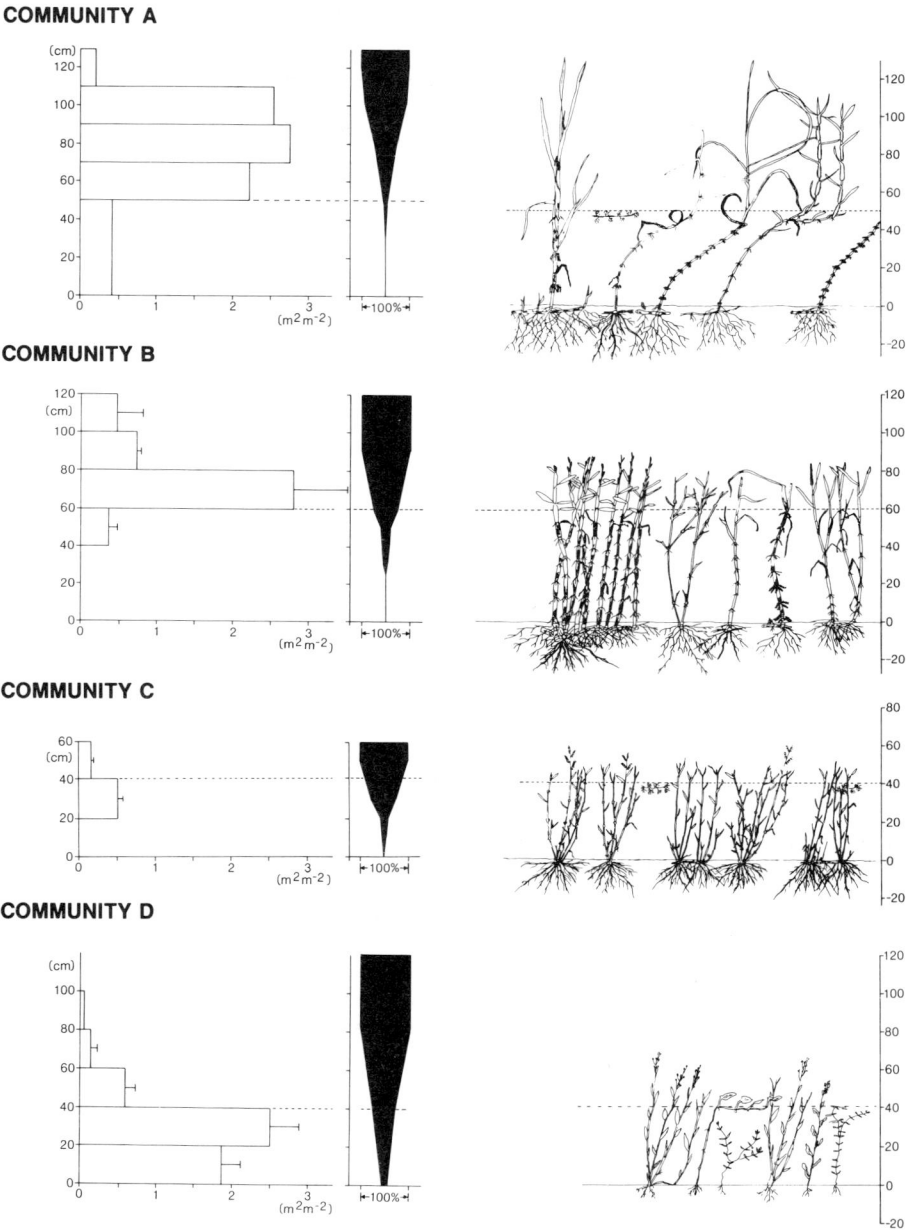

Fig. 2. Stratification of leaf area in exclosures in four floodplain communities, penetration of PAR in the leaf canopies, and schematic vegetation profiles. The dashed lines indicate the water height at the time the illustrated stand was sampled. Community A: floodplain grassland with *Vossia cuspidata* and *Echinochloa scabra*; Community B: water meadow with *Paspalidium obtusifolium*; Community C: water meadow with *Panicum repens* and *Leersia denudata*; Community D: water meadow with *Acroceras macrum*.

Normally Community C is permanently grazed. The lechwe and other animals remove most plant material above and at the water level. Also in the exclosure the community consists of plants with predominantly submerged leaves during the main growing season which is the period of flooding. About 80% of PAR penetrates into the water, but investment of dry matter in leaf area (LAR) is low and so is LAI. Under-water leaves are protected from grazing, but photosynthesis is hampered and, as total LAI is low, NAP is low.

Also Community D is permanently grazed. Flooding occurs for most of the growing season and most of the emergent canopy is removed by grazing. In the exclosure also this community consists largely of species with mainly submerged leaves during the period of flooding at peak standing crop, though there are also some emerged leaves. Just over 50% of PAR reaches the water level. Investment of dry matter in leaf area (LAR) is high. Total LAI and NAP are rather low, though much higher than in Community C.

The canopy profiles of the stands in the exclosures in Communities A and C are very similar to those of unprotected stands. Total leaf area of the stands in the exclosures in Communities B and D is larger than in unprotected stands but the stratification of leaf area is highly similar (*cf.* Ellenbroek 1987). It is thus clear that the intensive grazing of lechwe and some other large herbivores has had a strong selective effect on plant form and plant strategy in Community B and perhaps even stronger in Communities C and D. This resulted in canopy structures of the communities that seem adapted to grazing: in the case of Community B plants have their leaves very near, at, or even below the water surface to limit grazing efficiency, while in the cases of Communities C and D plants have their leaves largely or nearly entirely submerged to prevent defoliation by grazing. These canopy structures resulted either from co-evolution of plant form and grazers in evolutionary time, or from selection for dominance of some plant species adapted to the intensive grazing regime in ecological time. Our simple experiment cannot discriminate between these two processes. It is tempting, however, to speculate that the canopy profiles result from co-evolution, since the heavy grazing of the swards of these floodplain grasslands by the lechwe most likely has occurred since a very long period of time. This is suggested by geomorphological features of the area (Wedman 1980) and by the fact that by far the most important grazer at Kafue, the Kafue lechwe, is endemic to the area. Milchunas *et al.* (1988) pointed out that in wet grasslands adaptations to grazing diverge strongly from adaptations to successfully exploit the environmental resources, while in dry grasslands adaptations to either seem to require similar morphological and physiological features. If this reasoning is valid, it is to be expected that prolonged protection from grazing will lead to changes in canopy structure of the floodplain grasslands that have been under severe grazing regimes as a result of replacement of the present species by species with a taller and lusher growth form.

High production can result from a large leaf area as well as from a high net assimilation rate. In the four grassland communities studied by us LAI is a good predictor of NAP (NAP = 473.6 LAI $- 316.9$; $r = 0.9622$). It is possible that the higher NAP of Community D as compared to Community C may not only result from a more favorable canopy structure with a higher LAI and LAR. Part of the higher NAP may result from higher net assimilation rates by the species of Community D: in Community C half of the main species are C_4-plants (as are all grass species in Communities A and B), the other half C_3-plants, with the C_4-species becoming somewhat more prominent towards the very end of the growing season when the floods recede. In Community D all main species are C_3-plants. C_4-species have higher photosynthetic yields at high temperatures and C_3-species have higher yields at relatively low temperatures. Communities C and D are flooded during most of the growing season and their canopies are submerged. Therefore temperature conditions of the leaves are relatively low, and the strong predominance of C_3-plants in Community D may contribute to a higher photosynthetic yield and production in that community.

Conclusions

1. The floodplain grasslands of the Kafue Flats are subject to different grazing intensities as a result of depth of flooding during the growing season.
2. In response to the different prevailing grazing regimes under normal conditions the communities have developed differently structured canopies.
3. Intensively and prolongedly grazed communities develop largely submerged canopies.
4. The shorter the period a community is grazed, the more its canopy is developed above the water surface, and the higher its net annual above-ground production.
5. Among the floodplain communities studied leaf area index varies by a factor of 13, net annual above-ground production by a factor of 10, and leaf area ratio by a factor of 3. Net above-ground annual production correlates highly with leaf area index.
6. The amount of net annual above-ground production may not only be dependent on the canopy structure of the community, in particular the leaf area index or leaf area ratio. In the submerged communities it may also be partly determined by the photosynthetic pathways of the main species in the communities.

References

Douthwaite, R.J. and Van Lavieren, L.P. 1977. A description of the vegetation of Lochinvar National Park, Zambia. Techn. Report 34. Nat. Council for Sci. Res. Lusaka, Zambia.
Ellenbroek, G.A. 1987. Ecology and productivity of an African wetland system. Dr W. Junk Publishers, Dordrecht, Boston.
McNaughton, S.J. 1984. Grazing lawns: animals in herds, plant form, and co-evolution. Amer. Nat. 124: 863–886.
McNaughton, S.J. 1985. Ecology of a grazing ecosystem: the Serengeti. Ecol. Monogr. 55: 259–294.
Milchunas, D.G., Sala, O.E. and Lauenroth, W.K. 1988. A generalized model of the effects of grazing by large herbivores on grassland community structure. Amer. Nat. 132: 87–106.
Rees, W.A. 1978. The ecology of the Kafue lechwe. J. Appl. Ecol. 15: 163–217.
Thompson, K. 1985. Emergent plants of permanent and seasonal wetlands. In: P. Denny (ed.), The Ecology and Management of African Wetland Vegetation, pp. 43–107. Dr W. Junk Publishers, Dordrecht, Boston.
Wedman, E. 1980. Hydrogeology. DHV: Kafue Flats Hydrological Studies Final Report, pp. 133–148. DHV, Amersfoort.
Werger, M.J.A. and Ellenbroek, G.A. 1980. Water resource management and floodplain ecology: An example from Zambia. In: J.I. Furtado (ed.), Tropical Ecology and Development, pp. 693–702. Int. Soc. Trop. Ecol., Kuala Lumpur, Malaysia.

LARGE AFRICAN MAMMALS AS REGULATORS OF VEGETATION STRUCTURE

S.J. McNAUGHTON and G.A. SABUNI
Biological Research Laboratories, Syracuse University, Syracuse NY 13244-1220, USA

Abstract

Large mammals in savannas and grasslands have direct, indirect, and interactive effects upon vegetation structure so profound that it is difficult to regard grazing-browsing ecosystems as donor-controlled, *i.e.*, as systems solely determined by climatic and geological factors. Rather, the importance of animals suggests that component control, by trophic participants themselves, is an important determinant of ecosystem properties. The effects of animals on vegetation structure in Africa range from the interactive effects of browsers and fires controlling the balance of woody and herbaceous vegetation to the effect of animals upon the evolution of grass growth forms. But neither can the animals be regarded as sole ecosystem controllers. Instead, African ecosystems are clear examples of (a) how the physical setting controls the bounds within which ecosystem structure develops, and (b) how biotic evolution within those constraints leads to structurally-defined interactions within the trophic web.

Introduction

Vegetation form and function is constrained by and has evolved under the great global patterns of climate and geology (Walter 1979). But in a paper entitled 'Elephants as agents of habitat and landscape change in East Africa,' Laws (1970) argued that 'After man himself, probably no other animal has had as great an effect on African habitats as the African bush elephant, *Loxodonta a. africana* Blumenbach.' It is now generally recognized that the large mammal populations of Africa have profound effects upon both the structural and the functional properties of the vegetation that supports them (McNaughton 1976, 1984, Cumming 1982, Menaut *et al.* 1985). Savannas, indeed, rangelands generally, are characterized by an evidently fragile balance between woody and herbaceous plants that has been subject to severe disruption, characterized by bush encroachment, throughout Earth from the latter part of the 19th century to the present (Walter 1979). This encroachment seems to be particularly severe when periodic drought leads to animal population densities well beyond carrying capacities in both agricultural rangelands (Herbel 1986) and game reserves (Scholes 1985).

Perspectives

By structure we commonly mean the spatial arrangement of parts in some complex entity. An early, definitive approach to analyzing vegetation structure was Beard's (1946) analytical diagrams of foliage and bole organization in the vegetation of Trinidad, which was widely applied to rainforest (Richards 1957), and led to an architectural classification of wet tropical forest trees based on their branching and reproductive patterns (Hallé and Oldeman 1970). Another well developed approach to structure is applied to the vertical zonation of marine and freshwater biomes.

In ecosystem terms, structure is a state variable, a variable fixed at time of observation, but subject to modification through time by the processes taking place in an ecosystem. Vertical zonation of aquatic ecosystems, for example, changes due to diurnal migration of organisms, and the canopy structure of forests changes in ecological time with canopy growth and in evolutionary time with the evolution of growth and branching patterns in response to both the physical and the biotic environments. Vegetation supporting substantial grazer populations may be subject to more varying structural modification in ecological time than any other vegetation since those grazers modify canopy geometry in major ways.

In selecting the participants for this section of the Utrecht Symposium on Vegetational Structure, a difficult choice had to be confronted, the traditional choice between the extensive or the intensive. That is, should participants consider a broad range of vegetation types in which the types and effects of herbivores varied dramatically, or concentrate on a single vegetation type where the effects of herbivory are both significant and reasonably well characterized? The latter alternative was selected, assembling scientists who study the effects of grazers upon grasslands. Recognizing that state is both a consequence of, and a factor influencing, processes, attention is given to both in this symposium section. Since our own research is centered in African ecosystems where large, mammalian herbivores dominate ecological processes, that is what we emphasize in this paper.

The physical setting

African savanna grasslands occur over a broad range of climates, geological formations, and hydrologic settings (Werger 1977, 1983; Werger et al. 1978, 1979). So, also, vegetation physiognomy ranges from open grasslands of largely C_4 grasses (Werger and Ellis 1981) to the closed miombo woodland. There is a major continuum from dystrophic savannas on sandy, heavily leached, infertile soils, often derived from the ancient African shield, to eutrophic savannas on clayey, unweathered, fertile, volcanic soils of the Rift Valleys (MacVicar 1977; Huntley 1982; Bell 1982). The herbivores' food environments in dystrophic savannas are characterized by high standing crops of high fiber, less digestible plant tissues and the herbivores are either large-bodied, unselective grazer-browsers, or small bodied, highly selective browsers (Jarman 1974; Bell 1982; Demment and van Soest 1985). The eutrophic systems often support open savanna to grassland with low standing crops of nutritious, easily digestible, low fiber plant tissues and high densities of animals of small to intermediate body sizes. These are the typical game parks of eastern and southern Africa. It is this continuum from dystrophic to eutrophic ecosystems that characterizes the physical setting within which African plants and herbivores evolved.

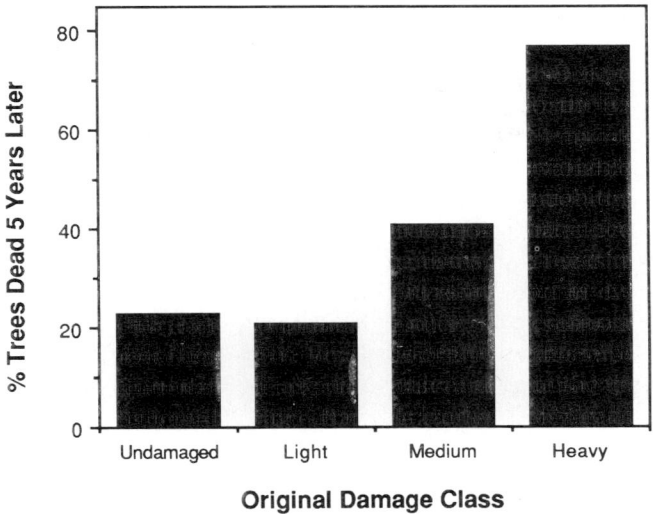

Fig. 1. Relationship between tree damage class and survival after five years (redrawn after Eltringham 1980).

The balance between trees and herbs

The balance between trees and underlying grass canopies in savannas is the most distinctive structural feature of arid to subhumid tropical and subtropical vegetation throughout Earth (Walter 1979; Huntley and Walker 1982; Tothill and Mott 1985). Savannas have an especially complex architecture involving two canopy levels and the balances between and within those levels (Sarmiento 1984). The mixture of life forms at the producer level has important consequences for other trophic levels. There can be little doubt that climate and geomorphology interact to provide the physical setting within which such structure develops (Cole 1986). Critical balances may involve soil fertility and the degree to which precipitation penetrates into deep soil layers, allowing a vertical partitioning of available soil moisture and nutrients between shallow rooted herbaceous plants, principally grasses, and deeply rooted trees (Walter 1971; Walker et al. 1981). The trees often have an extensive fine root system just below the major grass rooting zone and a deep taproot system that penetrates to the water table or deep water storage levels of the profile (Sarmiento 1984). Nevertheless, grasses can deplete topsoil water sufficiently to reduce penetration into lower edaphic layers, a type of preemptive competition that can reduce tree growth significantly (Knoop and Walker 1985).

The effects of animals, particularly elephants, upon the adult tree canopy indicate that African savannas are not merely donor-controlled ecosystems existing passively within a certain climatic-geological complex. Instead, the animals themselves influence the structure of the vegetation that supports them. As Laws (1970) observed, elephants have been a major agent regulating the structure of African vegetation. Eltringham (1980) studied the relationship between degree of elephant impact on trees and their survivial after five years had passed (Fig. 1).

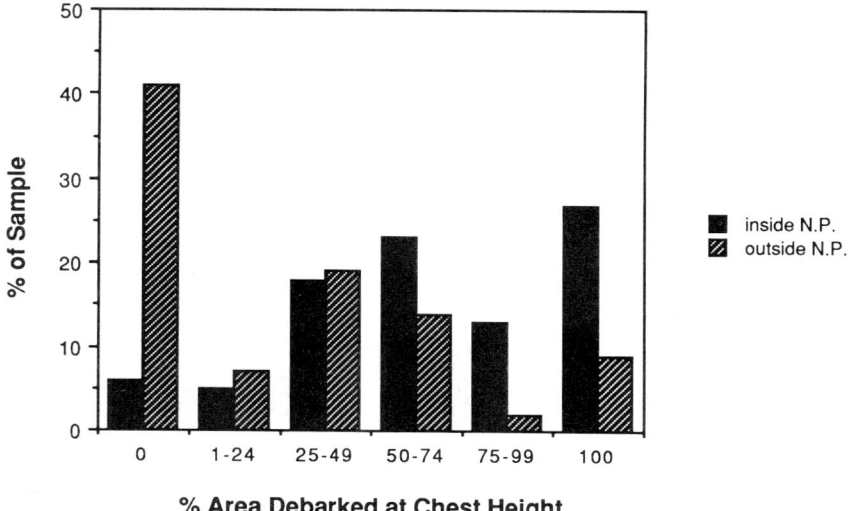

Fig. 2. Tree (*Adansonia digitata* L.) damage classes in relation to location within or without the South Luangwa National Park, Zambia (redrawn after Lewis 1986).

Trees that experienced light levels of elephant damage had survivorships equivalent to control trees, about 20% dying during the five years. Trees that were moderately damaged had mortalities about double the undamaged and lightly damaged individuals. Those suffering heavy damage, often involving extensive branch-breaking and ring-barking, had very high levels of mortality, with about 80% dying over the five years, a four-fold increase in mortality compared to undamaged or lightly damaged trees.

The effects of elephants upon trees vary with both habitat factors and tree properties. The compression (Laws *et al.* 1975) of elephants into national parks and other game reserves is an inevitable consequence of the incompatibility of utilization of landscapes by both humans and elephants, which are in strong conflict for surface water (Matzke 1975). In Zambia's Luangwa Valley, tree damage is markedly related to national park boundaries (Lewis 1986). Within the parks, the most frequent damage class is 100% debarking with most of the trees showing greater than 25% debarking and less than 10% escaping debarking entirely (Fig. 2). In the multiple use areas outside the park, in contrast, where human activities of various sorts are allowed, more than 40% of the trees had no evidence of ringbarking, and fewer than 10% were completely ringbarked. The mere presence of humans, quite apart from the recent massive increase in poaching both within and outside of game reserves (Douglas-Hamilton 1983, 1984), acts to deter utilization of areas by elephants (Matzke 1975), while the poaching has a drastic effect upon elephant behavior as well as numbers.

Elephant impact upon vegetation also varies seasonally (Croze 1974; Laws *et al.* 1975). During the wet season, grasses are the major component of the diet, and the impact upon woody vegetation is minor (Fig. 3). During the dry season, however, when grass becomes of low quality, the utilization of woody material increases substantially, in this example increasing to 90% of the diet indexed from volume percentage in dung.

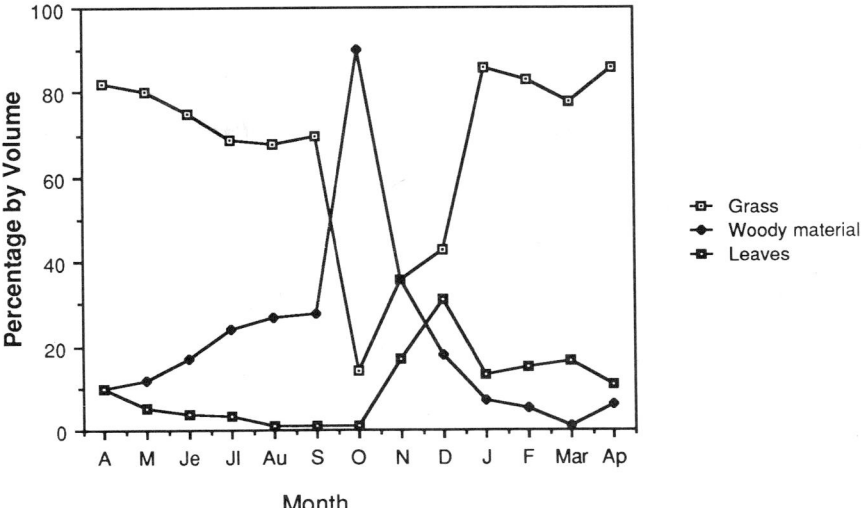

Fig. 3. Seasonal variation in elephant diet in the Luangwa Valley, Zambia based on dung volume class (redrawn after Lewis 1986).

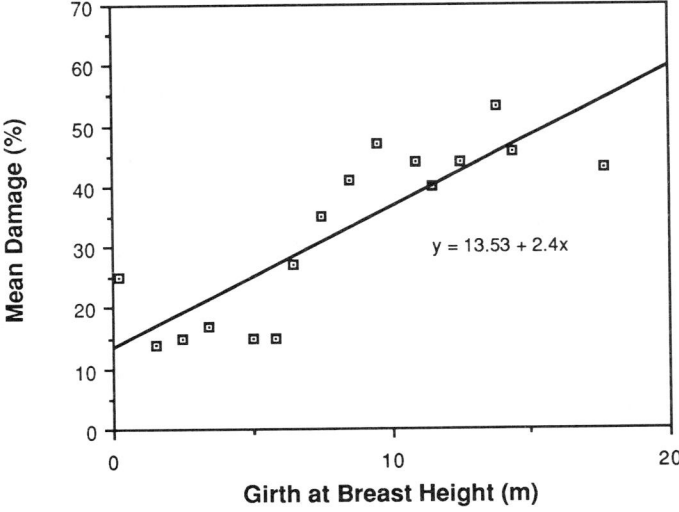

Fig. 4. Damage to baobab (*Adansonia digitata*) trees in Arusha National Park, Tanzania, in relation to tree girth at breast height. For the linear fit, $r = 0.73$, $p \leq 0.01$ (redrawn after Weyerhauser 1985).

Elephant impact is also related to tree size (Weyerhauser 1985). Large trees often suffer an inordinate degree of damage (Fig. 4). Although a straight line is drawn in this figure, the data appear to us to more closely fit a sigmoid function. Trees with a girth of less than 8 m sustained low, uniform levels of damage, there was a transition to high levels of damage between girths of 8 and 12 m, with trees above this level sustaining about 40% damage for this study.

It is generally, but perhaps not universally, agreed that elephants alone are in-

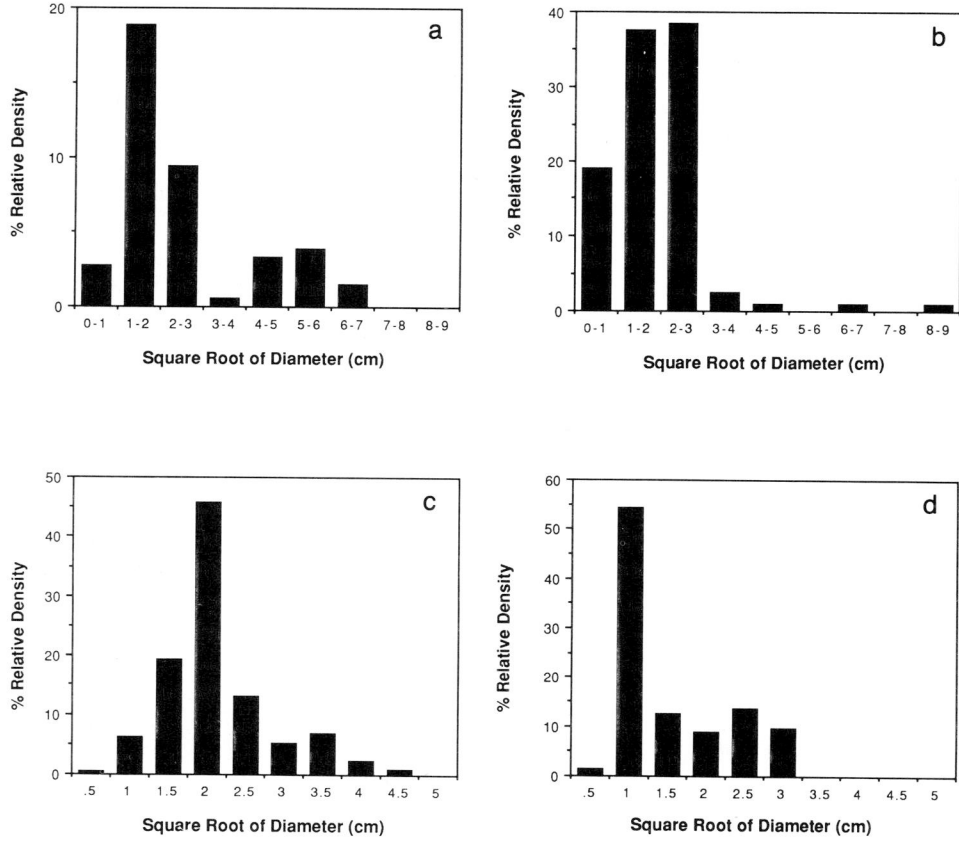

Fig. 5. Frequency distributions of trees of different diameter classes in Serengeti National Park, Tanzania, in relation to prevalence of elephant activity, fire, and browsing: (a) *Acacia tortilis* in a low impact area, (b) *A. tortilis* in a high impact area, (c) *A. drepanolobium* in a low impact area, (d) *A. drepanolobium* in a high impact area (redrawn after Sabuni 1977).

capable of converting woodland into savanna, or savanna into open grasslands (Norton-Griffiths 1979). Rather, it seems, combinations of elephant destruction of adult trees, browsing which retards the growth of seedlings and saplings into the adult canopy out of the fire-sensitive size classes, and recurrent fire must act in concert for those vegetation structural conversions to occur. Surveys of populations of *Acacia tortilis* (Forsk.) Hayne and *A. drepanolobium* Harms ex Sjöstedt in locations within Tanzania's Serengeti National Park (Sabuni 1977) where the balance of elephants, browsers, and fire differed conspicuously revealed radically different population size structures in different habitats and in the two species (Fig. 5). *A. tortilis*, a large canopy tree, had a biomodal size distribution in the locality with low fire, elephant, and browser impact. There were significant numbers of large, adult trees of sufficient size to be impervious to the effects of fire and a large reservoir of smaller, sapling-sized trees capable of emerging into the upper canopy. The open spacing of adults in such savannas appears to be due to competition, with self-thinning involved in the development of the canopy overstory and competition suppressing recruitment until that canopy is opened by death of adults (Smith and Walker 1983). *A. drepanolobium* is a short, spindly species that is less drastically affected by elephants than adults of *A. tortilis*. Its population size structure was less noticably biomodal in the locality with lower fire, browser, and elephant impacts, but there was a pronounced enrichment of smaller size classes in the locality where these impacts were more pronounced (Fig. 5). These results suggest that African savanna structure will be a complicated consequence of tree growth form and both the biotic and abiotic environments that trees experience.

In addition to the direct effects of animals upon the balance between arborescent and herbaceous vegetation through mortality and growth rate modification, that balance might also be affected by the effects of trampling upon soil properties. Soil bulk density, mechanical resistance to penetration, and moisture runoff are increased by trampling while soil aeration and moisture infiltration rate are reduced (Eavis and Payne 1969; Watkin and Clements 1978). Root penetration at a given water potential decreases as bulk density increases (Taylor and Gradiner 1963). Decreased plant growth due to trampling can be caused, in part, by decreased root growth and soil oxygen concentrations (Shielaw and Alston 1984; Scholefield and Hall 1985). Restricted root growth results in reduced nutrient absorbtion from compacted compared to uncompacted soils (Boone and Veen 1982), and differential resistance of species to trampling (Edmond 1964; High *et al.* 1965) by herbivores can influence vegetation structure, and the balance between woody and herbaceous vegetation, by affecting soil water and nutrient balances and the nutrient harvesting abilities and resultant plant growth rates.

Grass sward canopy structure

Sward canopies of African grasslands supporting large grazers are often distinctively heterogenous both spatially and temporally (Vesey-Fitzgerald 1974). During the wet season in Tanzania's Arusha National Park, *Cynodon dactylon* (L.) Pers. stands has a strong modal canopy height at 15 cm but by the end of the dry season that modal height was 5 cm. This pattern indicates that the vegetation grew more rapidly than the grazing rate during the wet season while the grazers, principally African buffalo (*Syncerus caffer* Sparrman), grazed the vegetation down to the limits of their ability to forage during the dry season. There also was a pronounced spatial grazing mosaic during the wet season with a definite dichotomy evident between grazed *C. dactylon* patches and ungrazed *Sporobolus macranthelus* Chiov. (syn. *S. greenwayi* Napper) tussocks (Fig. 6). This tendency for grazers to create a heterogenous canopy height structure is evident throughout African grasslands during the wet season, reaching extreme structural heterogeneity when such palatable species as *C. dactylon* are intermixed with such very unpalatable species as *Eleusine jaegeri* Pilger (Vesey-Fitzgerald 1974). Because high sward height hinders grazing (Arnold 1964), rank patches can become self-perpetuating once established.

The temporal modification of grass canopy structure can be very abrupt where migratory grazers move into mature grass and consume it in a matter of minutes

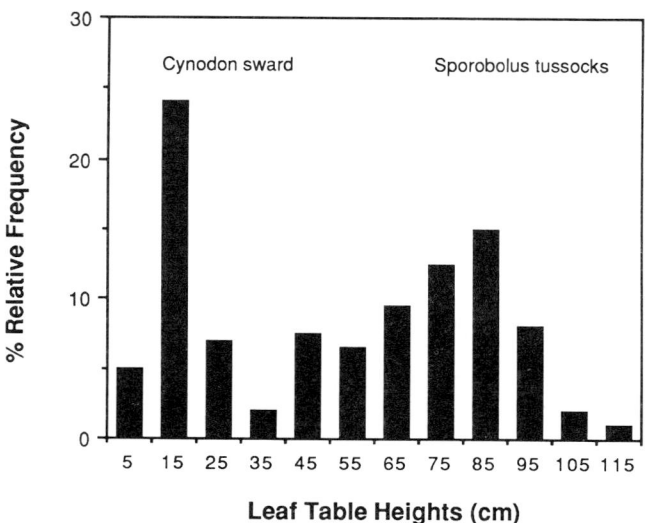

Fig. 6. Grass sward canopy height in a mosaic grassland in Arusha National Park, Tanzania, consisting of a palatable (*C. dactylon*) and an unpalatable (*S. macranthelus*) species (redrawn after Vesey-Fitzgerald 1974).

(McNaughton 1976). Closed canopy mid-height grasslands with much of the foliage between 50 and 100 cm above the soil surface and very little basal foliage can be rapidly transformed into a grassland with only about 5% of the initial foliage standing crop remaining. These grasslands then undergo basal tillering to produce a short, dense canopy of very compact structure that is subsequently exploited by smaller-bodied grazers such as Thomson's gazelles (*Gazella thomsonii* Gunther).

Grazing systems and sward structure

There are three types of seasonally-distinct grazing systems in the Serengeti ecosystem (McNaughton 1976, 1983, 1985): (a) sustained yield systems during the wet season when the nomadic grazers concentrate on the SE Serengeti Plains and the resident herds concentrate on upper catena locations throughout the rest of the region (Bell 1970): during this period the animals crop the vegetation continuously so long as there is sufficient rain to sustain plant growth; (b) passage systems during the early dry season, or when rains are interrupted during the wet season, when the large nomadic herds pass through grasslands that are largely unexploited, graze them heavily, and then do not return until much later, or in the following year (McNaughton 1976, 1983); and (c) rotational systems during the dry season when passage stands receive sufficient rain to generate regrowth that either migrants or residents utilize to re-exploit the same stand (McNaughton 1985). These grazing systems have distinctly different temporal patterns of sward structure. Sustained yield systems are arrested at short canopy heights throughout the growing season. The grazed vegetation is much shorter than fenced plots in vege-

Large African mammals as regulators of vegetation structure   347

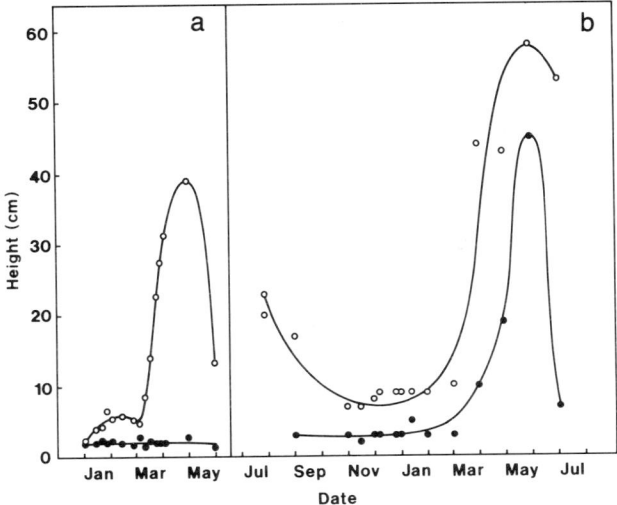

Fig. 7. Canopy height through an annual cycle in the Serengeti National Park, Tanzania, in (a) a location with a sustained yield grazing system, and (b) a passage/rotation grazing system. Circles are fenced vegetation and dots are unfenced vegetation (redrawn after McNaughton 1984).

tation subject to this grazing system (Fig. 7a). While fenced vegetation reaches canopy heights near a half meter, the grazed vegetation is always less than 5 cm in height. In passage and rotational systems, in contrast, the fenced vegetation tracks the height of unfenced vegetation, although it commonly is a few centimeters shorter due to canopy opening by resident animals (Fig. 7b). These stands reach terminal canopy heights between 50 and 200 centimeters, depending upon whether they are mid-height or tall grasses (McNaughton 1983).

A major sward structural property that summarizes canopy organization is biomass concentration (McNaughton 1976) or forage bulk density (Stobbs 1973), expressed as units of mass per unit foliage volume. This sward parameter reflects the balance between biomass and height and will be both influenced by and an influence upon grazing. Sward bulk density can be a major parameter influencing bite size of grazers (Stobbs 1973; McNaughton 1976, 1984, 1985; Laca and Demment 1988). Generally, the higher the grazing intensity, the higher the bulk density; and the higher the bulk density, the greater the intake for a given bite volume. Biomass concentration of grazed swards in the Serengeti declined drastically with increased canopy height up to about 15 cm, then it declined more gradually as canopy height increased beyond this (Fig. 8). In fenced vegetation, in contrast, bulk density declined in a linear fashion with canopy height so that ungrazed vegetation, on average, had a higher bulk density over the height range between 10 and 70 cm, while grazed vegetation had decidedly higher bulk densities at canopy heights below about 10 cm.

The arrested canopy height of short, sustained yield grasslands is evident in the relationship of maximum height reached inside and outside fences during the growing season (Fig. 9). Many passage and rotational grasslands approach the canopy height equality line while short grasslands are almost totally arrested at

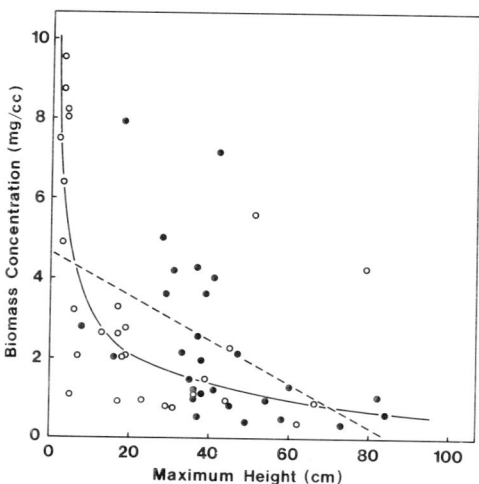

Fig. 8. Relationship between maximum height and maximum biomass concentration in fenced (dots) and unfenced (circles) vegetation in the Serengeti National Park, Tanzania. The solid line is the best fit line for grazed vegetation, the dashed line is best fit for plots outside fences (redrawn after McNaughton 1984).

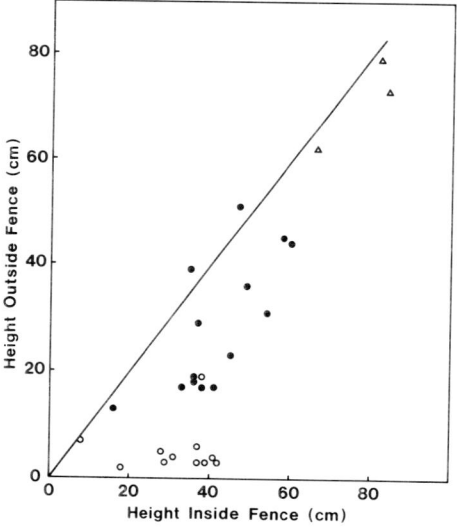

Fig. 9. Relationship between maximum height reached by paired unfenced and fenced plots across a range of grassland types in the Serengeti National Park. Circles are short grasslands, dots are mid-height grasslands, and triangles are tall grasslands. The solid line is the equality line (redrawn after McNaughton 1984).

the very short canopy heights leading to high bulk densities. In addition to the greater potential mass yield per bite volume, arresting the grasses in a short, rapidly growing state, with grazers cropping the growth increment above their minimum bite height, undoubtedly results in the harvest of forage with a much higher quality, i.e., nutrient density, than can be obtained by grazing taller, more mature swards (McNaughton 1979, 1986).

Sward structure and grass evolution

Sward structure can be due to the proximate effect of grazers on grass growth patterns, removing the upper canopy levels and stimulating basal tillering, but the developmental ability of grasses to respond in this fashion is evolutionarily constrained (McNaughton 1979, 1984). Among the earliest direct demonstrations of the effects of natural selection upon plant populations were studies examining grass growth form in population samples from localities where herbivory differed due to natural factors or human cultural practices (Gregor and Sansome 1926; Stapledon 1928; Turresson 1929). Subsequent research revealed that the differentiation of short, prostrate growth forms under grazing and tall, erect forms in hay meadows could take place within three years (Kemp 1937), suggesting that selection coefficients are large in these two distinctive habitats.

Clones of *C. dactylon* were sampled from localities within the Serengeti ecosystem where grazing intensity varied dramatically and transplanted them into a uniform garden to determine the extent to which growth form, and resultant canopy structure, might be genetically determined (McNaughton 1984). Clone traits differed conspicuously according to grazing intensity at the native site, with grazing intensity defined as the annual average of $G = 1 - (g/ug)$, where g was the biomass of unprotected vegetation and ug was the biomass inside fences. The principal plant trait accounting for much of the variation in plant structure was internode length, which was negatively correlated with grazing intensity of the site of origin (Fig. 10). Internode length varied from 10 cm in plants from lightly grazed locations to less than 2 cm for plants from the most heavily grazed locations. This single trait had major ramifications for plant form and resulting vegetational structure, with short internodes producing a reduced canopy height, a smaller area colonized per clone, and a greater ramet density within colonized areas, all traits that would contribute to a higher forage bulk density. In addition to the stature-dwarfing associated with shorter internodes, leaf blade length varied from over 20 cm for plants from lightly grazed habitats to less than 5 cm for plants from heavily grazed locations.

These major structural changes can occur quite rapidly because they are due to the combined effects of simple genetic changes, high heritabilities, and strong selection. Agronomic grain breeding experiments with wheat (*Triticum aestivum* L. emend Foir.) indicate that dwarfing can be due to a single dominant gene controlling internode length (Qualset et al. 1970). This would lead to the syndrome of structural properties associated with internode length in *C. dactylon*. Plant height, erectness, and leafiness of *Sorghastrum nutans* (L.) Nash. has narrow sense heritabilities of 0.75, 0.48, and 0.36, respectively (Vogel et al. 1981). Moreover, coefficients of selection against prostrate plants in unmowed areas were 0.77, and against tall plants in mowed areas ranged from 0.53 and 0.68 (Warwick and Briggs 1978). Therefore, grass growth form and resultant vegetation structure may be expected to respond rapidly to environmental changes in-

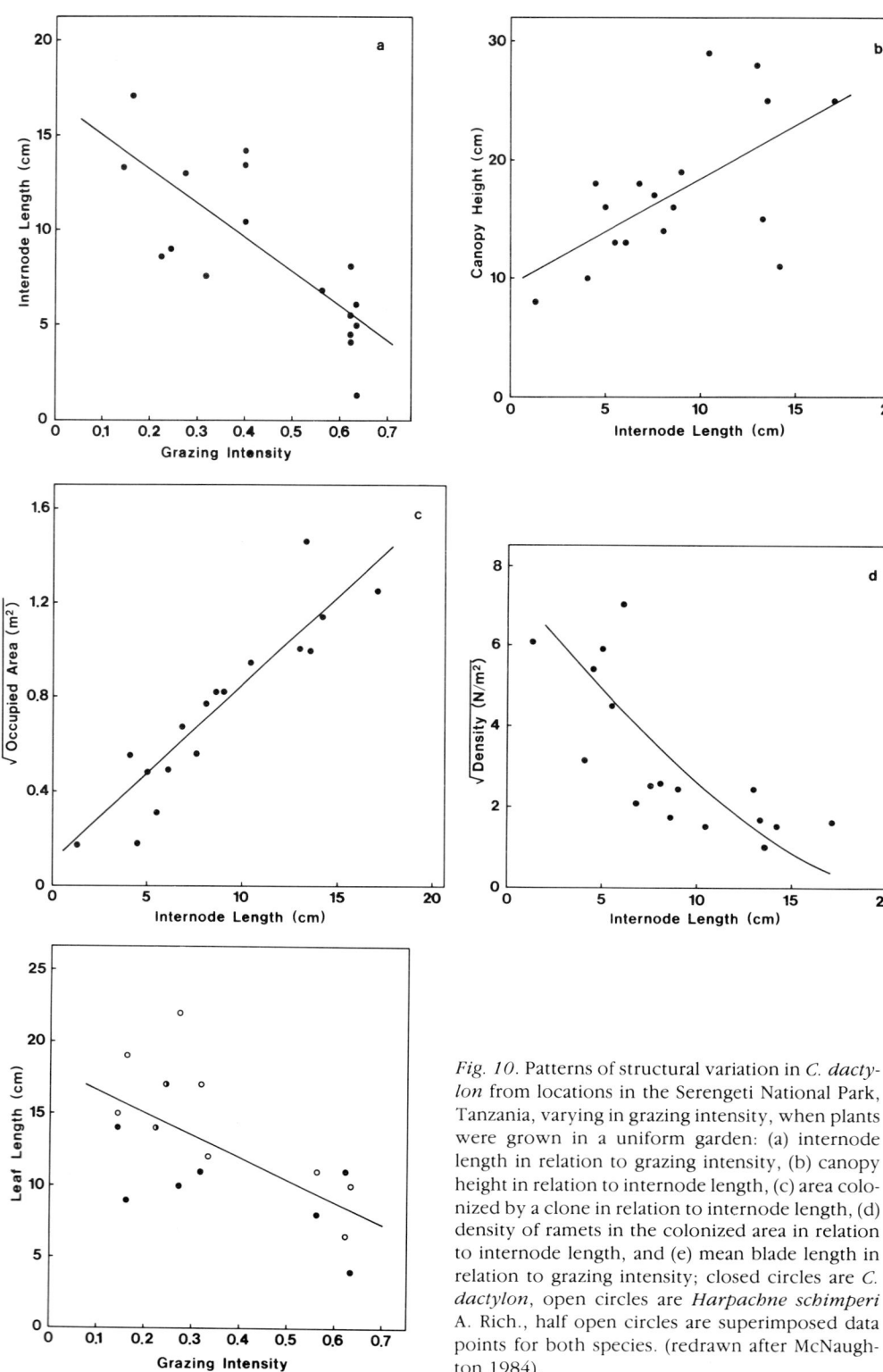

Fig. 10. Patterns of structural variation in C. dactylon from locations in the Serengeti National Park, Tanzania, varying in grazing intensity, when plants were grown in a uniform garden: (a) internode length in relation to grazing intensity, (b) canopy height in relation to internode length, (c) area colonized by a clone in relation to internode length, (d) density of ramets in the colonized area in relation to internode length, and (e) mean blade length in relation to grazing intensity; closed circles are C. dactylon, open circles are Harpachne schimperi A. Rich., half open circles are superimposed data points for both species. (redrawn after McNaughton 1984).

volving, on the one hand, increased grazing, and, on the other, the taller canopies that occur under ungrazed conditions (Detling and Painter 1983). This is, we believe, the clearest evidence available of how strong selection by herbivores, and by competition in the absence of herbivores, can lead to plant evolution having a profound effect upon vegetational structure.

It has been known since long before the emergence of ecology as a science that grass stature gradients are commonly related to rainfall gradients, with short grasslands in low rainfall areas and tall grasslands in high rainfall areas (Weaver 1954). These stature gradients, within and between species, can provide the genetic raw material allowing grasses to respond to the complex gradients of natural selection associated with varying grazing intensities (Coughenour 1984). Low grazing intensities release plants from the selective force of cropping and introduce a new force of light competition. High grazing intensities make competition for light unlikely, but can also impose competitive interactions related to the nutrient demands that grazing imposes (McNaughton et al. 1983; Ruess et al. 1983; McNaughton and Chapin 1985).

The ability of Serengeti grasses from the short grass plains to rapidly elaborate more erect, leafier, taller genotypes when released from grazing by fencing (McNaughton 1979) suggests that there is substantial intraspecific genetic variation within populations that natural selection can act upon. Some of that variation may be revealed in natural microsites, e.g., around bushes, where taller variants may find refuge from grazing. In addition, the apparently minor genetic difference governing internode length can lead to rapid form conversions with major structural ramifications (Qualset et al. 1970).

Conclusion: Belowground vegetational structure

The discussion above about the role of differential exploitation of soil depths by trees and grasses indicates that belowground structure is fully as important as aboveground structure as a regulator of, and consequence of, community and ecosystem organization. Models and empirical evidence both indicate that plant nutrient harvesting ability is influenced strongly by structural organization of the root system (Barber 1984). Competition for nutrients due to the higher nutritional demands of grazed plants (McNaughton et al. 1983, Ruess et al. 1983; Coughenour et al. 1985a, 1985b) might, in turn, impose selective forces for root structural organizations particularly efficient at exploiting the available soil volume (Barber 1984).

This symposium reveals that there is a growing knowledge of aboveground structure; understanding belowground structure is, we believe, an important goal for future research on vegetational structure. This research in savannas might profitably be guided by the objectives of (a) understanding the importance of the vertical root distributions of the arborescent and herbaceous layers as a mechanism partitioning nutrient and water harvesting (Knoop and Walker 1985), and (b) determining the importance of root proliferation within various soil layers as a nutrient harvesting mechanism (Barber 1984) potentially alleviating the nutrient demands placed upon plants by herbivory (McNaughton et al. 1983; Ruess et al. 1983; McNaughton and Chapin 1985).

Acknowledgment

Preparation of this paper was facilitated by NSF BSR 8505862 from the Ecosystem Studies Program to Syracuse University.

References

Arnold, G.W. 1964. Factors within plant associations affecting the behaviour and performance of grazing animals. In: D.J. Crisp (ed.), Grazing in Terrestrial and Marine Environments, pp. 133–154. Blackwell, Oxford.
Barber, S.A. 1984. Soil Nutrient Bioavailability. A Mechanistic Approach. Wiley/Interscience, New York.
Beard, J.S. 1946. The Natural Vegetation of Trinidad. Oxf. For. Mem. No. 21., Oxford.
Bell, R.H.V. 1970. The use of the herb layer by grazing ungulates in the Serengeti. In: A. Watson (ed.), Animal Populations in Relation to Their Food Resources, pp. 111–123. Blackwell, Oxford.
Bell, R.H.V. 1982. The effect of soil nutrient availability on community structure in African ecosystems. In: B.J. Huntley and B.H. Walker (ed.), Ecology of Tropical Savannas, pp. 193–245. Springer-Verlag, Berlin/New York.
Boone, F.R. and Veen, B.W. 1982. The influence of mechanical resistance and phosphate supply on morphology and function of maize roots. Neth. J. Agric. Sci. 30: 179–192.
Cole, M.M. 1986. The Savannas. Biogeography and Geobotany. Academic Press, New York.
Coughenour, M.B. 1984. Graminoid responses to grazing by large herbivores: adaptations, exaptations and interacting processes. Ann. Mo. Bot. Gard. 72: 852–863.
Coughenour, M.B., McNaughton, S.J. and Wallace, L.L. 1985a. Responses of an African tall-grass (*Hyparrhenia filipendula* Stapf.) to defoliation and limitations of water and nitrogen. Oecologia (Berl.) 68: 80–86.
Coughenour, M.B., McNaughton, S.J. and Wallace, L.L. 1985b. Responses of an African graminoid (*Themeda triandra* Forsk.) to frequent defoliation, nitrogen, and water: a limit of adaption to herbivory. Oecologia (Berl.) 68: 105–110.
Croze, H. 1974. The Seronera bull problem. II. The trees. E. Afr. Wildl. J. 12: 29–47.
Cumming, D.H.M. 1982. The influence of large herbivores on savanna structure in Africa. In: B.J. Huntley and B.H. Walker (ed.), Ecology of Tropical Savannas, pp. 217–245. Springer-Verlag, Berlin/New York.
Demment, M.W. and van Soest, P.J. 1985. A nutritional explanation for body-size patterns of ruminant and non-ruminant herbivores. Am. Nat. 125: 641–672.
Detling, J.K. and Painter, E.L. 1983. Defoliation responses of western wheatgrass population with diverse histories of prairie dog grazing. Oecologia (Berl.) 57: 65–71.
Douglas-Hamilton, I. 1983. Elephants hit by arms race. African Elephant and Rhino Group Newletter 2: 11–13.
Douglas-Hamilton, I. 1984. Trends in key African elephant populations. Pachyderm 4: 7–9.
Eavis, B.W. and Payne, D. 1969. Soil physical conditions affecting seedling growth. In: W.J. Whittington (ed.), Root Growth, pp. 315–321. Butterworth's, London.
Edmond, D.B. 1964. Some aspects of sheep treading on growth of ten pasture species. N. Z. J. Agric. Res. 7: 1–16.
Eltringham, S.K. 1980. A quantitative assessment of range usage by large African mammals with particular reference to the effects of elephants on trees. Afr. J. Ecol. 18: 53–71.
Gregor, J.W. and Sansome, F.W. 1926. Experiments on the genetics of wild populations. I. Grasses. Genetics 17: 349–364.
Halle, F. and Oldeman, R.A.A. 1970. Essai sur L'architecture et la Dynamique de Croissance des Arbres Tropicaux. Macon et Cie, Paris.
Herbel, C.H. 1986. Vegetation changes on arid rangelands of the southwestern United States. In: P.J. Joss, P.W. Lynch and O.B. Williams (eds), Rangelands: A Resource Under Siege, pp. 8–10. Aust. Acad. Sci., Canberra.
High, T.W., Chapman, E.J., Wittenburg, B.C. and High, T.W. 1965. Fescue pastures under different management systems and orchardgrass, clover for yearling slaughter steer production. Bull. Tenn. Agric. Exp. Sta. 385.

Huntley, B.J. 1982. Southern African savannas. In: B.J. Huntley and B.H. Walker (eds), Ecology of Tropical Savannas, pp. 101–119. Springer-Verlag, Berlin/New York.
Huntley, B.J. and Walker, B.H. (eds). 1982. Ecology of Tropical Savannas. Springer-Verlag, Berlin/New York.
Jarman, P.J. 1974. The social organization of antelope in relation to their ecology. Behaviour 48: 215–266.
Kemp, W.B. 1937. Natural selection within plant species as exemplified in a permanent pasture. Hered. 38: 329–333.
Knoop, W.T. and Walker, B.H. 1985. Interactions of woody and herbaceous vegetation in a southern African ecosystem. J. Ecol. 73: 235–253.
Laca, E.A. and Demment, M.W. 1988. A mechanistic model of a grazing ruminant: harvesting limitations to food intake. J. Appl. Ecol. 25: in press.
Laws, R.M. 1970. Elephants as agents of habitat and Landscape change in East Africa. Oikos 21: 1–15.
Laws, R.M., Parker, I.S.C. and Johnstone, R.C.B. 1975. Elephants and their Habitats. Clarendon, Oxford.
Lewis, D.M. 1986. Luangwa Valley Elephants: Toward Developing a Management Policy. Lupande Res. Proj. Publ. No. 2, Lupande, Zambia.
MacVicar, C.N. 1977. Soil Classification: A Bionomial System for South Africa. Agr. Tech. Serv., Pretoria.
Matzke, G. 1975. Large animals, small settlements, and big problems: a study of overlapping space preference in southern Tanzania. Syracuse U., Ph.D. diss.
McNaughton, S.J. 1976. Serengeti migratory wildebeest: facilitation of energy flow by grazing. Science 191: 92–94.
McNaughton, S.J. 1979. Grassland-herbivore dynamics. In: A.R.E. Sinclair and M. Norton-Griffiths (eds), Serengeti. Dynamics of an Ecosystem, pp. 46–81. U. Chicago Press, Chicago.
McNaughton, S.J. 1983. Serengeti grassland ecology: the role of composite environmental factors and contingency in community organization. Ecol. Monogr. 53: 291–320.
McNaughton, S.J. 1984. Grazing lawns: animals in herds, plant form, and coevolution. Am. Nat. 124: 863–886.
McNaughton, S.J. 1985. Ecology of a grazing ecosystem: the Serengeti. Ecol. Monogr. 55: 259–294.
McNaughton, S.J. 1986. Grazing lawns: on domesticated and wild herbivores. Am. Nat. 128: 937–939.
McNaughton, S.J. and Chapin III, F.S. 1985. Effects of phosphorus nutrition and defoliation on C4 graminoids from the Serengeti Plains. Ecology 66: 1617–1629.
McNaughton, S.J., Wallace, LL. and Coughenour, M.B. 1983. Plant adaptation in an ecosystem context: effects of defoliation, nitrogen, and water on growth of an African C4 sedge. Ecology 64: 307–318.
Menaut, J.C., Barbault, R., Lavelle, P. and Lepaze, M. 1985. African savannas: biological systems of humification and mineralization. In: J.C. Tothill and J.J. Mott (eds), Ecology and Management of the World's Savannas, pp. 14–33. Aust. Acad. Sci., Canberra.
Norton-Griffiths, M. 1979. The influence of grazing, browsing, and fire on the vegetation dynamics of the Serengeti. In: A.R.E. Sinclair and M. Norton-Griffiths (eds), Serengeti. Dynamics of an Ecosystem, pp. 310–352. U. Chicago Press, Chicago.
Qualset, C.O., Fick, G.N., Constantin, M.J. and Osborne, T.S. 1970. Mutation in internode length affects wheat plant-type. Science 169: 1090–1091.
Richards, P.W. 1957. The Tropical Rain Forest. Cambridge U. Press, Cambridge.
Ruess, R.W., McNaughton, S.J. and Coughenour, M.B. 1983. The effects of clipping, nitrogen source, and nitrogen concentration on the growth responses and nitrogen uptake of an East African sedge. Oecologia (Berl.) 59: 253–261.
Sabuni, G.A. 1977. A comparative study of the autecology of selected Acacia species in the Serengeti National Park. U. Dar-es-Salaam, M.S. thesis.
Sarmiento, G. 1984. The Ecology of Neotropical Savannas. Trans. O. Solbrig. Harvard U. Press, Cambridge, Massachusetts.
Scholefield, D. and Hall, D.M. 1985. Constricted growth of grass roots though rigid pores. Plant Soil 85: 153–162.
Scholes, R.J. 1985. Drought related grass, tree and herbivore mortality in a southern African savanna. In: J.C. Tothill and J.J. Mott (eds), Ecology and Management of the World's Savannas, pp. 350–353. Aust. Acad. Sci., Canberra.

Smith, T.M. and Walker, B.H. 1983. The role of competition in the spacing of savanna trees. Proc. Grassland Soc. Southern Afr. 18: 159–164.

Stapledon, R.G. 1928. Cocksfoot grass (Dactylis glomerata L.) ecotypes in relation to the biotic factor. J. Ecol. 16: 71–104.

Stobbs, T.H. 1973. The effect of plant structure on the intake of tropical pasture. I. Variation in the bite size of grazing cattle. Aust. J. Agric. Res. 24: 809–819.

Taylor, H.M. and Gardiner, H.R. 1963. Penetration of cotton seedling taproots as influenced by bulk density, moisture content, and strength of soil. Soil Sci. 96: 153–156.

Tothill, J.C. and Mott, J.J.(eds) 1985. Ecology and Management of the World's Savannas. Aust. Acad. Sci., Canberra.

Turreson, G. 1929. Ecotypical selection in Siberian Dactylis glomerata L. Hereditas 12: 225–351.

Vesey-Fitzgerald, D.F. 1974. Utilization of the grazing resources by buffaloes in the Arusha National Park, Tanzania. E. Afr. Wildl. J. 12: 107–134.

Vogel, K.P., Gorz, H.J. and Haskins, F.A. 1981. Heritability estimates for height, color, erectness, leafiness, and vigor in indiangrass. Crop Sci. 21: 734–736.

Walker, B.H., Ludwig, D., Holling, C.S. and Peterman, R.M. 1981. Stability of semi-arid savanna systems. J. Ecol. 69: 473–498.

Walter, H. 1971. Ecology of Tropical and Subtropical Vegetation (D. Mueller-Dumbois, trans.). Oliver and Boyd, Edinburgh.

Walter, H. 1979. Vegetation of the Earth. Springer-Verlag, Berlin/New York.

Warwick, S.I. and Briggs, D. 1978. The genecology of lawn weeds. II. Evidence for disruptive selection in Poa annua L. in a mosaic environment of bowling green lawns and flower beds. New Phytol. 81: 725–737.

Watkin, B.R. and Clements, R.J. 1978. The effects of grazing animals on pastures. In: J.R. Wilson (ed.), Plant Relations in Pastures, pp. 273–289. CSIRO, E. Melbourne.

Weaver, J.E. 1954. North American Prairie. Johnson Publ. Co., Lincoln, Nebraska.

Werger, M.J.A. 1977. Gradients in vegetation structure and soil types in an African marginal savanna. Acta del IV Symp. Intern. de Ecologia Tropical : 476–498.

Werger, M.J.A. 1983. Tropical grasslands, savannas, woodlands: natural and man-made. In: W. Holzner, M.J.A. Werger and I. Ikusima (eds), Man's Impact on Vegetation, pp. 107–137. Dr. W. Junk, The Hague.

Werger, M.J.A. and Ellis, R.P. 1981. Photosynthetic pathways in the arid regions of South Africa. Flora 171: 64–75.

Werger, M.J.A., Morris, J.W. and Louppen, J.M.W. 1979. Vegetation-soil relationships in the southern Kalahari. Doc. Phytosociologiques 4: 967–981.

Werger, M.J.A., Wild, H. and Drummond, B.R. 1978. Vegetation structure and substrate of the northern part of the Great Dyke, Rhodesia: gradient analysis and dominance-diversity relationships. Vegetatio 37: 151–161.

Weyerhauser, F.J. 1985. Survey of elephant damage to baobabs in Tanzania's Lake Manyara National Park. Afr. J. Ecol. 23: 235–243.

INDEX OF KEYWORDS

(Page numbers refer to the first page in a section in which the subject is discussed. Subsequent pages are not indicated.)

-3/2 power model 127, 162
A/P plant 233
aerodynamic resistance 202
allometric equations 183
anatomy 1, 193
animal disturbance 47, 227, 243, 281, 301, 317, 331, 355
annual 13, 211

bark relief 5
branch density 168
browsers 339

C3-plants 225, 307, 331
C4-plants 225, 307, 331
canopy photosynthesis 153, 171
carbon-nitrogen ratio 292
chamaephyte 13, 49, 184, 211, 307
clonal growth 73, 121, 222, 282
convergence 192

deciduous forest 245
desert 211
diversity 263, 322, 334
drought avoidance 98, 211
drought tolerance 72, 211

ecotypic differentiation 305
energy balance 202
epiphyte 5
exotic species 321

facilitation model 94
feed-back processes 292
flowering stage 172
fog interception 96
foliage distribution 105, 154, 171, 183, 276, 331
foraging 122

gene flow 110
genetic differentiation 109
germination 72, 78
grass 18, 55, 105, 214, 269, 290, 301, 331
grassland 77, 105, 111, 214, 263, 281, 301, 317, 331, 339
grazers 340
grazing 281, 301, 317, 331, 339
grazing history 325, 331, 339
growth forms 2, 9, 45, 87, 105, 109, 201, 211
grubbing 283
gynodioecy 129

heat transfer 202
hemicryptophyte 9, 45, 211, 243
herbivory 243, 253, 263, 281, 301, 317, 331, 339
high energy strategy 229
historical survey of growth forms and life forms 11
homoeostasis 110

inhibition model 94
insecticide 263
insects 253, 265
interference 73
isostatic uplift 282

keystone species 290
Krummholz 201

landscape units 45
leaf area index (LAI) 5, 105, 150, 153, 174, 318, 334
leaf area ratio (LAR) 109, 174, 334
leaf nitrogen concentration (LNC) 135, 148, 171
leaf size 2, 191
leaf weight/leaf + root weight 147
leaf weight ratio (LWR) 136, 179
Lesser Snow Goose 282
lichens 10
life expectancy of branches 162
light climate 2, 106, 149, 153, 172, 191, 323, 331
limit-cycle hypothesis 296
Little Ice Age 282
low vs high energy strategy 231

meristems 203
microclimate 7, 87, 206
models of succession 93
modular growth 161
mortality 91, 121, 162, 243, 254
moss layer 6, 294

net assimilation rate (NAR) 135
nitrogen availability 136, 171
nitrogen concentration 135, 148, 171, 309
nitrogen partitioning 135, 174
nitrogen productivity 136
non-equilibrium theory (of vegetational structure) 282

organism-centered approach 88

P/A plant 234

partitioning of dry matter 80, 105, 109, 135, 147, 171, 183
peat 'barrens' 295
permafrost 46, 290
phenological plant types 9
phenotypic response 109, 121, 147, 192
photosynthesis 148, 153, 171, 183
physiological integration 124
pioneer species 93
plant communities 5, 47, 183, 213, 281, 302, 325, 331
population maturity continuum 92

R:FR ratio 323
radial increment 206, 256
relative growth rate (RGR) 72, 77, 136
root consumption 310
root length 77, 91
root mass 72, 79, 91, 222, 310
root:shoot ratio 109, 135, 147
roughness length 202

salt marsh 283
sedge meadow 283
seedling establishment 77, 98, 250, 258, 295
seedling rooting strategies 88

self-thinning 127, 162
seral species 93
shade leaves 3, 192
shoot pulling 286
shrub 88, 183, 201, 211
spatial scales 296, 327
specific absorption rate (SAR) 137
specific leaf area (SLA) 147
specific leaf weight (SLW) 4, 135, 172, 191
spruce budworm 253
steady-state exponential growth 141
stem weight/whole plant weight 147
stolons 122
succession 88, 105, 260, 263, 317
succulent 216
sun leaves 2, 192

time of invasion 87
time scales 296, 325
tolerance model 94
tree crown 154, 162, 253
tropical rain forest 192

within-population variability 109
woodland 243
woodland herb 243